COMPUTER-AIDED STATISTICAL PHYSICS

AIP CONFERENCE PROCEEDINGS 248

COMPUTER-AIDED STATISTICAL PHYSICS

TAIPEI, TAIWAN 1991

EDITOR:
CHIN-KUN HU
INSTITUTE OF PHYSICS
ACADEMIA SINICA, TAIPEI

American Institute of Physics **New York**

L.C. Catalog Card No. 91-78378
ISBN 0-88318-942-9
DOE CONF-9106313

Printed in the United States of America.

CONTENTS

PREFACE

Taiwan has a comparatively short history of doing research on basic sciences and an even shorter history of doing research on statistical physics. To improve this situation, the Institute of Physics of Academia Sinica in Taipei has held conferences on statistical physics almost every year since 1984. These included the 1984 Workshop on Statistical Physics, the 1986 Summer School on Statistical Mechanics, the 1987 Summer Symposium on Statistical Mechanics, the 1988 Workshop on Statistical Mechanics, the 1989 Workshop on Statistical Mechanics, and the 1990 Taipei Workshop on Chaos and Critical Phenomena. In every conference, some foreign speakers were invited. The proceedings of the 1986 Summer School on Statistical Mechanics and the 1988 Symposium on Statistical Mechanics were edited by Chin-Kun Hu in the book *Progress in Statistical Mechanics*, which was published by the World Scientific Publishing Company, in 1988. Since all these conferences were intended mainly for researchers in Taiwan, they were not announced internationally.

In 1990, the building "Center for International Academic Activities" was completed. This Center is quite convenient for visitors to the Academia Sinica. With such nice facilities, we held the 1991 International Symposium on Statistical Physics. We also chose a specific subject: Computer-Aided Studies in Statistical Physics. We chose this subject because workstations and personal computers have become quite popular and inexpensive in recent years. It is no longer difficult for researchers to acquire very powerful computing facilities. Furthermore, there are many interesting problems in statistical physics that may be studied with the aid of computers. The conference was held on June 20–25, 1991. It was sponsored by the Institute of Physics and the Computing Center of Academia Sinica (Taipei), the National Science Council of the Republic of China (Taiwan), the Physics Research Promotion Center, and the Physical Society of the Republic of China (Taiwan). Besides expressing our thanks to the speakers and participants, we are also grateful to the following people for their contributions to the conference: Members of the International Advisory Committee: Professors K. Binder, B. Hu, L. P. Kadanoff, D. Kim, M. Suzuki, and F. Y. Wu; and members of the Local Organizing Committee: Professors H. H. Chen, L. J. Chen, P. M. Hui, S. C. Lin, S. P. Li, H. C. Tzeng and Drs. C.-N. Chen and W.-J. Tzeng.

The book *Computer-Aided Studies in Statistical Physics* contains invited and review papers presented at the 1991 Taipei International Symposium on Statistical Physics: Computer-Aided Studies in Statistical Physics. These papers are classified into the following categories: (1) lattice models; (2) phase transitions: statics and dynamics; (3) chaos, neural network, and self-organized criticality; and (4) quantum systems. We hope that this book will be useful for people working on statistical physics.

Chin-Ku Hu
Academia Sinica, Taipei,
September 1991

LATTICE MODELS

KNOT THEORY AND STATISTICAL MECHANICS

F. Y. Wu
Department of Physics, Northeastern University, Boston, Massachusetts 02115

ABSTRACT

Recent development in the mathematical theory of knots using the method of statistical mechanics is examined. We show that knot invariants can be obtained by considering statistical-mechanical models on a lattice. Particulary, we establish that the Kauffman's bracket polynomial is the partition function of a q-state vertex model previously considered by Perk and Wu, and that the Jones polynomial is generated by a q^2-state Potts model partition function. The generation of further new knot and link invariants may very well rely on computed-aided studies of solutions of certain Yang-Baxter equations.

INTRODUCTION

It is rare that frontiers of researches in two unrelated fields suddenly find themselves focussing on the same topic. But in recent years we have witnessed precisely one such happening. This is the exciting new development in knot theory in mathematics inspired by results in exactly solvable models in statistical physics[1]. Overviews of aspects of this new approach are now available,[2,3] but these papers have been written with the mathematics community in mind. In this lecture I shall give a brief description of the connection between knot theory and statistical mechanics using a language that is more familiar to physicists. I shall also draw attention to earlier works in statistical mechanics relevant to the new approach in knot theory.

We begin with a brief description of some terms in knot theory. For our purposes we define a knot as a set of nonintersecting loops in the three-dimensional space, which may be linked. Two knots are equivalent if they can be transformed into each other by a continuous deformation. Knots can always be projected onto a plane and represented by a planar diagram. Diagrams of several knots are shown in Fig. 1.

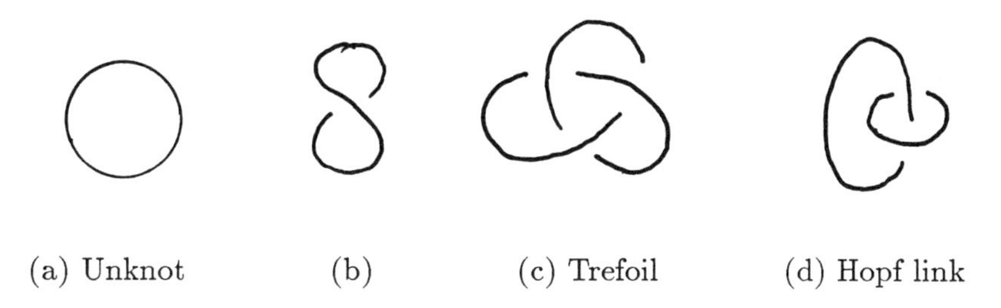

(a) Unknot (b) (c) Trefoil (d) Hopf link

Fig. 1. Examples of unoriented knots.

Two equivalent knots may have very different projections. For example, knots in (a) and (b) of Fig. 1 are both a single loop and hence are equivalent. Generally, it is not easy to tell whether two knots are equivalent by just looking at their projections. One central problem in the mathematical theory of knots has been the construction of algebraic quantities that are the same for equivalent knots. This leads to the notion of polynomial invariants.

It has been found that certain Laurent polynomials can be constructed for knots such that two equivalent knots share the same polynomial. Of fundamental importance is the polynomial discovered by Jones in 1984, the Jones polyno- mial.[4] Traditionally, the knot invariants are obtained via a recursion relation, the Skein relation,[5] using which one obtains a polynomial by starting from those of simpler knots. Particularly, one starts from the normalization that the invariant for an unknot of Fig. 1a is 1. The route of construction of invariants using the Skein relation is a complicated one, and one must also prove that the polynomial so obtained is well-defined and unique. The new approach using statistical mechanics circumvents these problems in a surprisingly simple fashion. For a given knot one constructs a vertex (or spin) model, and the desired knot invariant is given by its partition function! The construction of the polynomial then becomes straightforward, and its existence is by definition unique. Furthermore, different knot invariants can be generated from different statistical mechanical models.

A knot is oriented if its loops are directed. Thus, starting from a given unoriented knot, one can generate 2^ℓ distinct oriented knots, where ℓ is the number of loops of the knot. Since the Skein relation involves knots whose loops are oriented, one generally consider oriented knots. However, it turns out that in our considerations the orientation of a knot gives rise to only an overall multiplication factor to its invariant.[2,6] Therefore, we confine first to unoriented knots and return to the Skein relation and oriented knots below.

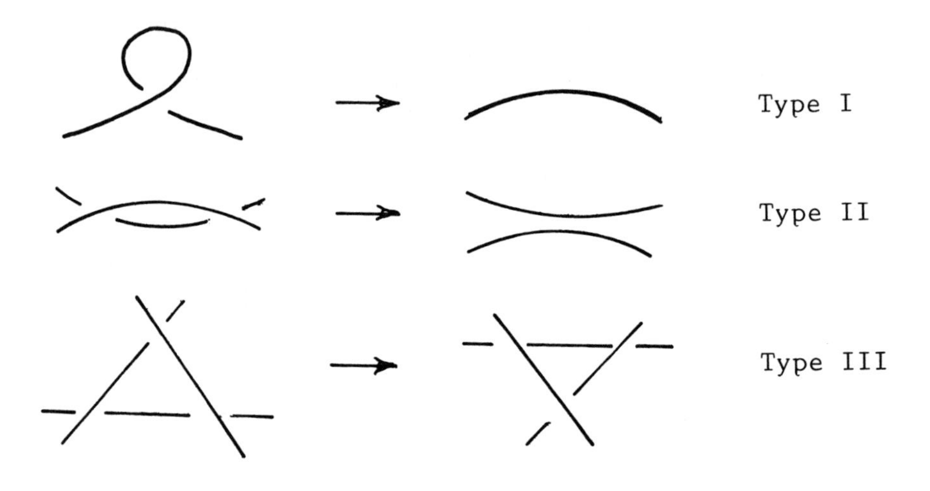

Fig. 2. Three types of Reidemeister moves.

Two equivalent knots can be deformed into each other by executing a sequence of well-defined operations known as the Reidemeister moves.[7] It has been found that three Reidemeister moves, shown in Fig. 2, are needed to generate equivalent knots. In the statistical mechanics approach, the requirement of knots invariance under the three Reidemeister moves leads to the Yang-Baxter equations.

MAPPING OF A KNOT ONTO A VERTEX MODEL

The main idea of the statistical mechanical approach is to formulate a one-one mapping of a knot onto a vertex (or spin) model. Regarding the loops of a knot as lattice edges, we obtain for each knot a lattice $\mathcal{L}$. We next introduce a vertex model on $\mathcal{L}$. For our purposes we consider a particular vertex model studied by Perk and Wu.[9,10] This is a nonintersecting string (NIS) model which, as we shall see, leads to the Jones polynomial.

The NIS model is a vertex model whose lattice edge can be in q different states. Denote the state of the four edges at a site by $\{a, b; c, d\} = \{1, 2, ..., q\}$ in the order shown in Fig. 3, and the vertex weight by $\omega(a, b; c, d) = \omega(d, c; b, a)$. Note that in Fig. 3 we have attached two dots to each site to remind us of the way the two lines interlace. Then the vertex weight of the NIS model is given by

$$\omega(a, b; c, d) = A\, \delta(a, c)\delta(b, d) + B\, \delta(a, b)\delta(c, d). \tag{1}$$

This leads to a $q(2q-1)$-vertex model whose allowed vertex configurations are shown in Fig. 4. We remark that the NIS model (1) is a subclass of a more general q-state vertex models studied previously in the context of obtaining exact solutions for the square lattice.[11–15] However, Perk and Wu considere *arbitrary* lattices and *arbitrary* weights[9] (thus making the model applicable to knots), and showed that the NIS model is equivalent to a Potts model. The NIS model partition function (2) was discovered independently by Kauffman[2,6] who called it the bracket polynomial, and applied by him to obtain the Jones polynomial.[6] The NIS model (1) has also been re-discovered by Lipson.[16]

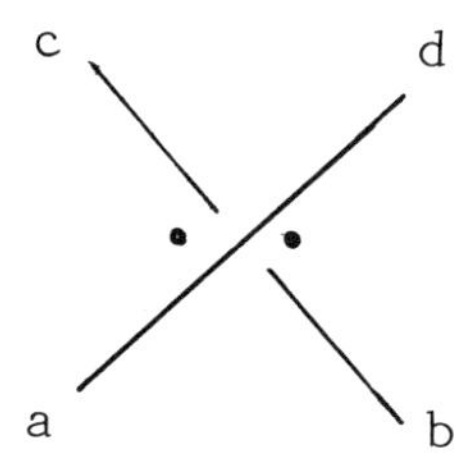

Fig. 3. Labelling of lattice edges.

The main idea is to rewrite the partition function $Z_{\rm NIS}$ from a sum-over-state into a global graphical summation. It is not difficult to see that $Z_{\rm NIS}$ is a polynomial in q, A, and B in the form of (Cf. Eq. (9) of Ref. 9)

$$Z_{\rm NIS}(q,A,B)=\sum_{a_i=1}^{q}\prod \omega(a_i,a_j;a_k,a_l)=\sum_P q^{p(P)}\prod_i \mathrm{W}_i(P). \tag{2}$$

Here, the summation is taken over all nonintersecting polygonal decompositions P of $\mathcal{L}$, $p(P)$ the number of polygons (loops) in P, and $\mathrm{W}_i(P)$ the weight of the i-th site, which is either A or B, depending on the particular decomposition taken at the site.

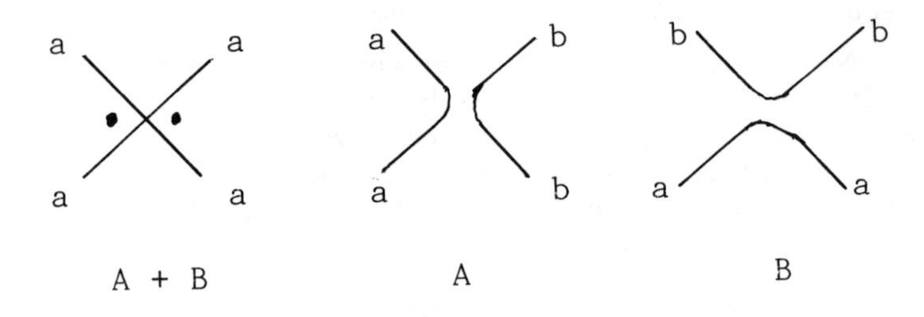

Fig. 4. Vertex configurations and weights of the NIS model ($a \neq b$).

Let $Z_{\rm NIS}(D)$ be the partition function of the NIS model for which a particular lattice site is fixed to be the configuration $D=\{D_+,D_-,D_0,D_\infty\}$ shown in Fig. 5. We then have the identities

$$\begin{aligned} Z_{\rm NIS}(D_+) &= AZ_{\rm NIS}(D_0)+BZ_{\rm NIS}(D_\infty)\\ Z_{\rm NIS}(D_-) &= BZ_{\rm NIS}(D_0)+AZ_{\rm NIS}(D_\infty). \end{aligned} \tag{3}$$

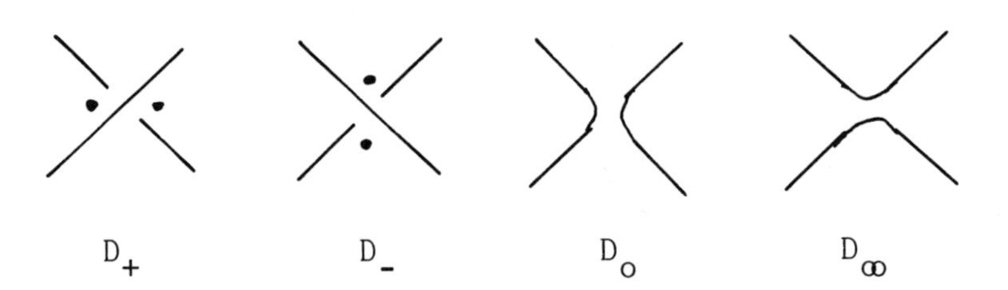

Fig. 5. Four configurations at a site for unoriented knots.

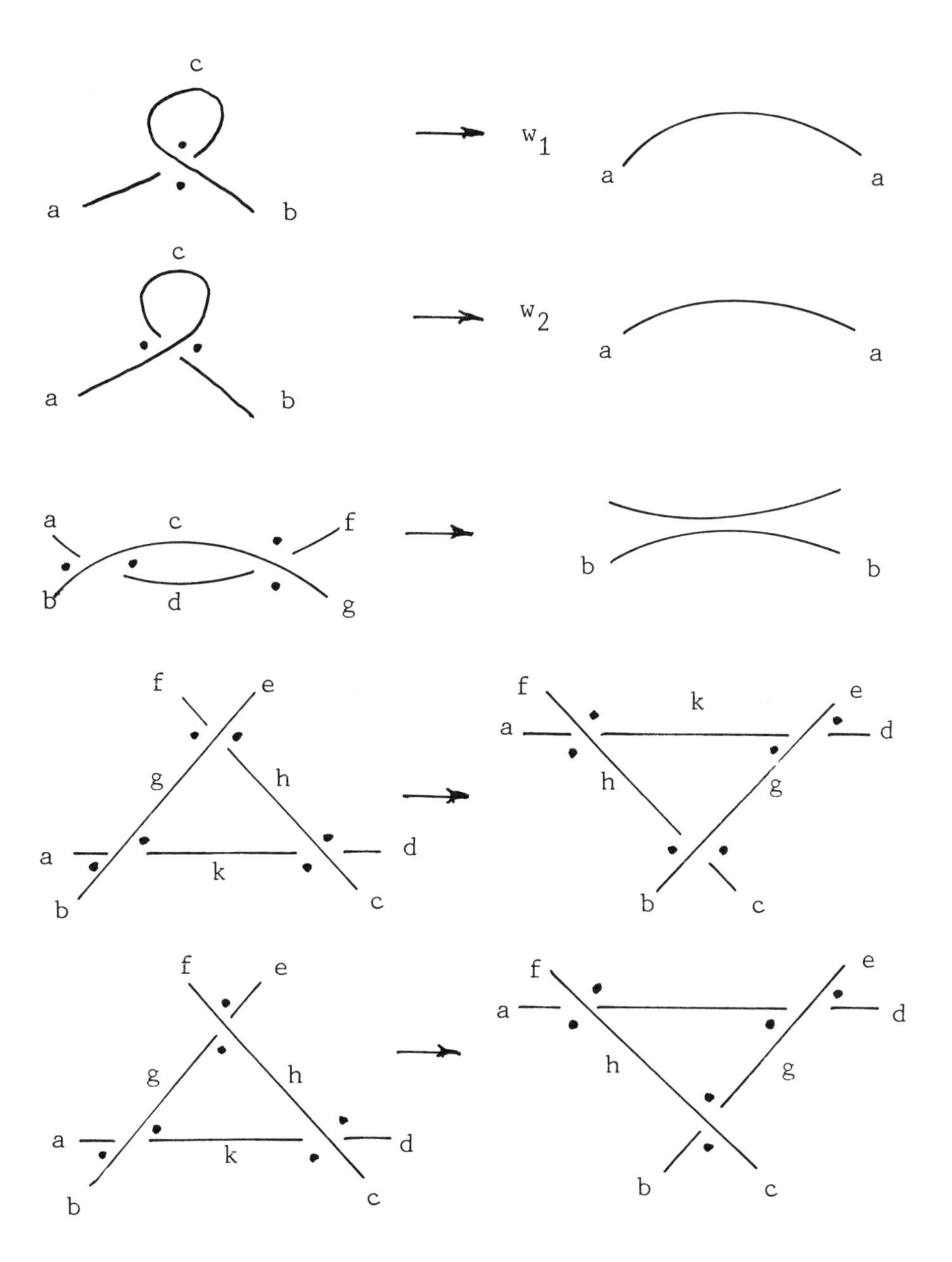

Fig. 6. Reidemeister moves for the NIS model.

INVARIANTS FOR UNORIENTED KNOTS

Our next task is to see whether it is possible to choose q, A and B so as to render the polynomial (2) invariant under the Reidemeister moves. While for the NIS model (1) this requirement, the Yang-Baxter equations, is most conveniently considered using graphical means[2,6,10], here we give a formulation of the Yang-Baxter equations for general vertex weights $\omega(a,b;c,d)$, thus allowing the possibility of extending to other vertex weights.

The three types of possible Reidemeister moves are shown in Fig. 6 with appropriate labellings of the lattice edges. Note that in the most general cases there are two type I moves, one type II move, and two type III moves. Explicitly, the Yang-Baxter equations for these moves can be read off from Fig. 6. They are:

$$\begin{aligned} &\sum_c \omega(c,a;c,b) = \mathrm{w}_1\delta(a,b) \\ &\sum_c \omega(a,b;c,c) = \mathrm{w}_2\delta(a,b), \qquad \text{Type I moves.} \end{aligned} \tag{4}$$

$$\sum_{cd} \omega(b,d;a,c)\omega(c,d;f,g) = \delta(a,f)\delta(b,g), \qquad \text{Type II move} \tag{5}$$

$$\begin{aligned} &\sum_{hgk} \omega(b,k;a,g)\omega(c,d;k,h)\omega(g,h;f,e) = \omega(b,c;h,g)\omega(g,d;k,e)\omega(h,k;a,f) \\ &\sum_{hgk} \omega(b,k;a,g)\omega(c,d;k,h)\omega(h,e;g,f) = \omega(h,b;g,c)\omega(g,d;k,e)\omega(h,k;a,f) \\ &\qquad \text{Type III moves.} \end{aligned} \tag{6}$$

Note that we have allowed the possibility that the two type I Reidemeister moves gives rise to multiplication factors w_1 and w_2. In addition, the application of a type I move to the knot in Fig. 1b transforms it to an unknot. Applying (4) to this move, *i.e.*, setting $a = b$ in (4) and summing over a, we find the 'partition function' of an unknot to be q.

The most general solution of (4) - (6) is not yet known. While computer-aided studies may very well play a role in the searching of general solutions, for our purposes it is sufficient to show that q, A, and B can be chosen so that (1) is a solution. Consider first (5). The substitution of (1) into (5) leads to

$$(A^2 + B^2 + qAB)\,\delta(f,g)\delta(a,b) + AB\,\delta(a,f)\delta(b,g) = \delta(a,f)\delta(b,g) \tag{7}$$

and hence we require

$$B = A^{-1}, \qquad q = -(A^2 + A^{-2}). \tag{8}$$

It is straightforward verify that, using (8), the Yang-Baxter equations (6) are also satisfied. Furthermore, we find from (4)

$$\begin{aligned} \mathrm{w}_1 &= qA + B = -A^3 \\ \mathrm{w}_2 &= qB + A = -A^{-3} = 1/\mathrm{w}_1. \end{aligned} \tag{9}$$

Thus, for each unoriented knot, we have a Laurent series in $A^{\pm 1}$

$$f(A) = \frac{1}{q} Z_{\mathrm{NIS}}(-A^2 - A^{-2}, A, A^{-1}) \tag{10}$$

which is invariant under type II and type III Reidemeister moves. Here we have included a factor $1/q$ in the right hand side of (10) so that the invariant of an unknot is mormalized to 1. However, the type I Reidemeister moves introduce an extra factor $\mathrm{w}_1^{\pm 1}$ to $f(A)$.

THE JONES POLYNOMIAL

We now return to oriented knots. Invariants for oriented knots are generated by using the Skein relation, starting from the invariant of an unknot normalized to 1. Jones[4] showed that there exists a Laurent polynomial $V(t)$ of $t^{\pm 1/2}$ satisfying the Skein relation

$$t^{-1} V_{L_+}(t) - t V_{L_-}(t) + (t^{-1/2} - t^{1/2}) V_{L_0}(t) = 0. \tag{11}$$

Here, $V_L(t)$ is the polynomial invariant of the knot when a particular intersection of the knot is fixed by the structure $L = \{L_+, L_-, L_0\}$ shown in Fig. 7. Let $n = n_+ - n_-$ where $n_\pm$ is the number of $L_\pm$ interactions in the knot. Then, following Kauffman,[2,6] we have the identities $A = t^{-1/4}$ and

$$V(t) = (-t^{3/4})^n f(t^{-1/4}). \tag{12}$$

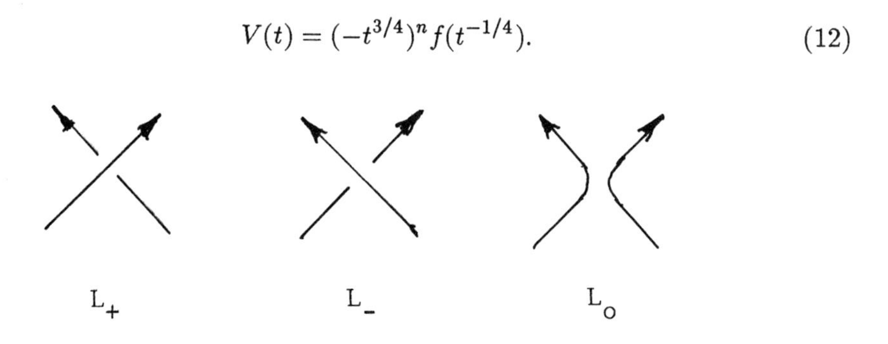

Fig. 7. Three configurations for oriented knots.

To see this, we observe from (3) and $B = A^{-1}$ that we have

$$AZ_{\mathrm{NIS}}(D_+) - A^{-1}Z_{\mathrm{NIS}}(D_-) = (A^2 - A^{-2})Z_{\mathrm{NIS}}(D_0). \tag{13}$$

Substituting with

$$Z_{\mathrm{NIS}}(D_s) = q(-t^{3/4})^{-n+s\times 1}V_{D_s}(t) \tag{14}$$

where $s = \pm, 0$, then (13) leads to the Skein relation (11) upon replacing $\{D_\pm, D_0\}$ by $\{L_\pm, L_0\}$. The Jones polynomial $V(t)$ for any given knot is now constructed from the partition function (10) and is thus well-defined.

EQUIVALENCE WITH A POTTS MODEL

The Jones polynomial $V(t)$ can also be generated by the partition function of a Q-state Potts model.[3] This connection follows from a complete equivalence between the q-state NIS model (1) and a q^2-state Potts model.[9] A convenient way to visualize this equivalence is to shade every other face of $\mathcal{L}$, a device introduced by Baxter, Kelland and Wu[17] in a consideration of the Potts model18 for arbitrary lattices. Then the Potts spins are located in the shaded faces and the Potts interactions go through sites of $\mathcal{L}$. The complete equivalence is given by (Cf. Eq. (5) of Ref. 8)

$$Z_{\mathrm{NIS}} = q^{-N} \sum_{\sigma_i=1}^{q^2} \prod_{<ij>} B(\sigma_i, \sigma_j), \tag{14}$$

where N is the number of Potts spins, and $B(\sigma_i, \sigma_j)$ is the Boltzmann factor

$$\begin{aligned} B(\sigma, \sigma') &= B_i + qA_i \qquad && \sigma = \sigma' \\ &= B_i && \sigma \neq \sigma'. \end{aligned} \tag{15}$$

Here $q = -(A^2 + A^{-2})$, $\{A_i, B_i\} = \{A^{-1}, A\}$ if the two dots at a site are in the shaded faces, and $\{A_i, B_i\} = \{A, A^{-1}\}$ if the two dots are in unshaded faces. Hence, there are two kinds of Potts interactions depending upon the relative positionings of the shaded faces and the dots, the latter denoting the way the two lines interlace at the site. It also clear that we have

$$Q = q^2 = (A^2 + A^{-2})^2 = 2 + t + t^{-1}. \tag{16}$$

These results are the same as those of Ref. 3.

This work is support in part by NSF Grants DMR-9015489 and INT-8902033.

REFERENCES

1. For a readable account of this development, see V. F. R. Jones, *Scientific American* (November 1990), p. 98.
2. L. H. Kauffman, *Am. Math. Monthly* **95**, 195 (1988).
3. V. F. R. Jones, *Pacific J. Math.* **137**, 311 (1989).
4. V. F. R. Jones, *Bull. Am. Math. Soc.* **12**, 103 (1985).
5. J. H. Conway, *An enumeration of knots and links*, Computational Problems in Abstract Algebra (Pergamon, N. Y. 1970) p. 329.
6. L. H. Kauffman, *Topolygy* **26**, 395 (1987).
7. K. Reidemeister, *Knotentheorie* (Chelsea, New York, 1948); English translation: *Knot Theory* (BSC Associate, Moscow, Idaho 1983).
8. V. G. Turaev, *Invent. Math.* **92**, 527 (1988).
9. J. H. H. Perk and F. Y. Wu, *J. Stat. Phys.* **42**, 727 (1986).
10. J. H. H. Perk and F. Y. Wu, *Physica* **138A**, 100 (1986).
11. Yu. G. Stroganov, *Phys. Lett.* **74A**, 116 (1979).
12. C. L. Schultz, *Phys. Rev. Lett.* **46**, 629 (1981).
13. J. H. H. Perk and C. L. Schultz, *Phys. Lett.* **84A**, 407 (1981).
14. J. H. H. Perk and C. L. Schultz, *Physica* **122A**, 50 (1983).
15. J. H. H. Perk and C. L. Schultz, in *Nonlinear Integrable Systems - Classical Theory and Quantum Theory*, Proc. RIMS Symposium, M. Jimbo and T. Miwa, eds. (World Scientific, Singapore, 1983).
16. A. S. Lipson, *Pacific J. Math.* to appear.
17. R. J. Baxter, S. B. Kelland, and F. Y. Wu, *J. Phys.* *A***9**, 397 (1976).
18. For a review on the Potts model and its physical significances see F. Y. Wu, *Rev. Mod. Phys.* **54**, 235 (1982).

PLANAR POLYGONS: REGULAR, CONVEX, ALMOST CONVEX, STAIRCASE AND ROW CONVEX.

A.J. Guttmann
Department of Mathematics, The University of Melbourne,
Parkville, Vic. 3052, Australia.

ABSTRACT

In recent years there has been renewed interest in a variety of models of planar polygons. These are combinatorial models of interest in their own right. In addition, they have application to such diverse areas as Computer Science and Polymer Chemistry. They also represent increasingly close approximations to the important unsolved problem of self-avoiding polygons. We review the known exact solutions, and discuss the known numerical results for the self-avoiding polygon problem. In addition we discuss the enumeration by area, instead of the more usual expansion parameter of perimeter, and discuss the two-variable generating function, expanded in terms of both area and perimeter.

1. INTRODUCTION

The enumeration of self-avoiding walks and polygons on regular lattices has been recognized as an extremely difficult problem for many years. Virtually the only exact (but not rigorous) numerical result is that of Nienhuis[1,2] who showed that the perimeter generating function for walks and polygons on the honeycomb lattice is singular at the point $x_c = (2 + \sqrt{2})^{-1/2}$, at which point the generating function behaves as $A(x_c - x)^{\lambda}$, where $\lambda = -43/32$ (walks) and $\lambda = 3/2$ (polygons). The difficulty of this problem has prompted the study of various restricted classes of walks and polygons in an effort to gain some insight into the structure of the unrestricted problem. These restricted models include convex polygons studied by Delest and Viennot[3], Guttmann and Enting[4], Lin and Chang[5], Kim[6], staircase and row-convex polygons by Brak and Guttmann[7], almost convex polygons by Enting *et al.*[8], spiral self-avoiding walks studied by Guttmann and

Wormald[9] and by Szekeres and Guttmann[10], as well as various directed walk models, discussed by Redner and Majid[11].

This study will review the known solved planar polygon problems, as well as the current status of the numerical work on the self-avoiding polygon problem.

2. STAIRCASE POLYGONS

Staircase polygons were first studied by Pólya[12] in 1969. Staircase polygons on the square lattice, with unit lattice spacing, may be most conveniently viewed as polygons convex with respect to a line at -45^{0} to the x-axis. This implies that such a line passing through the polygon has only one segment within the polygon. Pólya showed that, if a_{2n} denotes the number of such polygons with perimeter 2n, and $b_{m,2n}$ denotes the number of such polygons with area m and perimeter 2n, then

$$a_{2n} = \binom{2n}{n} / (4n - 2) \tag{2.1}$$

a result obtained by Delest and Viennot[3] in a particularly elegant form:

$$a_{2n+2} = C_n = \binom{2n}{n} / (n+1) \tag{2.2}$$

where C_n is the nth Catalan number. The generating function $P(x) = \sum a_{2n} x^n$ is therefore given by

$$P(x) = -\frac{1}{2} [(1-4x)^{1/2} - 1 + 2x].$$

The two variable generating function is

$$\sum_{n=1}^{\infty} \sum_{m} b_{mn} q^m x^n = \ldots + q^{-1}x^2 + 2x + qx^2 + 2q^2x^3 + (q^4 + 4q^3)x^4 + \ldots$$

$$= 1 - \frac{1}{1 + P_1(q)x + P_2(q)x^2 + P_3(q)x^3 + \ldots}, \tag{2.3}$$

where, by definition,

$$b_{01} = 2, \qquad b_{-mn} = b_{mn},$$

$$P_n(q) = \sum_{r=0}^{n} \begin{bmatrix} n \\ r \end{bmatrix}^2 q^{-r(n-r)}, \tag{2.4}$$

and

$$\begin{bmatrix} n \\ r \end{bmatrix} = \frac{(1 - q^n)(1 - q^{n-1})\ldots(1 - q^{n-r+1})}{(1 - q)(1 - q^2)\ \ldots\ (1 - q^r)} \tag{2.5}$$

Pólya pointed out that while the binomial coefficient $\binom{n}{r}$ is well-known to be the number of distinct shortest staircase paths on the lattice joining (0,0) to (r, n–r), it is less well-known that the corresponding Gaussian binomial coefficients $\begin{bmatrix} n \\ r \end{bmatrix}$, which are polynomials in q, generate the area under such staircase paths. The polynomial coefficient mq^n denotes that there are m distinct paths enclosing an area n.

An alternative derivation was given by Brak and Guttmann[7], who used a method due to Temperley[13] to obtain the two-variable area and perimeter generating function for staircase polygons, as follows:

Let $G(y,z) = \sum_{n\geq 1} \sum_{m\geq 1} c_{n,m} y^n z^m$ be the generating function for the $c_{n,m}$ staircase polygons with perimeter n and area m.

Let h_n be the generating function of staircase polygons, on the square lattice, whose first row contains exactly n squares. Then h_n satisfies the following equations

$$h_1 = zy^4+y^2z[h_1+h_2+h_3+h_4+\ldots]$$

$$h_2 = z^2y^6+y^2z^2[y^2h_1+(1+y^2)h_2+(1+y^2)h_3+(1+y^2)h_4+\ldots]$$

$$h_3 = z^3y^8+y^2z^3[y^4h_1+(y^2+y^4)h_2+(1+y^2+y^4)h_3+(1+y^2+y^4)h_4+\ldots] \tag{2.6}$$

$$h_4 = z^4y^{10}+y^2z^4[y^6h_1+(y^4+y^6)h_2+(y^2+y^4+y^6)h_3+(1+y^2+y^4+y^6)h_4+...]$$

from which the following recurrence relation may be derived:

$$h_{n+2} - z(1 + y^2 - y^2z^{n+1})h_{n+1} + y^2z^2h_n = 0. \tag{2.7}$$

This recurrence relation is not a difference equation with constant (i.e. n independent) coefficients and thus not amenable to the 'text book' methods of solution. We try the following ansatz[28]:

$$h_n = \lambda^n \sum_{m=0}^{\infty} r_m(z)z^{mn} \tag{2.8}$$

where $r_m(z)$ is an arbitrary function of z for $m > 0$ and $r_0 = 1$ (in order to obtain the correct $z \to 0$ limit). Substituting (2.8) into (2.4) and rearranging gives

$$\lambda^2 - z(1 + y^2)\lambda + z^2y^2$$
$$+ \sum_{m=1}^{\infty} [y^2\lambda z^{1+m}r_{m-1} + r_m(\lambda^2z^{2m} - \lambda z(1+y^2)z^m + z^2y^2)]z^{mn} = 0. \tag{2.9}$$

We then choose λ and r_m such that

$$\lambda^2 - z(1+y^2)\lambda + z^2y^2 = 0 \tag{2.10}$$

and

$$y^2\lambda z^{1+m}r_{m-1} + r_m(\lambda^2z^{2m} - \lambda z(1+y^2)z^m + z^2y^2) = 0. \tag{2.11}$$

Equation (2.10) has two solutions $\lambda_1 = z$ and $\lambda_2 = zy^2$. As (2.11) is just a first-order difference equation for r_m, it is readily solved to give

$$r_m = \prod_{k=1}^{m} \frac{-y^2\lambda z^{k+1}}{\lambda^2z^{2k}-\lambda z(1+y^2)z^k+y^2z^2} \qquad z \neq 1$$

$$= \frac{(-zy^2\lambda)^m z^{m(m+1)/2}}{\prod_{k=1}^{m} (\lambda z^k - zy^2)(\lambda z^k - z)}. \tag{2.12}$$

Thus the general solution is $h_n = A_1 h_n^{(1)} + A_2 h_n^{(2)}$, where

$$h_n^{(i)} = \lambda_i^n + \lambda_i^n \sum_{m=1}^{\infty} \frac{(-zy^2\lambda_i)^m z^{m(m+1+2n)/2}}{\prod_{k=1}^{m} (\lambda_i z^k - zy^2)(\lambda_i z^k - z)} \tag{2.13}$$

and when A_1 and A_2 are arbitrary functions of y and z determined by the initial conditions.

The generating function G(y,z) is given by

$$G(y,z) = \sum_{n=1}^{\infty} (A_1 h_n^{(1)} + A_2 h_n^{(2)}). \tag{2.14}$$

In the limit $y \to 0$, $z = 1$ we have $\lambda_1 \sim O(1)$ and $\lambda_2 \sim O(y^2)$ whilst $h_n \sim O(y^{2n+2})$. This implies that $A_1 = 0$. The second constant A_2 is obtained by substituting h_2 and h_1 obtained from (2.13) into the recurrence relation (2.7) (with $n = 0$, $h_0 \equiv y^2$) and solving for A_2. If this is done and the result substituted into (2.14) then, after some rearranging the following result is obtained

$$G(y,z) = \frac{y^2}{\Delta}\left(\frac{y^2 z}{1 - y^2 z} + \sum_{m=1}^{\infty} \frac{y^2 z^{m+1}}{1 - y^2 z^{m+1}} R_m z^{m(m+1)/2}\right) \tag{2.15}$$

where

$$\Delta = 1 - y^2 z + \sum_{m=1}^{\infty} (1 - y^2 z^m + y^2(1-z))R_m z^{m(m+3)/2} \tag{2.16}$$

and

$$R_m = \frac{(-y^2)^m}{\prod_{k=1}^{m} (1 - z^k)(1 - y^2 z^k)}. \tag{2.17}$$

We note that the form of (2.15) is very different to that given by Pólya's result. His result is also invariant under the transformation $z \to 1/z$ (i.e. it generates negative powers of z). It would be of some interest to relate the two results, particularly as the Pólya result contains the more 'natural' Gaussian polynomials which generate areas under 'staircase' paths.

The perimeter generating function behaves like

$$P(x) = \sum a_{2n} x^n \sim \frac{-1}{2} (1 - 4x)^{1/2}$$

where a_{2n} is the number of polygons with perimeter 2n. The "critical point" is seen to be at $x_c = 1/4$, with "critical exponent" equal to 1/2, corresponding to a cusp-like singularity.

The area generating function A(y) is found to be

$$A(y) = \sum b_n y^n \sim A(1 - y/y_c)^{-1}$$

where b_n is the number of staircase polygons of area n, and $y_c = 0.4330619...$.

3. CONVEX POLYGONS

Convex polygons may be defined as self-avoiding polygons whose perimeter is equal to the perimeter of the minimum bounding rectangle. As a consequence, any vertical or horizontal line that perpendicularly bisects any edge must bisect exactly one other edge. We first consider the perimeter distribution function. This problem was first solved by Delest and Viennot[3], who used the method of Schützenberger[14] to express the perimeter generating function as an algebraic language.

In that method one first encodes the objects to be enumerated by the words of an algebraic language. Next, one writes down a non-ambiguous grammar to generate the language. This grammar is expressible as an algebraic system, which is then solved to give an explicit or implicit generating function for the words of the grammar.

Finally, one obtains an exact or asymptotic expansion for the number of objects in the set.

In this way it was shown[3] that

$$P(x) = \sum p_{2n} x^n = x^2(1 - 6x + 11x^2 - 4x^3)/(1 - 4x)^2 - 4x^4/(1 - 4x)^{3/2} \quad (3.1)$$

and hence that

$$p_{2n+8} = (2n + 11).2^{2n} - 4(2n + 1)\binom{2n}{n} \quad n \geq 0. \quad (3.2)$$

where p_{2n} is the number of distinct 2n-step polygons. Note that this shows that the "critical point" is $x_c = 1/4$ and the "critical exponent" is -2, corresponding to the double pole at $x = x_c$. This result was independently obtained by three different methods[4,5,6] in 1988.

Enting and Guttmann[4] enumerated the convex polygons to 64 steps, and found a recurrence relation among the coefficients, from the coefficients up to 24 steps. This recurrence relation was then found to generate all the other known coefficients, and so was conjectured to be exact. The generating function (3.1) was thereby obtained.

Lin and Chang[5], who were also unaware of the earlier work[3], proved that this was so. In their method, they showed that any convex polygon can be divided into two polygons with a horizontal cutting line through two vertices. The line is so chosen that the top polygon has width *at the bottom* equal to the width of the bounding rectangle of the top polygon. The bottom polygon has the property that its north-easterly vertex is also the north-easterly vertex of its bounding rectangle. Lin and Chang found the generating functions for these two classes of polygons, and then calculated the number of distinct ways a member of each class could be joined along a common cutting line. In this way they verified the above result, and also extended it to distinguish between the number of horizontal and vertical steps in the polygon. That is to say, they obtained a two-variable generating function.

Kim[6] took a different approach. Drawing the minimum bounding rectangle around any convex polygon, it is clear that there is precisely one segment of the perimeter coincident with each of the four edges of the bounding rectangle. These four segments are joined by "staircase-like" paths. Now the unrestricted number of such staircase paths is given by a binomial coefficient. Kim showed that the construction of a convex polygon by the concatenation of the four linear segments and the four staircase segments could be simply expressed by binomial coefficients, and it then remained to subtract all possible overlaps of the staircase segments. This proved to be possible using the earlier results of Gessel and Viennot[15]. Simplification of the various binomial sums then gave the results (3.1) and (3.2).

Using the same numerical approach as for square lattice convex polygons described above, Enting and Guttmann[16] obtained the generating function for convex polygons on the "brickwork" lattice (isomorphic to the honeycomb lattice, though "convexity" has no natural interpretation on the honeycomb lattice[16]). They found that the exactly enumerated coefficients satisfied the generating function

$$P(x) = \sum p_{2n} x^n = x^6(1 - 2x + x^2 - x^4)/[(1+x)^2(1 - 2x)^2]$$

$$- x^8(1 + 2x)^{1/2}/[(1 + x)^2(1 - 2x)^{3/2}] \qquad (3.3)$$

Note that this shows that the "critical point" is $x_c = 1/2$ and the "critical exponent" is again -2, corresponding to a double pole at $x = x_c$, as for the square lattice case. There is also a non-physical double pole at $x = -1$. This result was also proved by Lin and Chang[5].

An alternative generating function, in terms of area rather than perimeter, was first considered by Klarner and Rivest[17], who proved that $\lim_{n\to\infty} (c(n))^{1/n} = 2.309138..$, where c_n is the number of distinct convex polygons with area n. This result was improved by Bender[16], who showed that $c_n \sim 2.67564..(2.30913859330)^n$. Remarkably[16], this is the same exponential growth factor as for staircase polygons by area. Both these authors based their calculation on a subdivision of the polygon into three pieces by two horizontal lines. The central piece of the polygon is a staircase polygon, while the top and bottom pieces are "stack" or "pyramid" polygons, in which each successive row is is contained within the width of the row underneath (or on top of) the preceding row. This three section subdivision was also used by Lin[23] in his calculation of the three-variable generating function in terms of area and perimeter, as discussed below.

Finally, in 1991 two authors[21-23] have solved the full, two-variable generating function for convex polygons by both area and perimeter. Earlier partial solutions to this problem included the calculation of the first and second area-weighted moments of the perimeter generating function, and were given by Enting and Guttmann[19], whose conjectured results were proved by Lin[20], who gave an algorithm for producing the rth such moment, and who explicitly gave the first 10 such moments. M. Bousequet-Melou[21,22] and Lin[23] in fact obtained the three variable generating function that includes the distinction between horizontal and vertical steps.

M. Bousequet-Melou's results may be stated as follows, where as usual

$$(a)_n = (1 - a)(1 - aq)...(1 - aq^{n-1}), \quad (a;q)_\infty = \prod_{n=0}^{\infty} (1 - aq^n),$$

$(a;q)_n = (a;q)_\infty/(aq^n;q)_\infty$ and $\begin{bmatrix} n \\ k \end{bmatrix}_q = (q;q)_n/[(q;q)_k(q;q)_{n-k}]$ denotes the Gaussian q-binomial coefficient. Let the number of horizontal bonds be 2n, the number of vertical bonds 2m, and the area k. Then the generating function

$$P(x,y,q) = \sum_{n\ge 0} \sum_{m\ge 0} \sum_{k\ge 0} c_{n,m,k} x^{2n} y^{2m} q^k = 2yV(R - \hat{N})/N - 2yM - B$$

where

$$N = \sum_{n\ge 0} \frac{(-1)^n x^n q^{\binom{n+1}{2}}}{(q)_n (yq)_n}, \qquad \hat{N} = \sum_{n\ge 1} \frac{(-1)^n x^n q^{\binom{n+1}{2}}}{(q)_{n-1} (yq)_n}$$

$$R = y \sum_{n\geq 2} \frac{x^n q^n}{(yq)_n} \sum_{m=0}^{n-2} \frac{(-1)^m q^{\binom{m+2}{2}}}{(q)_m (yq^{m+1})_{n-m-1}} ,$$

$$V = \sum_{m\geq 0} xq^{m+1} \frac{T_m N_{m+1}}{(xq)_m}, \quad M = \sum_{m\geq 0} \sum_{0\leq n\leq m} N^{n+1}_{m+1} \frac{T_m T_n}{(xq)_m (xq)_n},$$

$$B = \sum_{n\geq 1} \frac{xy^n q^n (T_n)^2}{(xq)_{n-1}(xq)_n}$$

$$T_n(x) = 1 + \sum_{k=1}^{n/2} x^k q^{k^2} \left\{ \sum_{j=0}^{n-2k} \binom{n-k-1}{k-1}_q \binom{k+j}{k}_q \right\} ,$$

$$N_n(x) = \sum_{k=0}^{n} (-x)^k q^{k(k+1)/2} \left\{ \sum_{s=0}^{n-1-k} \binom{k+s}{k}_q \binom{n-s-1}{k}_q y^s \right\} .$$

The N^n_m are deduced from the $N_n(x)$ as follows: $N^n_m = 0$ if $m < n$;

$$N^n_n = -xy^{n-1}q^n, \quad N^n_m = x^2 y^{n-1} q^{n+m} N_{n-m}(xq^n) \quad \text{if } m > n.$$

Lin's approach is to break the polygon up into three pieces[17,18], calculate the generating function for each piece, and then calculate the combinatorics of the number of ways of putting the pieces together. Following Lin[23], let

$$G = x^2 \sum_{n=1}^{\infty} \frac{(y^2 z)^n (1 - x^2 z^n)}{\prod_{k=1}^{n} (1 - x^2 z^k)^2}$$

and

$$g_m = x^{2m} \sum_{n=1}^{\infty} (y^2 z)^n q_n^{-2} [t_{m-1,n} - (2+z^n) t_{m-2,n} + (1+2z^n) t_{m-3,n} - z^n t_{m-4,n}]$$

where

$$t_{k,n} = \sum_{r=0}^{k} q_{n+r}(z)q_{n+k-r}(z)/q_r(z)q_{k-r}(z) \qquad k \geq 0$$

$$= 0 \qquad k < 0$$

and $$q_n(z) = \prod_{k=1}^{n} (1 - z^k) \qquad n > 0$$

$$= 1 \qquad n = 0.$$

Let $S_m = \sum_{n=1}^{m-2} g_n x^{-2n}(m-n-1)$, with $S_2 = 0$

and $$h_n = y^2(x^2z)^n \, [1 + \sum_{m=1}^{\infty} R_m z^{m(m+1+2n)/2}]/[1 + \sum_{m=1}^{\infty} R_m z^{m(m+1)/2}]$$

where

$$R_m = \frac{(-y^2)^m}{\prod_{r=1}^{m} (1 - z^r)(1 - x^2z^r)} \qquad \text{with } R_0 = 1.$$

Also, let $h_{m,n} = h_{n,m} = A[h_m(h'_n - h_n)] + \delta_{m,n} y^2 z^n x^{2n} \qquad m \geq n$

where $$h'_n = y^2z^n[1 + \sum_{m=1}^{\infty} R'_m z^{m(m+1+2n)/2}]/[1 + \sum_{m=1}^{\infty} R'_m z^{m(m+1)/2}]$$

and $$R'_m = (-y^2)^m / \prod_{r=1}^{m} [(1-z^r)(x^2-z^r)]$$

and $A = y^2z/(h'_1 - h'_1)$.

Finally let

$$T_m = \sum_{n=1}^{m-1} x^{-2n} \sum_{p=0}^{\infty} h_{n+p}$$

$$= \frac{y^2[\sum_{k=0}^{\infty} R_k z^{k(k+1)/2}(z^{k+1} - z^{m(k+1)})/(1-x^2 z^{k+1})(1-z^{k+1})]}{\sum_{k=0}^{\infty} R_k z^{k(k+1)/2}}$$

then

$$P(x,y,z) = G + \sum_{m=3}^{\infty} g_m S_m + 2y^2 \sum_{m=2}^{\infty} g_m \sum_{n=1}^{m-1} x^{-2n} \sum_{r=n}^{\infty} z^r x^{2r} S_{r+1}$$

$$+ 2(\sum_{m=2}^{\infty} g_m T_m)[1 + A \sum_{r=2}^{\infty} S_{r+1}(h'_r - h_r)]$$

$$+ 2A \sum_{m=2}^{\infty} g_m \sum_{n=1}^{m-1} x^{-2n} \sum_{r=n+1}^{\infty} S_{r+1} \sum_{p=0}^{r-n-1} (h_r h'_{n+p} - h'_r h_{n+p}).$$

4. ROW CONVEX POLYGONS

Row convex polygons are polygons convex with respect to only one axis. They appear to have been first discussed by Temperley[13], who referred to them as "Model Q". Temperley used the method given above for staircase polygons, and obtained a recurrence relation for the three-variable generating functions $g_n(x,y,z)$, where g_r is the generating function of row-convex polygons whose first column contains exactly r squares. The variables x,y,z are conjugate to the number of horizontal steps, the number of vertical steps, and the area respectively. Temperley showed that the problem simplified greatly if one sets y=1, so that the generating function by area and number of columns is obtained. Simplifying this further by setting x=1 gives the generating function by area alone. This is

$$G(z) = z(1 - z)^3/[1 - 5z + 7z^2 - 4z^3] \qquad (4.1)$$

This generating function diverges at $z_c = 0.3119570553$, at which point it has a simple pole. This result was independently re-derived by Klarner[24] in 1965, and by Klarner[25] again in 1967 using a different method, by Polya[12] in 1969, and extended by Delest[26] in 1988, who obtained the generating function by perimeter, a result previously given in implicit form by Temperley[13]. Delest's explicit result may be stated as follows:

Let $P(t) = \sum_{n>0} p_n t^n$ be the generating function for the number of column-convex polygons having perimeter 2n+2. Then

$$P(t) = [-(AC^{1/3} + D + EC^{-1/3}) - F]^{1/2}/[2(AH)^{1/2}] - H/[2(2A)^{1/2}] - G \tag{4.2}$$

where

$$A = 18t^4(2t^3 - 23t^2 + 38t - 18)^2,$$

$$B = (t^2 - 38t + 1)(t - 1) - 6(t^2 - 6t + 1)\sqrt{-3t},$$

$$C = \frac{2(t-1)^9(t+1)^3 B}{3At^2},$$

$$D = -3t^2(t-1)^4(t+1)(11t^3 + 49t^2 - 439t + 171), \tag{4.3}$$

$$E = 2(t-1)^6(t+1)^2(t^2 + 10t + 1),$$

$$F = 81t^3(t-1)^5(t+1)^2(t^2 - 6t + 1)(t^3 - 79t^2 + 163t - 81),$$

$$G = -\frac{(t-1)(5t^3 - 25t^2 + 47t - 21)}{4t(2t^3 - 23t^2 + 38t - 18)},$$

$$H = \sqrt{\frac{2AC^{2/3} - DC^{1/3} + 2E}{C^{1/3}}}.$$

In 1990 Brak *et al.*[27] solved Temperley's implicit recurrence explicitly, and obtained for $G(y) = y^2P(y^2)$ the result

$$G(y) = \frac{1}{\Delta}\Big\{ (y^2 - 1)(-21 + 47y^2 - 35y^4 + 5y^6)$$

$$- 3(y^2 - 1)^2(1 + y^2)(1 - 6y^2 + y^4)^{1/2}$$

$$- 9\sqrt{2}(y^2 - 1)^2[(y^4 - 1)(y^2 - 1) - (y^4 - 1)(1 - 6y^2 + y^4)^{1/2}]^{1/2}$$

$$- \sqrt{2}\, y(y^4 - 1)[(y^4 - 1)(y^2 - 1) + (y^4 - 1)(1 - 6y^2 + y^4)^{1/2}]^{1/2} \Big\} \tag{4.4}$$

where

$$\Delta = 4(18 - 38y^2 + 23y^4 - 2y^6). \tag{4.5}$$

from which they showed that

$$p_{n-1} \sim (3 + 2\sqrt{2})^{n-1/2} n^{-3/2} [c_0 + c_1/n + O(n^{-2})] \tag{4.6}$$

and gave explicit expressions for c_0 and c_1. In a subsequent calculation, Brak and Guttmann[7] obtained the two variable generating function, using Temperley's method, and an *ansatz* of Gautschi[28] to solve the recurrence relations. If $c_{n,m}$ is the number of row-convex polygons with perimeter n and area m, the generating function $G(y,z) = \sum c_{n,m} x^{2n} y^m$ is

$$G(y,z) = \sum_{m=0}^{\infty} R_m z^{m(m+2)} \frac{y^2 z}{1 - y^2 z^{m+1}} [A_1 + A_2((1 - y^2 z^{m+1})^{-1} + 2S_m)] \tag{4.7}$$

where

$$A_1 = \frac{1}{\Delta} \sum_{m=0}^{\infty} R_m z^{m(m+2)} [y^2(y^2 - 1)\, Q_m^{(4)} - y^2 Q_m^{(2)}]$$

$$A_2 = \frac{1}{\Delta} \sum_{m=0}^{\infty} R_m z^{m(m+2)} [y^2 Q_m^{(1)} - y^2(y^2 - 1)\, Q_m^{(3)}]$$

$$\Delta = \sum_{m=0}^{\infty} \sum_{l=0}^{\infty} R_m R_l [Q_m^{(1)} Q_l^{(4)} - Q_m^{(2)} Q_l^{(3)}]\, z^{m(m+2)+l(l+2)}$$

$$Q_m^{(1)} = y^2 z^m - 1 + y^2 z(1 - 2y^2)/T_m$$

$$Q_m^{(2)} = 2S_m Q_m^{(1)} + 2y^2 z^m - 1 + y^2 z(1 - 2y^2)/T_m^2 \tag{4.8}$$

$$Q^{(3)}_m = 1 - y^2 z/T^2_m$$

$$Q^{(4)}_m = 2S_m(1 - y^2 z/T^2_m) + 1 - y^2 z(1 + y^2 z^{m+1})/T^3_m$$

$$T_m = 1 - y^2 z^{m+1}$$

$$S_m = \sum_{k=1}^{m} \frac{(1 - y^2 z^{2k})}{(1 - z^k)(1 - y^2 z^k)}$$

$$R_m = \frac{y^{2m}(1 - y^2)^{2m}}{\prod_{k=1}^{m} (1 - y^2 z^k)^2 \prod_{k=1}^{m} (1 - z^k)^2}.$$

This result was generalized to the rectangular lattice by Lin and Tzeng[34]. The honeycomb lattice version of this problem was solved by Lin and Wu[29], using an adaptation of Temperley's method. If x,y,z denotes the three directions on the honeycomb lattice, and $p_{l,m,n}$ denotes the number of polygons with l,m,n edges in the x,y,z, direction respectively, convex with respect to an axis perpendicular to the x-direction, then they found the generating function of such polygons G(x,y,z). In the symmetric case x=y=z their result simplifies to

$$G(x) = \sum c_n x^{2n} = (1 - x)^4 N/4x^2(2 - x^2 + x^4)D \qquad (4.9)$$

where

$$N = a + bS^{1/2} + cT^{1/2} + d(ST)^{1/2}$$

$$D = 2 + 8x^2 + 5x^4 - 16x^6 - 18x^8 + 8x^{10} + 12x^{12}$$

$$S = (1 + x^2)(1 - 3x^2)$$

$$T = (1 - x^4)[(1 - x^2)(2 + 6x^2 + x^4 + x^6) + 2(1 + x^2)S^{1/2}]$$

$$a = 2 + 15x^2 + 17x^4 - 36x^6 - 38x^8 + 33x^{10} + 7x^{12} - 16x^{14} + 24x^{16}$$

$$b = 2 + x^2 - 10x^4 - 4x^6 + 8x^8 - x^{10} + 4x^{14} \qquad (4.10)$$

$c = -1 - 7x^2 - 7x^4 + 7x^6 + 8x^8$

$d = -1 + 3x^4 - 4x^8.$

The singularity structure is similar to that for row-convex polygons on the square lattice. The critical point is at $x_c^2 = 1/3$, with a "cusp-like" square root singularity. The asymptotic form of the coefficients was also given as

$$a_n \sim 3^n n^{-3/2}[A + B/n + O(n^{-2})],$$

where A and B are constants expressible in terms of square roots of integers and of π.

5. ALMOST-CONVEX POLYGONS

Very recently, Enting *et al.*[8] introduced a class of polygons referred to as "almost-convex". As well as their intrinsic interest, they appear to offer a systematic route to go from convex polygons to the unrestricted polygon problem, by systematically increasing the concavity measure. They are also of practical value in providing "correction terms" that enable us to extend the finite lattice enumerations of the unrestricted problem[30]

We define a concavity index for polygons on the square lattice by associating with each 2n-step polygon a minimal bounding rectangle of side $i \times j$. For each rectangle of dimension (i,j) we denote the number of 2n-step polygons by $u_{n;i,j}$. This will be zero for $n < i + j$. The concavity measure is $m = n - i - j$ and the number of 2n-step polygons with concavity measure m is denoted by $N_{m,n}$. Enting *et al.* obtained expansions of the generating function for polygons with perimeter up to about 60 steps for m=1,...,10. They also proved the following two theorems:

Theorem 1.

If $m=o(n^{2/3})$, then

$$N_{n,m} \sim 2^{2n-m-7} n^{m+1} \exp(m^2/2n)/m! \qquad (5.1)$$

Theorem 2.

If m=o(n), then the radius of convergence of the generating function is 1/4.

The proofs of the theorems are quite elaborate, requiring some sixteen printed pages. The underlying method is to first apply a sequence of rotations and reflections to the polygon to transform it into a self-avoiding walk. The "non-convex" steps are treated as perturbations, and the number of ways m perturbations can be inserted without the walk undergoing self-interesction was calculated. There is a close formal similarity between this method and the algebraic language method of Delest and Viennot[3].

From a careful study of the exact enumerative data and the exact result for the case m=0 (which corresponds to convex polygons), it was conjectured that

$$N_{n,m} \sim 2^{2n-m-7} n^{m+1} \exp(m^2/2n)/m! \; [1 - 4/(n\pi)^{1/2} + O(1/n)]. \qquad (5.2)$$

Thus we see that the radius of convergence is unchanged from that of convex polygons, through the critical exponent increases by 1 for each increase in concavity measure. Of course, if m was not restricted to o(n), the radius of convergence must approach that of the "true" self-avoiding polygon problem. No results are known for this case however.

6. SELF-AVOIDING POLYGONS

For self-avoiding polygons, the only exact (though non-rigorous) results are those of Nienhuis[1,2], who showed that the perimeter generating function for self-avoiding polygons on the honeycomb lattice has radius of convergence $x_c = (2 + \sqrt{2})^{-1/2}$, and that the number of 2n step polygons on the honeycomb lattice, $p_{2n} \sim n^{-5/2}/x_c^{2n}$. The best numerical results for other lattices come from series analysis studies. For the triangular, square and honeycomb lattices the number of polygons is known up to perimeter 35, 56 and 82 respectively[31,30,19]. Analysis of these series by the method of differential approximants permits us to conjecture that $p_{2n} \sim n^{-5/2}/x_c^{2n}$ for all three lattices, where x_c(triangular) $\approx$ 0.2409185 and x_c(square) $\approx$ 0.3790523. The numerical results are in complete agreement with Nienhuis's results for the honeycomb lattice.

The generating function by area is known to 21 terms[32], and series analysis suggests that the generating function has radius of convergence $y_c \approx 0.251834$, and that the number of polygons of area n, $a_n \sim n^{-1}/y_c^{\,n}$. The two-variable generating function is also given by Enting and Guttmann[32], and an analysis of this generating function is reported both there and in more detail by Fisher *et al.*[33]

7. SUMMARY

It has recently been shown[35] that most exactly solvable two-dimensional lattice models satisfy an *algebraic* equation with polynomial coefficients. That is, the various generating functions satisfy an equation of the form

$$\sum_{n=0}^{k} P_n(x)\, f(x)^n = 0 \qquad (7.1)$$

where the $P_n(x)$ are polynomials, and f(x) is the generating function for the system. If the degree of the polynomial $P_i(x)$ is α_i, we denote the algebraic equation by $[\alpha_k, \alpha_{k-1}, \dots, \alpha_1, \alpha_0]$. For example, the generating function of staircase polygons by perimeter,

$$P(x) = [(1 - 4x)^{1/2} - 1 + 2x]/2 \qquad (7.2)$$

satisfies the [0,1,2] algebraic equation

$$P(x)^2 - (1 - 2x)\,P(x) + x^2 = 0. \tag{7.3}$$

This seems to be a convenient and illuminating way to summarise the known exact results for polygon generating functions. From the theory of algebraic equations, it is clear that only branch-point singularities can be represented by such equations. Given that the unsolved self-avoiding polygon generating function has a square root singularity, and shows no evidence of a confluent term with an exponent other than an integer multiple of 1/2, it is entirely possible that this generating function might also satisfy such an algebraic equation. A project to search for such a solution is currently in progress.
In the following table we summarise the known critical points, critical exponents and algebraic equations for the solved problems, and give the best currently known numerical estimates for the unsolved problems.

Model	Area/ Perimeter	Lattice	Critical point	Exponent	Algebraic Equation
Staircase	(p)	square	1 / 4	1/2	[0,1,2]
Staircase	(a)	square	0.433062•	-1	?
Convex	(p)	honey.	1 / 2	-2	[8,11,14]
Convex	(p)	square	1 / 4	-2	[4,7,10]
Conv., r^{th} mom.	(p)	square	1 / 4	-2(r+1)	[4+4r,7+5r,10+6r]
Convex	(a)	square	0.433062•	-1	?
Row convex	(p)	honey.	1 / 3	1/2	8th order
Row convex	(p)	square	$3-2\sqrt{2}$	1/2	[3,4,5,6,6]
Row convex	(a)	square	0.311957•	-1	[3,4]
Almost convex	(p)	square	1 / 4	-2-m	?
Self avoiding	(p)	honey.	$(2+\sqrt{2})^{-1/4}$	3/2*	?
Self avoiding	(p)	square	$0.1436806^{\lozenge}$	3/2*	?
Self avoiding	(a)	square	$0.251834^{\lozenge}$	0 (log)#	?
Self avoiding	(p)	triang.	$0.2409185^{\lozenge}$	3/2*	?
Self avoiding	(a)	triang.	$0.33937^{\lozenge}$	0 (log)#	?

* Exact but not rigorous

Conjectured from numerical study

• Obtainable to arbitrary numerical accuracy

$^{\lozenge}$ Best numerical estimate

ACKNOWLEDGMENTS

I have benefitted greatly from discussions and collaborations with Richard Brak, Ian Enting and Stuart Whittington in the course of this study. This work has been supported by the Australian Research Council.

REFERENCES

1. B. Nienhuis, Phys. Rev. Letts., **49**, 1062 (1982)
2. B. Nienhuis, J. Stat. Phys. **34**, 731 (1984)
3. M.P. Delest and G. Viennot, Theor. Comp. Sci. **34**, 169 (1984)
4. A.J. Guttmann and I.G. Enting, J. Phys. A, **21**, L467 (1988)
5. K.Y. Lin and S.J. Chang, J. Phys. A, **21**, 2635(1988)
6. D. Kim, Disc. Math. **70**, 47, (1988)
7. R. Brak and A.J. Guttmann, J. Phys. A, **23**, 4581 (1990)
8. I.G. Enting, A.J. Guttmann, L.B. Richmond and N.C. Wormald, Random Struct. & Algor. (to appear) (1991)
9. A.J. Guttmann and N.C. Wormald, J. Phys. A, **17**, 271(1984)
10. G. Szekeres and A.J. Guttmann, J. Phys. A, **20**, 481 (1987)
11. S. Redner and I. Majid, J. Phys. A., **16**, L307, (1983)
12. G. Pólya, J. Comb. Theory **6**, 102 (1969)
13. H.N.V. Temperley, Phys. Rev. **103**, 1, (1956)
14. M.P. Schützenberger, Inform. and Control, **6**, 246, (1963)
15. I. Gessel and G. Viennot, Adv. in Math., **58**, 300 (1985)
16. I.G. Enting and A.J. Guttmann , J. Phys. A, **22**, 1371 (1989)
17. D.L. Klarner and R.L. Rivest, Disc. Math. **8**, 31 (1974)
18. E.A. Bender, Disc. Math. **8**, 219 (1974)
19. I.G. Enting and A.J. Guttmann , J. Phys. A, **22**, 2639 (1989)
20. K.Y. Lin, Int. J. of Mod. Phys. **4B**, 1717 (1990)
21. M. Bousequet-Melou, Proc. Collq. Formal Power Series and Algebraic Combinatorics, Bordeaux, France, May 1991
22. M. Bousequet-Melou, Ph. D. Thesis, Univ. Bordeaux, (1991)
23. K.Y. Lin, J. Phys.A, (to appear)
24. D.L. Klarner, Fibbonaci Q., **3**, 9 (1965)
25. D.L. Klarner, Canad. J. Math. **19**, 851, (1967)
26. M.P. Delest, J. Comb. Theory, Ser. A, **48**, 12, (1988)
27 R. Brak I.G. Enting and A.J. Guttmann, J. Phys. A, **23**, 2319 (1990)

28. W. Gautschi, SIAM Review, **9**, 24, (1967)
29. K.Y. Lin and F.Y. Wu, J. Phys. A, **23**, 5003 (1990)
30. A.J. Guttmann and I.G. Enting, J. Phys. A, **21**, L165 (1988)
31. A.J. Guttmann and I.G. Enting, (in preparation).
32. I.G. Enting and A.J. Guttmann, J. Stat. Phys. **58**, 475 (1990)
33. M.E. Fisher, A.J. Guttmann and S.G. Whittington, J. Phys. A, **24** (in press) (1991)
34. K.Y. Lin and W.J. Tzeng, Int. J. Mod. Phys. B (submitted) (1991)
35. R. Brak and A.J. Guttmann, J. Phys. A, **23**, L1331, (1991)

INTERACTING SELF-AVOIDING WALK AND POLYGON SYSTEMS: MODELS OF COLLAPSE.

A.J. Guttmann
Department of Mathematics, The University of Melbourne,
Parkville, Vic. 3052 Australia,

ABSTRACT
The behaviour of a variety of models of collapse transition in polymer systems is discussed. Both rigorous results and numerical results are presented. Some of the work described is still in progress, so the results are incomplete. Nevertheless, the overall picture demonstrates how the association of an attractive fugacity between nearby monomers can give rise to a collapse transition in a wide variety of models. These then qualitatively describe the transition undergone by collapsing polymeric systems.

I INTRODUCTION

A number of recent studies of models of collapse in polymer systems, as modelled by self-avoiding walks and polygons, are discussed. Both rigorous results in the form of existence proofs and bounds on thermodynamic quantities and numerical results are presented. The numerical results are primarily used to locate the position and identify the nature of the collapse transition. Exact results for a simplified model that nevertheless retains the essential features of a more complicated model are also discussed. The effects of a geometrical constraint on a particular system are also discussed.

A linear polymer molecule in dilute solution in a good solvent can be modelled by a self-avoiding walk (saw) on a regular lattice. By suitably weighting near-neighbour interactions with an appropriate fugacity, the (infinite) walk is believed to undergo a transition which adequately models the transition in real polymers brought about by the dominance of attractive forces at low temperatures or a poor solvent. When the fugacity is neutral, or favours near-neighbour repulsion, the walk is in a swollen phase, corresponding to the behaviour in a good solvent. As the fugacity is weighted towards attraction, the walk adopts a more compact configuration. Between these two regimes is an intermediate regime, corresponding to a tricritical point. This model, based on weighting saws has been studied theoretically by many authors[1-7]. A related model in which the collapse transition in suitably weighted self-avoiding polygons has been studied is reported in Maes and Vanderzande[8].

Just as saws model linear polymers, so do lattice animals model the configurational properties of randomly branched polymers in dilute solution in a good solvent. As the solvent quality deteriorates, branched polymers are expected to become more compact and a collapse transition analogous to that discussed above is expected to occur. In these models one associates a fugacity with every cycle, so that the energy is proportional to the cyclomatic index. Making this energy suitably attractive then is expected to lead to a collapse phenomenon, since animals with many cycles will be more compact than those with few cycles. This model has also attracted considerable attention[9-14]. For a directed animal problem it can be proved that there is a collapse transition[15].

In subsequent sections we describe our work on several models of collapse in linear polymeric systems and report on the results obtained to date. In some cases this is ongoing work, and additional results are expected to be published subsequently. In most of these problems there is no natural volume, thus the thermodynamic limit in the usual sense cannot be taken. However if n is the number of monomers in the walk, and we choose the entropy S and internal energy U as our two extensive thermodynamic variables, then the appropriate procedure is to use the conventional canonical formalism in which the generating function $Z_n(\beta)$, defined by

$$Z_n(\beta) = \sum_m c_n(m)\, e^{\beta m} \tag{1.1}$$

is the canonical partition function parameterized by n. Here $c_n(m)$ is the number of n monomer walks or polygons, with m interactions. These interactions might be nearest neighbours, or bends, or any other appropriate quantity. The appropriate thermodynamic limit is then the limit $n \to \infty$.
A general scaling theory for such models has been provided by Whittington and De'Bell[16]. The corresponding "free-energy" is

$$A_n(\beta) \equiv n^{-1} \log Z_n(\beta) \tag{1.2}$$

where β can be conveniently considered as $-\varepsilon/kT$, where ε is an energy per interaction. In the large-n limit the free-energy becomes

$$A(\beta) \equiv \lim_{n\to\infty} n^{-1}\log Z_n(\beta). \tag{1.3}$$

Here β is an effective inverse temperature. Defining $t = (T - T_c)/T_c$, we expect that, near $t = 0$, $A(\beta) \sim t^{2-\alpha}$. For finite n we take the scaling form to be

$$A_n(\beta) \sim t^{2-\alpha}\Theta(nt^{1/\phi}) \sim n^{-(2-\alpha)\phi}\psi(n^{\phi}t). \tag{1.4}$$

Differentiating twice with respect to temperature gives the *intensive* heat capacity

$$C_n = (\langle m^2\rangle - \langle m\rangle^2)/n \sim n^{\alpha\phi}\Psi(n^{\phi}t). \tag{1.5}$$

If t_n and h_n denote the value of t at which C_n takes its maximum, and the value of C_n at that maximum respectively, then differentiating (1.5) to determine the position of that maximum yields

$$t_n \sim n^{-\phi} \quad \text{and} \quad h_n \sim n^{\alpha\phi}. \tag{1.6}$$

Thus a study of the fluctuations, $(\langle m^2\rangle - \langle m\rangle^2)/n$ allows one to calculate the exponents α and ϕ.

An alternative approach is based on the study of the weighted mean-square radius of gyration (or mean-square end-to-end distance)

$$R_n^2(\beta) = \sum_m c_n(m)\, R_G^2\, e^{\beta m} \Big/ \sum_m c_n(m)\, e^{\beta m}. \tag{1.7}$$

For large n, $R_n^2 \sim n^{2\nu}$, where $\nu = \nu$(non-interacting) for β small, $\nu = \nu$(strongly - interacting) for β large, and in the intermediate, collapse region, $\nu = \nu_t$ for $\beta = \beta_t$. Near β_t, the following scaling law is expected to hold:

$$R_n^2 \sim n^{2\nu} F(\Delta\beta\, n^{\phi}) \tag{1.8}$$

where $\Delta\beta = \beta - \beta_t$. From ν and ϕ, the other exponents can be calculated by scaling.

II A DIRECTED WALK MODEL

In this work we consider a variant of the interacting saw model which is exactly solvable. On the square lattice, we consider saws in which no steps in the negative x direction are allowed. In the non-interacting case, this is a straightforward model to solve. The generating function for n-step walks is singular at $(1+\sqrt{2})^{-1}$, with a simple pole at that point. However if we consider the number n-step walks with m nearest-neighbour contacts, $c_n(m)$, the problem is more difficult, but can still be exactly solved[17-20]. To simplify certain calculations, we impose the additional technical condition that the first step of the walk is in the positive x-direction.

A number of rigorous results can be obtained by concatenation arguments. Consider the generating function

$$Z_n(x) = \sum_m c_n(m)\ x^n \tag{2.1}$$

and the corresponding quantity

$$\kappa(x) \equiv \lim_{n\to\infty} n^{-1}\log Z_n(x) \tag{2.2}$$

then, as noted above $\kappa(1) = \log(1+\sqrt{2})$. Concatenation arguments, plus the observation that $Z_n(x)^{1/n}$ is bounded above for $x < \infty$ are sufficient to establish the existence of the limit (2.2), and its finiteness for $x < \infty$. Further, the functional inequality derived by concatenation implies that $\kappa(x)$ is log-convex and hence continuous. From the observation of monotonicity of $\kappa(x)$ in both the regimes $x < 1$ and $x > 1$, plus the fact that $\max(m) = n + o(n)$, we can prove that

$$\lim_{x\to 0+} \kappa(x)/\log x = 0$$

$$\lim_{x\to +\infty} \kappa(x)/\log x = 1. \tag{2.3}$$

We shall find it useful to define the two-variable generating function

$$G(x,y) = \sum_{n,m} c_n(m)\, x^m y^n. \tag{2.4}$$

At fixed x, G(x,y) converges for $y < \exp(-\kappa(x))$, which defines the boundary $y=y(x)$ in the (x,y) plane. From (2.3) it follows that, for large x, $y \sim 1/x$.

We now use the technique of Temperley[21] to derive an explicit expression for G(x,y). Let $c_n(r,m)$ denote the number of n-step walks with m contacts, with the first step in the positive x direction and exactly r steps in the y direction. Define the generating function

$$g_r(x,y) = \sum c_n(r,m)\, x^m y^n \tag{2.5}$$

and then write down recurrences satisfied by g_r for $r \geq 0$. By eliminating terms between the g_r we obtain the recurrence relation

$$g_{n+1} - (1+x)yg_n - (1-x)x^n y^{n+2} g_n + xy^2 g_{n-1} = 0 \tag{2.6}$$

After some manipulation, we obtain

$$G = [2yg_1' - (2+y-xy)g_0']/[y^2(1+x+y-xy)g_0' - 2yg_1'] \tag{2.7}$$

where

$$g_n' = y^n + y^n \sum_{m=0}^{\infty} z^{mn}(y-z)^m y^m z^{m(m+1)/2}/[\prod_{k=1}^{m} (z^k-1)(y^k-z)] \tag{2.8}$$

and $z = xy$.

The boundary is the locus of singular points of G closest to the x - axis for positive y. An analysis of g_0' and g_1' shows that they are analytic except on the hyperbola $y=1/x$, and for values of y such that $(xy)^k = 1$, for k = 1 to infinity. From (2.7) we see that additional singularities of G arise from zeros of the denominator. The recurrence relation (2.6) can also be solved on the hyperbola $z = 1$, where

$$G\left(\frac{1}{y}, y\right) = -1 + [1 - 4y + 2y^2 + y^4]^{1/2}/[1 - 3y - y^2 - y^4] \qquad z=1 \tag{2.9}$$

We have numerically determined the zeros of (2.7) which lie below the hyperbola for $x < x^* = 9/[(17 + 3\sqrt{33})^{1/3} + (17 - 3\sqrt{33})^{1/3} - 1]^2$, while for $x \geq x^*$ the boundary coincides with the hyperbola $z = 1$. The argument that the locus of zeros meets the boundary at x^* and then coincides with the boundary for $x > x^*$ is quite involved[20]. The point $y^* = 1/x^*$ is also a pole of $G(\frac{1}{y}, y)$. The boundary is shown in Fig.1. Physically, the significance of the boundary can be seen by considering the average value of n at some point (x, y), defined by

$$\langle n(x,y)\rangle = \frac{\partial \log G(x,y)}{\partial \log y} \tag{2.10}$$

On the boundary $\langle n\rangle$ is infinite, while below the boundary it is finite. Above the boundary we enter the non-physical region. The point (x^*,y^*) corresponds to a collapse transition, and models the coil-ball transition undergone by real polymers. Full details of this model are reported in Brak *et al.*[20], while a number of results are already reported Binder *et al.*[18]. A related, simpler model in which all horizontal segments are of equal length was introduced by Zwanzig and Lauritzen and by Nordholm[22-24].

III WALKS CROSSING A SQUARE

In this problem[25] we consider self-avoiding walks on the square lattice which are confined to lie in or on the boundary of a square with vertices at (0,0), (0,L), (L,0), and (L,L). We first consider the number of such walks that begin at the origin and end at the vertex (L,L), especially in the large L limit. At fixed L we also associate a fugacity with the number of steps in the walk and ask how the system behaves as a function of this fugacity.

We have investigated both problems analytically and by series analysis. Denote by $c_n(L)$ the number of saws with n steps confined to lie in the L x L square and which start at the origin and finish at (L,L). Denote by c(L) the sum $\Sigma_n c_n(L)$. By considering the maximum and minimum value of n for fixed L we can prove that

$$\lim_{L\to\infty} \sup L^{-2} \log c(L) \leq \log \mu \tag{3.1}$$

where μ is the growth constant for saws on the square lattice. By partially covering the L x L square with smaller squares of side M + 2, it is possible to prove by concatenation arguments that the limit

$$\lim_{L\to\infty} L^{-2} \log c(L) = \log \lambda \tag{3.2}$$

exists. We obtain a lower bound on c(L) by observing that, for L even, walks with the maximum number of steps are Hamiltonian walks. A lower bound for the number of such Hamiltonian walks is obtained by a construction similar to that used by Gujrati[26]. The square is effectively covered by disjoint rectangles of size (1 x m). We consider all possible ways of joining these rectangles so as to form Hamiltonian polygons, and this gives the bound

$$\log \mu_H \geq \log m/[2(m + 1)] \tag{3.3}$$

This bound is most effective when m = 4, when it is equal to 0.1386... Finally, to convert the polygon to a walk crossing the square we remove and insert appropriate edges. This doesn't change the bound so that we have finally

$$\lim_{L\to\infty} L^{-2} \log c(L) \geq 0.1386.. \tag{3.4}$$

That is to say, we have proved that the number of distinct self-avoiding paths grows exponentially with the *area* of the lattice, rather than exponentially with its linear dimension. Similarly, the mean number of steps in such a path, $\langle n \rangle = \Sigma_n n.c_n(L)/ \Sigma_n c_n(L)$ must be of order L^2, as can be seen by the following argument. Firstly, observe that $c_n(L) \leq c_n = \mu^{n+o(n)}$. If $n = o(L^2)$, then $c_n(L) \leq \mu^{o(L)}$ so that all except exponentially few walks have of order L^2 steps. Hence the mean number of steps must also be of order L^2.

Numerically, we have derived exact values of $c_n(L)$ for $L \leq 6$. Summing over n we obtain c(L) values which, when extrapolated, permit us to estimate $\lambda = 1.756 \pm 0.01$.

For the second aspect of this problem, we introduce a step fugacity x, and consider the generating function

$$C_L(x) = \Sigma_n c_n(L).x^n \tag{3.5}$$

By a refinement of the proof discussed above, we can prove the existence of the limit

$$\lim_{L\to\infty} L^{-2} \log C_L(x) = \log \lambda(x) \tag{3.6}$$

To estimate $\lambda(x)$, we first note that for $x \leq 1$ an upper bound is obtainable by observing that every walk which crosses the square is "doubly unfolded" (that is, the end-points are the "top" and "bottom" points). Since such walks can be concatenated to give a supermultiplicative inequality, we have $c_n(L) \leq \mu^n$ and hence that

$$C_L(x) \leq \Sigma\, \mu^n.x^n = (\mu x)^{2L}(1 - (\mu x)^{n_{max} -2L +1})/(1 - \mu). \tag{3.7}$$

If $x > 1/\mu$ then (3.7) implies that $\log \lambda(x) \leq \log \mu + \log x$. If $x < 1/\mu$ we have that $\log \lambda(x) \leq 0$, and combined with the bound $C_L(x) \geq c_{2L}(L).x^{2L}$ this implies that $\log \lambda(x) = 0$ for all $x \leq 1/\mu$. Hence $\log \lambda(x)$ is non-analytic. To determine the point of non-analyticity we note that $C_L(x) \geq c_{n_{max}}(L).x^{n_{max}}$, and so $\log \lambda(x) \geq \log \mu_H + \log x$, where μ_H is defined above. Thus there must be a singular point x^* in the range $1/\mu \leq x^* \leq 1/\mu_H$. To investigate this numerically, we have studied the mean number as a function of fugacity. We define this quantity by

$$\langle n(x,L)\rangle = \Sigma_n n.c_n(L)x^n / \Sigma_n c_n(L)x^n \tag{3.8}$$

and expect that $\langle n(x,L)\rangle = A(x)L^2[1 + o(1)]$. We have estimated $A(x)$ for a range of values of x, and found that it vanishes between $x=0.3$ and $x=0.4$. From the above bounds on x^*, we have that $0.37905 \leq x^* \leq 0.1386$. It is thus tempting to suggest that $x^* = 1/\mu$ exactly, though we have been unable to prove this suggestion. Very recently Burkhardt[27] has shown that this result follows from scaling theory and the s.a.w. correspondence with the $n \to 0$ limit of the n-vector model. Following Burkhardt, the generating function $C_L(x) \sim L^{-\eta}F[L^{1/\nu}(x^* - x)]$, where $\eta = 5/2$ is a corner exponent[28] and $\nu = 3/4$. From the scaling form, it follows that $C_L(x^*) \sim L^{-\eta}$, $\langle n(x^*,L)\rangle \sim L^{1/\nu}$ and $x^* = 1/\mu$. Our numerical data are insufficient to estimate the critical exponents associated with the transition, but Burkhardt's scaling arguments are quite convincing. Further details of this calculation appear elsewhere[25].

IV THE COLLAPSE TRANSITION IN TWO DIMENSIONAL VESICLES

In this problem we[29] enumerated self-avoiding polygons on the square lattice by both perimeter and area. By associating a fugacity with both area and perimeter, we can induce a collapse transition along a line of critical points in the two variable fugacity plane. We consider the number $p_m^{(n)}$ of polygons on the square lattice with m edges enclosing area n. The generating function for such polygons is given by

$$P(x,y) = \sum_{m,n} p_m^{(n)} x^m y^n$$

$$= \sum_m P_m(y)\, x^m = \sum_n A_n(x)\, y^n \tag{4.1}$$

where $P_m(y)$ and $A_n(x)$ are, respectively, the generating functions for polygons of perimeter m and area n. The generating function P(x,y) can be considered a grand partition function.

By considering polygons with maximum area for fixed perimeter, we show that

$$\lim_{m\to\infty} m^{-2} \log P_m(y) = \log y/16 \text{ for all } y \geq 1. \tag{4.2}$$

For $0 < y < 1$ $\lim_{m\to\infty} m^{-1} \log P_m(y) = \kappa(y)$ exists and is finite. We note that $\kappa(1) = \kappa$, the usual connective constant for polygons grouped by perimeter.

From the observation that

$$P_m(y_1)\, P_m(y_2) \geq \left[P_m\left(\sqrt{y_1\, y_2}\right)\right]^2, \tag{4.3}$$

it follows that $\kappa(y)$ is log-convex.

By considering the polygons of minimum area with fixed perimeter we obtain the bound $\kappa(y) \leq \kappa + \frac{1}{2} \log y$. Observing that these are just the number of a class of site trees on the dual lattice, we can concatenate them and hence prove the existence of the thermodynamic limit κ_0 in that case. Hence

$$\kappa(y) \geq (\kappa_0 + \log y)/2 \tag{4.4}$$

The two bounds together imply

$$\lim_{y \to 0+} \kappa(y)/\log y = 1/2 \tag{4.5}$$

Now the grand partition function can be written as

$$P(x,y) = \sum_m e^{mk(y)+o(m)} x^m \tag{4.6}$$

and for fixed x this converges for $x < e^{-\kappa(y)}$, so that $x = x_c(y) = e^{-\kappa(y)}$ defines a phase boundary. As $y \to 0$, $x \to \infty$ as

$$x \sim y^{-1/2} \tag{4.7}$$

Further, as $\kappa(y)$ is monotone non-decreasing, $x(y)$ is monotone non increasing. Further, $x(y)$ is bounded below by $e^{-\kappa}$ for $y \leq 1$ and then jumps discontinuously to zero. The bounds established above enable us to write

$$e^{-\kappa_0} \geq x_c^2 y \geq e^{-2\kappa} \tag{4.8}$$

Similar results have been obtained for the generating function $A_n(x)$. The limit

$$\lim_{n \to \infty} n^{-1} \log A_n(x) = \chi(x) \tag{4.9}$$

exists, and we show that $\chi(x)$ is log-convex. Additional simple arguments allow us to obtain the bounds

$$\chi(1) + 2 \log x \geq \chi(x) \geq \lim_{n \to \infty} n^{-1} \log p_{2n+2}^{(n)} + 2 \log x \quad \text{for } x \geq 1. \tag{4.10}$$

Hence

$$\lim_{x \to \infty} \frac{\chi(x)}{\log x} = 2 . \tag{4.11}$$

For $x \leq 1$ we prove that $\chi(x) \leq 2\kappa + 2 \log x$ when $e^{-\kappa} \leq x \leq 0$, and $\chi(x) = 0$ when $x \leq e^{-\kappa}$. Similar results have been obtained for 3 and higher dimensions. In d dimensions we now treat surfaces in Z^d made up of elementary unit (d - 1) dimensional "hyper-squares". A similar phase boundary is obtained, and the analogous result to (4.7) is

$$x \sim y^{-1/(2d-2)} \quad \text{as } y \to 0^+ \tag{4.12}$$

In all dimensions the phase boundary x(y) is monotone non-increasing and jumps discontinuously to zero at y = 1.

In two dimension we have constructed an accurate numerical phase diagram by enumerating all polygons with perimeter ≤ 44 for all areas. We analysed the series for fixed y, and determined x(y) in this way. The resulting phase diagram is shown in Fig. 2. In the vicinity of the multicritical point, the grand partition function behaves, to leading order, as

$$P(x,y) \sim C_0 |\tilde{x}|^{2-\alpha_0} Z(\tilde{y}/|\tilde{x}|^{\phi}) + B(x,y) \tag{4.13}$$

where the scaling axes $\tilde{x}$, $\tilde{y}$ are given by

$$\tilde{x} = x_c - x + (y-1)/e_2$$

$$\tilde{y} = 1 - y,$$

$\alpha_0 = \frac{1}{2}$ is the usual polygon exponent, and Z(z) is a scaling function with normalisation Z(0) = 1. The crossover exponent ϕ is shown to be equal to $2\nu = 3/2$, where $\nu = 3/4$ is the usual s.a.w. correlation function exponent. Near y = 1, the critical line is given by

$$x_c(y) \approx x_c(1) + (\Delta y/z_c)^{2/3} + \Delta y/e_2 + \ldots \tag{4.14}$$

so that if $z_c < \infty$, we see that the critical line has vertical slope as $y \to 1^-$.

V THE ROD-COIL TRANSITION

The transition between linear, rod-like phases and compact coiled phases can be modelled by square lattice saws if an energy is associated with each 90° bend[30,31]. By weighting such bends appropriately, either phase may be favoured. Let $c_n(m)$ be the number (per lattice site) of saws of n steps and m right-angle bends. Clearly

$$\sum_{m} c_n(m) = c_n, \tag{5.1}$$

the total number of saws, and

$$\lim_{n\to\infty} n^{-1} \log c_n = \kappa \tag{5.2}$$

exists, where κ is the connective constant of the lattice. We can define the partition function

$$Z_n(\beta) = \sum_{m} c_n(m)\, e^{\beta m} \tag{5.3}$$

and we are interested in the properties of the corresponding limiting free energy. We[32] have determined the coefficients $c_n(m)$ for $n < 28$, extending by five terms the data of Privman and Redner[31], and analysed the corresponding "specific heat ". We find $\alpha\phi \approx 0$. Hence at least one of α, ϕ must be 0. Plotting ln n vs. ln t we find that the gradient, $-\phi$, is initially positive, and only becomes negative for $n \geq 24$. However it is clearly not zero, so we conclude that $\alpha \approx 0$, and hence that $\nu \approx 1$. Taking the scaling *ansatz* $R_n^2 \sim n^{2\nu} F(\Delta\beta\, n^{\phi})$, Halley *et al.*[30] have in fact proved that $\phi = 1$. The exponents are therefore the same as for random walks, hence the problem is not particularly interesting from the point of view of the transition itself, which is by now well understood, but rather because it offers the opportunity to study the scaling function $F(x)$. The s.a.w. model we have described therefore exhibits the same dominant behaviour as the random walk model, the differences only emerging in the study of the scaling function. With our extended data we are currently studying this aspect.

VI THE COLLAPSE OF SELF-AVOIDING POLYGONS

If we introduce an energy term associated with near-neighbour contacts we would expect a collapse transition in polygons just as in walks[8]. One technical reason for studying the polygon problem is that polygons can be concatenated without (substantially) changing the total number of contacts. A second reason for studying this problem is that it allows us to investigate the influence of architecture an the collapse transition.

We consider the square lattice. Let $p_n(m)$ be the number (per lattice site) of polygons with n edges and m near-neighbour contacts. Clearly

$$\sum_{m} p_n(m) = p_n, \tag{6.1}$$

the total number of polygons, and

$$\lim_{n\to\infty} n^{-1} \log p_n = \kappa \tag{6.2}$$

exists, where κ is the connective constant of the lattice. We can define the partition function

$$Z_n(\beta) = \sum_{m} p_n(m)\, e^{\beta m}, \tag{6.3}$$

where β can be conveniently considered as $-\varepsilon/kT$, where ε is an energy per contact, and we are interested in the properties of the corresponding limiting free energy.

Pairs of polygons can be concatenated by the addition of a "neck"[25] to give the functional inequality

$$p_n(m) = \sum_{m_1} p_n(m_1)\, p_{n-n_1}(m-m_1-2) \tag{6.4}$$

which easily gives the existence of the limit

$$\lim_{n\to\infty} n^{-1} \log Z_n(\beta) \equiv A(\beta) \tag{6.5}$$

Moreover $A(\beta)$ is convex and hence continuous.

Clearly $A(0) = \kappa$ and, by monotonicity, $A(\beta) \le \kappa$ for $\beta \le 0$ and $A(\beta) \ge \kappa$ for $\beta \ge 0$. For $\beta \le 0$,

$$Z_n(\beta) \ge p_n(0) \tag{6.6}$$

so

$$A(\beta) \ge \kappa_o,$$

where κ_o is the connective constant for neighbour - avoiding walks, and, for $\beta \ge 0$.

$$p_n e^{\beta m_{max}} \ge Z_n(\beta) \ge p_n(m_{max})\, e^{\beta m_{max}} \tag{6.7}$$

Since $m_{max} = \frac{1}{2} n + o(n)$, taking logarithms, dividing by n and letting $n \to \infty$, we have

$$\kappa + \beta/2 \ge A(\beta) \ \ge \ \max\,[\kappa, \kappa^* + \beta/2]. \tag{6.8}$$

for $\beta \ge 0$, where $\kappa^* = \lim n^{-1} \log p_n(m_{max})$, and hence $\lim_{\beta\to\infty} A(\beta)/\beta = 1/2$.

Although these results give useful information on the general behaviour of $A(\beta)$ they do not establish that the model shows a phase transition. To investigate this we have derived exact values of $p_n(m)$ for $n \le 34$, and computed $c_n(\beta) = \partial^2 A_n(\beta)/\partial\beta^2$, where

$$A_n(\beta) = n^{-1} \log Z_n(\beta),$$

for a range of values of β. The graph of $c_n(\beta)$ vs. β for $n = 14 - 34$ displays a fairly sharp peak whose position is only weakly dependent on n. By extrapolating the peak positions we estimate that the transition occurs at $\beta_0 \approx 0.658$.

Further series extension and Monte Carlo work is being undertaken[32] and will be reported elsewhere, along with our exponent estimates. An earlier study by Maes and Vanderzande[8] was based on the behaviour of the radius of gyration, for which they obtained series expansions to $n = 28$. In this way they estimated $\nu = 0.58 \pm 0.01$ and $\phi = 0.90 \pm 0.02$. Their result for $\nu \approx 4/7$ is not unexpected, but their value of the crossover exponent ϕ is more controversial. We expect that our extended data will enable a more precise estimate to be made.

VII CONCLUSION

We have described five different models of a collapse transition, and shown how in each case, a combination of rigorous and numerical results elucidate the nature of the collapse transition. In each case a fugacity is associated with a particular geometrical property of the model. By making this fugacity attractive or repulsive one moves from one state to another. This transition is seen in real polymeric systems in, for example, the coil-ball transition. It is hoped that the description here of a range of models will gather together a sufficiently broad range of techniques to permit subsequent models to be similarly studied, and will also make clear the common features of this range of models.

ACKNOWLEDGMENTS

The author has benefitted greatly from discussions and collaborations with with R. Brak, I.G. Enting and S.G.Whittington, with whom much of this work has been done. He acknowledges the support of the Australian Research Council.

REFERENCES

1. A. Baumgärtner, J. Phys. (Paris) 43, 1407 (1982)
2. V. Privman, J. Phys. A, 19, 3287, (1986)
3. B. Derrida & H. Saleur, J. Phys. A, 18, L1075 (1985)
4. T. Ishinabe, J. Phys. A, 20, 6435 (1987)
5. F. Seno & A. L. Stella, J. Phys. (Paris) 49, 739, (1988)
6. F. Seno, A. L. Stella, & C. Vanderzande, Phys. Rev. Letts. 61, 1520, (1988)
7. R. Mark Bradley, Phys. Rev. A, 39, 3738, (1989)
8. D. Maes & C. Vanderzande, Phys. Rev. A, 41, 3074 (1990)
9. T.C. Lubensky & J. Isaacson, Phys. Rev A, 20, 2130 (1979)

10. B. Derrida & H.J. Herrmann, J. Physique 44, 1365, (1983)
11. R. Dickman & W.C. Shieve, J. Stat. Phys. 44, 465 (1986)
12. P.M. Lam, Phys. Rev. B, 38, 2813, (1988)
13. I.S. Chang & Y Shapir, Phys. Rev. B, 38, 6736, (1988)
14. D.S. Gaunt & S. Flesia, Physica A, (to appear) (1990)
15. D. Dhar, J. Phys. A. Math. Gen. 20, L847 (1987)
16. S.G. Whittington and K. De'Bell - private communication (1990)
17. A.R. Veal, J.M. Yeomans & G. Jug, J. Phys. A: Math. Gen, 23, L109 (1990)
18. P.M. Binder, A.L. Owczarek, A.R. Veal & J.M Yeomans, J. Phys. A: Math. Gen, 23, (1990)
19. D. Foster, J. Phys. A: Math. Gen **23**, L1135, (1990)
20. R. Brak, A.J. Guttmann & S. G. Whittington, J. Phys. A (submitted)
21. H.N.V. Temperley, Phys. Rev. 103, 1, (1956)
22. R. Zwanzig and J.I. Lauritzen, Jr. J. Chem. Phys. 48, 3351, (1968)
23. J.I. Lauritzen Jr and R Zwanzig, J. Chem. Phys. 52, 3740, (1970)
24. K.S.J. Nordholm, J. Stat. Phys., 9, 235, (1973)
25. S.G. Whittington and A.J. Guttmann, J. Phys. A: Math Gen, **23**, 5601, (1990)
26. P. Gujirati J. Phys. A:Math. Gen 13, L437 (1980)
27. T. Burkhardt , private communication. (1991)
28. J.L. Cardy, Nucl. Phys. **B240**, 514, (1984)
29. M.E. Fisher, A.J. Guttmann & S.G. Whittington, J. Phys. A, **24**, (to appear)
30. V. Privman and S. Redner, Z. Phys. B, **67**, 129 (1987)
31. J.W. Halley, H. Nakanishi and R. Sundararajan, Phys. Rev. **31B**, 293 (1985)
32. I.G. Enting, A.J. Guttmann, J. Shilling & S.G. Whittington (in preparation)

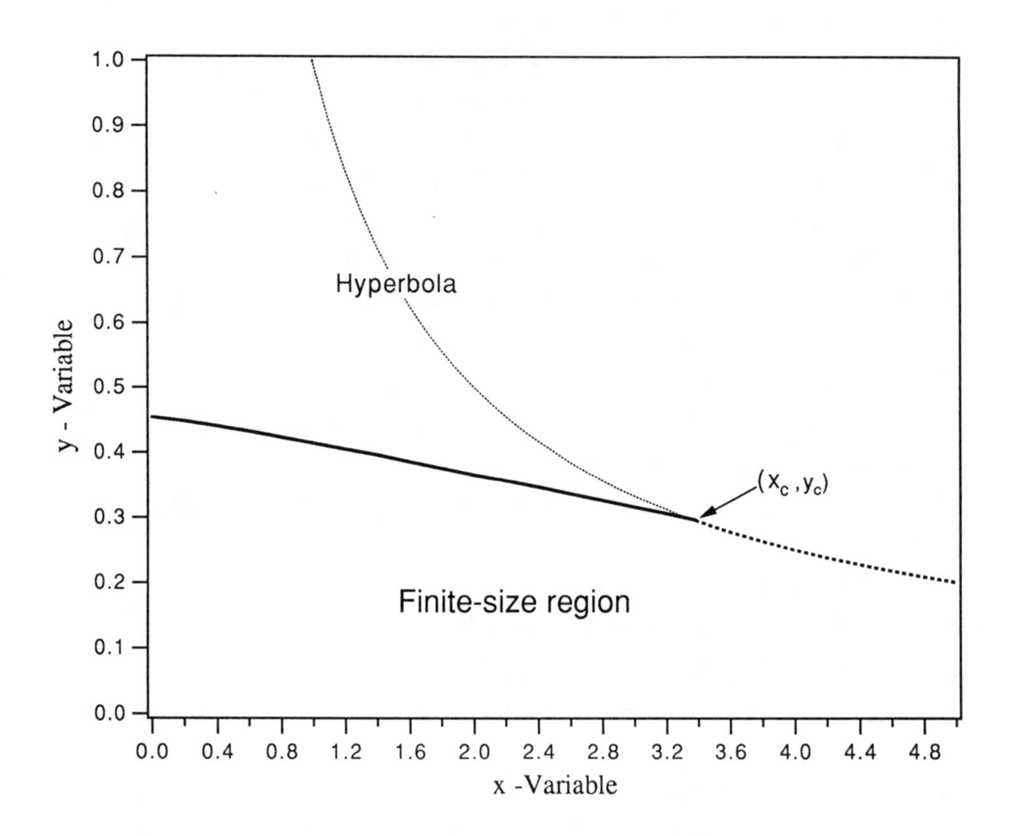

FIGURE 1

Boundary along which a collapse transition occurs for the directed walk model. Beyond (x_c, y_c) the boundary coincides with the hyperbola $xy = 1$.

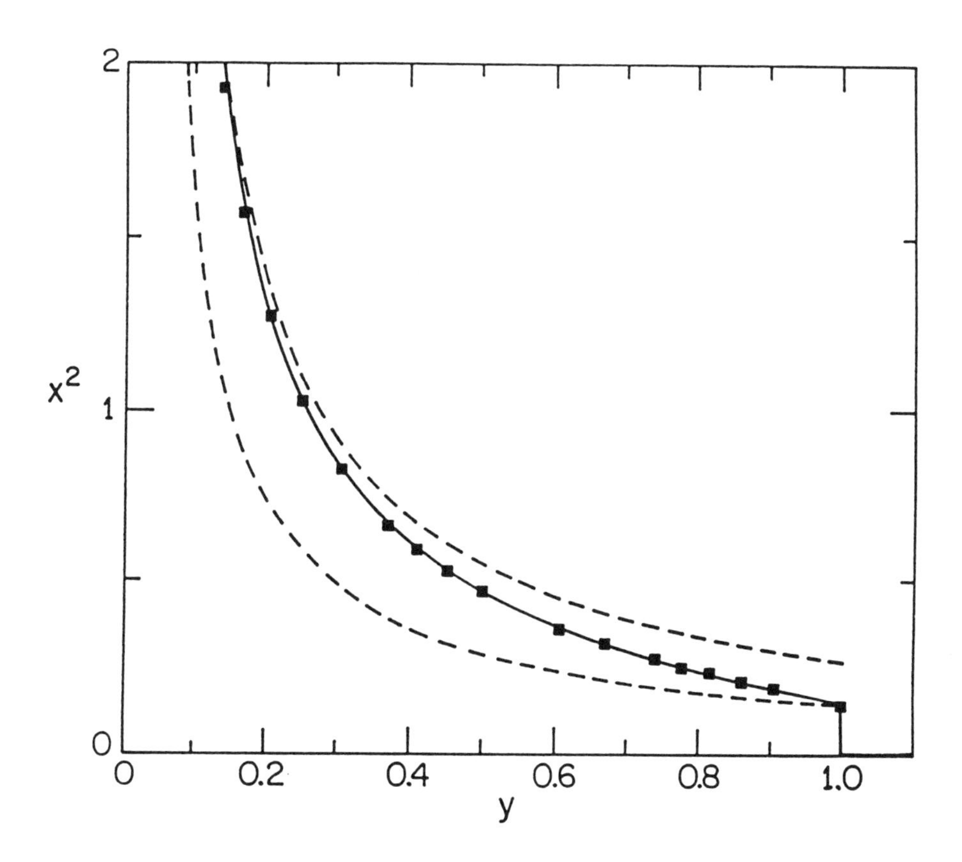

FIGURE 2

Phase diagram of two dimensional vesicles. The solid curves are the upper and lower bounds, the broken curve is the numerically determined phase boundary.

GENERATING FUNCTIONS OF SELF-AVOIDING POLYGONS ON TWO-DIMENSIONAL LATTICES

K. Y. Lin
Physics Department, National Tsing Hua University, Hsinchu, Taiwan 300, ROC.

ABSTRACT

Recent progress on the exact solutions of perimeter and area generating functions of self-avoiding polygons on two-dimensional lattices is reviewed.

1. INTRODUCTION

The self-avoiding polygon was considered as a model of crystal growth [1] or polymer. [2,3] The problem in two dimensions is to find the generating function for the number of polygons on a lattice with given perimeter and/or area. The generating function is defined by

$$G(X\ ,Z) = \sum_{N=1}^{\infty} \sum_{M=1}^{\infty} C_{N,M}\, X^N\, Z^M \tag{1}$$

where $C_{N,M}$ is the number of polygons with perimeter N and area M . An exact solution has not been found yet. However simpler problems can be solved. In particular there are four classes of self-avoiding polygons on the square lattice where exact solutions were obtained; these are the pyramid polygons, staircase polygons, convex and row-convex polygons. A large amount of results were derived during last two years. In this paper I shall review these recent developments.

2. PYRAMID POLYGONS

Pyramid polygon was first considered by Temperley [1] as a model of crystal growth on a plane substrate. An example is shown in figure 1.

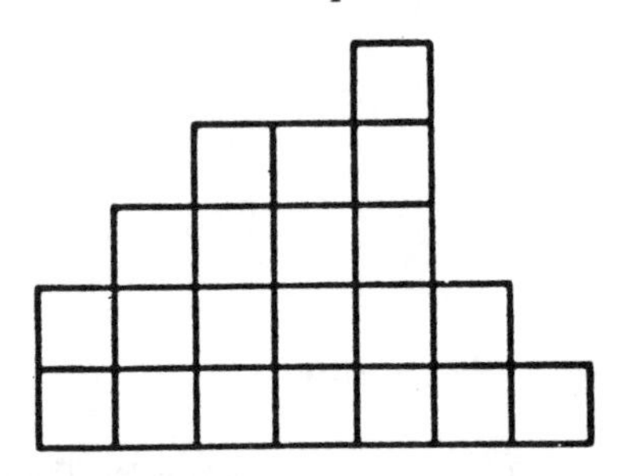

Fig. 1. A pyramid polygon on the square lattice.

The area generating functon was derived by Temperley and his result is

$$G(1\ , Z) = \sum_{N=1}^{\infty} Z^N\,(1-Z^N)\,/\,Q_N^2 = \sum_{N} C_N\, Z^N \tag{2}$$

$$= Z / (1 - Z) + Z^2 / (1 - Z)^2(1 - Z^2) + Z^3 / (1 - Z)^2(1 - Z^2)^2(1 - Z^3) + \dots$$

where

$$Q_N = \prod_{K=1}^{N} (1 - Z^K) \quad \text{for } N > 0$$

$$= 1 \qquad \text{for } N = 0 ,$$

and C_N is the number of pyramid polygons with area N. The generating function has an essential singularity at $Z = 1$. It was shown by Auluck [4] that for large N

$$C_N \approx 8^{-1}\, 3^{-3/4}\, N^{-5/4} \exp\,[2\pi(N/3)^{1/2}] . \tag{3}$$

Recently Lin [5] considered pyramid polygons on the rectangular lattice and derived the generating function

$$G(X,Y,Z) = \sum_{N=1}^{\infty} \sum_{M=1}^{\infty} \sum_{R=1}^{\infty} X^{2N}\, Y^{2M}\, Z^R\, C_{N,M,R} \tag{4}$$

where $C_{N,M,R}$ is the number of pyramid polygons with $2N$ horizontal steps, $2M$ vertical steps and area R. In the special case of $X = Y$, (4) reduces to (1). The generating function (4) can be derived by the method of Temperley as follows. We write

$$G(X , Y , Z) = \sum_{N=1}^{\infty} H_N(X , Y , Z) \tag{5}$$

where H_N generates all pyramid polygons whose height is N (see figure 2) and

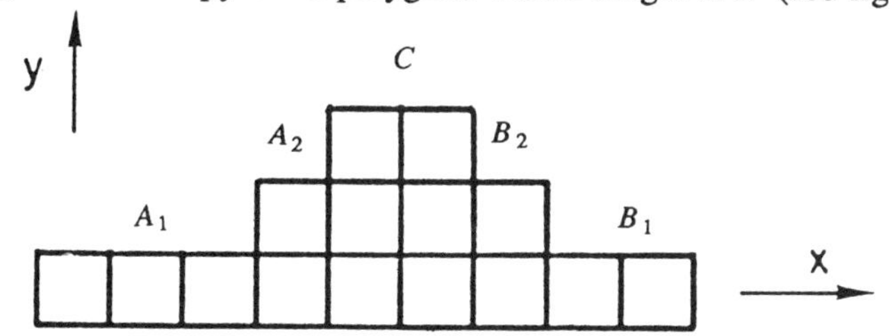

Fig. 2. A pyramid polygon with $N = 3, A_1 = 3, A_2 = 1, C = 2, B_2 = 1, B_1 = 2$.

$$H_1 = Y^2 \sum_{A=1}^{\infty} (X^2Z)^A = Y^2ZX^2 / (1 - X^2Z)$$

$$H_2 = Y^4 \,[\sum_{A=0}^{\infty} \sum_{B=0}^{\infty} (X^2Z)^{A+B}\,] \sum_{C=1}^{\infty} (X^2Z^2)^C = Y^4Z^2X^2 / (1 - X^2Z)^2(1 - X^2Z^2)$$

$$H_N = Y^{2N}\, [\sum_{A_1=0} \sum_{B_1=0} (X^2Z)^{A_1+B_1}\,] \dots [\sum_{A_{N-1}=0} \sum_{B_{N-1}=0} (X^2Z^{N-1})^{A_{N-1}+B_{N-1}}\,] \sum_{C=1} (X^2Z^N)^C$$

$$= (Y^2Z)^N\, X^2 / (1 - X^2Z^N) \prod_{K=1}^{N-1} (1 - X^2Z^K)^2 .$$

When $Z = 1$, we have

$$H_N = Y^{2N}X^2 / (1 - X^2)^{2N-1} . \tag{6}$$

In this special case the infinite series on the right hand side of (5) can be summed over N exactly to give

$$G(X\ ,Y\ ,1) = X^2(1-X^2)\sum_{N=1}^{\infty}[Y\ /\ (1-X^2)]^{2N} = X^2Y^2(1-X^2)\ /\ [\ (1-X^2)^2 - Y^2] \quad (7)$$

which was first derived by Lin and Chang. [6]

3. STAIRCASE POLYGONS

The perimeter and area generating function of staircase polygons (see figure 3) on the square lattice was first obtained by the famous mathematican Pólya, [7] and the result is

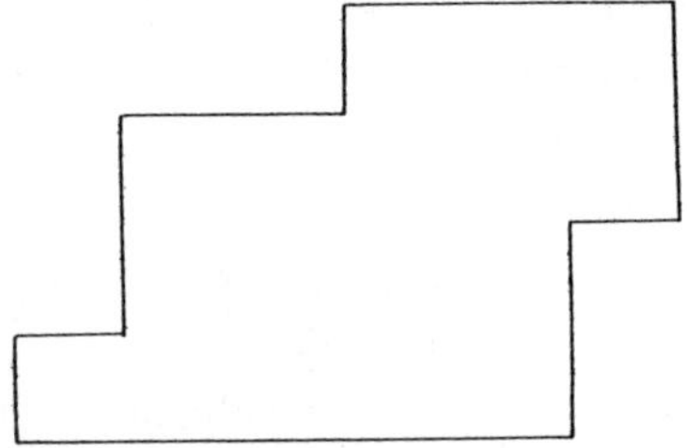

Fig. 3. A staircase polygon.

$$P(X\ ,Z) = \sum_{N=1}^{\infty}\sum_{M=-\infty}^{\infty} C_{N,M}\ X^N\ Z^M \quad (8)$$

$$= \ldots + Z^{-1}X^4 + 2X^2 + ZX^4 + 2Z^2X^6 + (Z^4 + 4Z^3)X^8 + \ldots$$

$$= 1 - [\ \sum_{N=0}^{\infty} P_N(Z)\ X^{2N}\ ]^{-1}$$

where, by definition,

$$C_{2,0} = 2\ , \qquad C_{N,-M} = C_{N,M}\ , \qquad P_0(Z) = 1\ , \qquad P_N(Z) = P_N(Z^{-1}) = \sum_{R=0}^{\infty}[N,R]^2\ Z^{-R(N-R)}\ ,$$

and

$$[N,R] = [N,N-R] = \frac{(1-Z^N)(1-Z^{N-1})\ldots(1-Z^{N-R+1})}{(1-Z)(1-Z^2)\ldots(1-Z^R)}$$

is the Gaussian binomial coefficient which is a polynomial in Z. The Gaussian binomial coefficient generates areas under zig-zag paths and reduces to $(N,R)=N!/R!(N-R)!$ for $Z = 1$. [7] The series converge for all complex values of Z and X subject to the conditions

$$|Z| = 1\ , \qquad |X| < 1/2\ . \quad (9)$$

His definition of generating function (8) is related to (1) by

$$P(X\ ,Z) = G(X\ ,Z) + G(X\ ,Z^{-1}) + 2X^2\ . \quad (10)$$

The perimeter generating function can be obtained by taking the limit $Z = 1$ and the result is

$$G(X\ ,1) = [\ P(X\ ,1) - 2X^2\ ]\ /\ 2 = [\ 1 - 2X^2 - (1-4X^2)^{1/2}\ ]\ /\ 2\ . \quad (11)$$

In a footnote of his 1969 paper, Pólya mentioned that the expansion (8) was entered into his diary in June 1938. It is amazing that he waited 30 years before announcing his result. He also promised to give a proof in a subsequent paper which never appears. Unfortunately nobody knows how to rederive his beautiful result. We conjectured recently [8] that his result can be generalized to the rectangular lattice as follows:

$$P(X\ ,Y\ ,Z)=\sum_{N=0}^{\infty}\sum_{M=0}^{\infty}\sum_{R=-\infty}^{\infty}C_{N,M,R}\,X^{2N}\,Y^{2M}\,Z^{R} \tag{12}$$

$$=1-\left[\sum_{N=0}^{\infty}\sum_{M=0}^{\infty}[N+M\ ,N]^2\,X^{2N}\,Y^{2M}\,/\,Z^{NM}\right]^{-1}$$

where

$$C_{2,0,0}=C_{0,2,0}=1\ , \qquad C_{N,M,R}=C_{N,M,-R}\ .$$

We have expanded (12) to $20th$ order in X and Y, and the coefficients of positive powers of Z agree with the exact counting of staircase polygons. The perimeter generating function,

$$G(X\ ,Y\ ,1)=[\,P(X\ ,Y\ ,1)-X^2-Y^2\,]\,/\,2=(\,1-X^2-Y^2-\Delta^{1/2}\,)\,/\,2 \tag{13}$$

where

$$\Delta=1-2X^2-2Y^2+(X^2-Y^2)^2=(1+X+Y)\,(1+X-Y)\,(1-X+Y)\,(1-X-Y)\ ,$$

also agrees with the exact result obtained earlier by Lin *et al.* [9]

Recently Brak and Guttmann [10] derived the perimeter and area generating function for staircase polygons and their expression is very different from the one obtained by Pólya. Their result has been generalized to the rectangular lattice by Lin and Tzeng [8] and the result is

$$G(X\ ,Y\ ,Z)=\sum_{M=1}^{\infty}H_M(X\ ,Y\ ,Z) \tag{14}$$

$$=X^2[1+\sum_{M=1}^{\infty}R_M\,Z^{M(M+3)/2}\,]\,[1+\sum_{M=1}^{\infty}R_M\,Z^{M(M+1)/2}\,]^{-1}-X^2$$

where H_M is the generating function for all staircase polygons whose top row contains M squares, and

$$H_N=Y^2(X^2Z)^N\,[\,1+\sum_{M=1}^{\infty}R_M\,Z^{M(M+1+2N)/2}\,]\,[\,1+\sum_{M=1}^{\infty}R_M\,Z^{M(M+1)/2}\,]^{-1}\ ,$$

$$R_M=\frac{(-Y^2)^M}{\prod_{R=1}^{M}(1-Z^R)\,(1-X^2Z^R)}\ .$$

4. CONVEX POLYGONS

A convex polygon on the square lattice has the property that a straight line on the bonds of the dual lattice cuts the bonds of the polygon at most twice. The pyramid polygon and the staircase polygon are special cases of the convex polygon. The perimeter generating function was first derived in 1984 by Delest and Viennot. [11] Their derivation is extremely complicated and the final result is very simple:

$$G(X\ ,1) = \sum_{N=2}^{\infty} C_{2N}\, X^{2N} \tag{15}$$

$$= X^4\,(1 - 6X^2 + 11X^4 - 4X^6)\,/\,(1 - 4X^2)^2 - 4X^8\,/\,(1 - 4X^2)^{3/2}$$

where

$$C_{2M+8} = (2M + 11)\,4^M - 4(2M + 1)\,(2M\ ,M)\,.$$

Their result has been rederived by simpler methods.[6,12,13] Lin and Chang[6] generalized (15) to the rectangular lattice and obtained

$$G(X\ ,Y\ ,1) = \sum_{R,S=1}^{\infty} C_{R,S}\, Y^{2R}\, X^{2S} \tag{16}$$

$$= X^2Y^2\,[1 - 3X^2 - 3Y^2 + 3X^4 + 3Y^4 + 5X^2Y^2 - X^6 - Y^6$$

$$- X^2Y^4 - X^4Y^2 - X^2Y^2(X^2 - Y^2)^2\,]\,/\,\Delta^2 - 4X^4Y^4\,/\,\Delta^{3/2}$$

where $C_{R,S}$ is the number of convex polygons with vertical height R and horizontal width S. The authors proved this formula by deriving recurrence relation for convex polygons with a given top-row width. Recently Gessel[14] proved that

$$C_{M+1,N+1} = \tag{17}$$

$$(M+N+MN)\,(2M+2N\ ,2M)\,/\,(M+N) - 2(M+N)\,(M+N-1\ ,M)\,(M+N-1\ ,N)\,.$$

The derivation follows from the identity

$$\Delta^{-R} = \sum_{I,J=0}^{\infty} X^{2I}\, Y^{2J}\,(R + 1/2)_{I+J}\,(2R)_{I+J}\,/\,I!\,J!\,(R + 1/2)_I\,(R + 1/2)_J \tag{18}$$

where we define

$$(R)_N = R\,(R + 1)\,.\,.\,.\,(R + N - 1)\,.$$

Enting and Guttmann[15] considered the generating functions for the first and second area-weighted moments of the number of convex polygons:

$$P_1(X) = \sum_N X^N \sum_K K\, C_{N,K}\ , \qquad P_2(X) = \sum_N X^N \sum_K [\,K(K - 1)\,/\,2\,]\, C_{N,K}\,. \tag{19}$$

Based on the exact series expansions up to $N = 64$, they obtained non-rigorously

$$P_1(X) = X^4\,[\,(1 - 12X^2 + 50X^4 - 76X^6 + 42X^8 - 48X^{10} + 32X^{12})\,/\,(1 - 4X^2)^4 \tag{20}$$

$$+ 4X^4\,/\,(1 - 4X^2)^{5/2}\,]\,,$$

$$P_2(X) = X^6\,[\,R(X)\,/\,(1 - 4X^2)^6 + S(X)\,/\,(1 - 4X^2)^{9/2}\,]\,,$$

where

$$R = 2 + 5X^2 - 224X^4 + 1306X^6 - 3352X^8 + 4536X^{10} - 3424X^{12} + 1664X^{14} - 512X^{16}\,,$$

$$S = -\,29X^2 + 172X^4 - 356X^6 + 312X^8 - 120X^{10}\,.$$

Their conjectures were verified later by Lin,[16,17] who gave a general method to derive the generating functions for the area-weighted moments and calculated the first ten moments.

Recently the perimeter and area generating function for the convex polygons on the rectangular lattice is derived by Lin. [18] The result is very complicated. The derivation is based on the observation that a convex polygon can be divided by two horizontal lines into three polygons such that the top polygon is a pyramid polygon, the middle one is a staircase polygon and the bottom one is an inverse pyramid polygon (see figure 4).

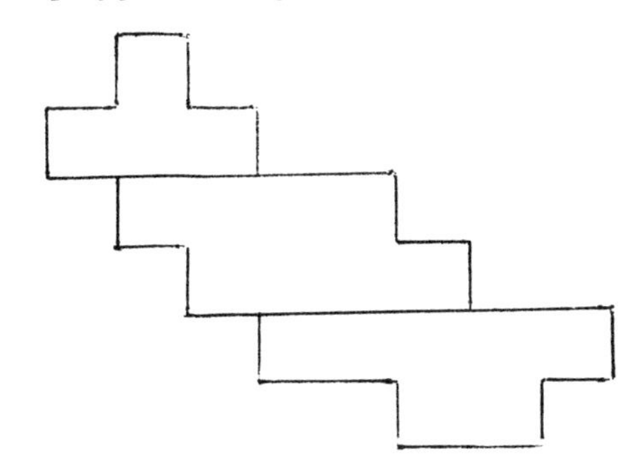

Fig. 4. A convex polygon divided into three polygons.

5. ROW-CONVEX POLYGONS

Recently Brak *et al.* [19] derived the perimeter generating function for row-convex polygons on the square lattice, which include the convex polygons as a subset. A vertical line on the bonds of the dual lattice cuts a row-convex polygon at most twice. Their result was generalized to the rectangular lattice by Lin. [20] The perimeter and area generating function for row-convex polygons on the square lattice was obtained by Brak and Guttmann [10] and their result was generalized to the rectangular lattice by Lin and Tzeng. [8]

Row-convex polygon was first considered by Temperley [2] and he wrote the generating function in the form

$$G(X\ ,Y\ ,Z) = \sum_{R,S,K=1}^{\infty} Y^{2R}\ X^{2S}\ Z^{K}\ P_{R,S,K} = \sum_{R=1}^{\infty} G_R \tag{21}$$

where G_R is the generating function for polygons whose first vertical row contains R squares, and $P_{R,S,K}$ is the number of row-convex polygons with $2R$ vertical steps, $2S$ horizontal steps and area K. It was shown by Temperley that G_R satisfies the following equations

$$G_1 = X^2Y^2Z + X^2Z\ [G_1 + 2G_2 + 3G_3 + 4G_4 + \ldots] \tag{22}$$

$$G_2 = X^2Y^4Z^2 + X^2Z^2\ [2Y^2G_1 + (1+2Y^2)G_2 + (2+2Y^2)G_3 + (3+2Y^2)G_4 + \ldots]$$

$$G_3 = X^2Y^6Z^3 + X^2Z^3\ [3Y^4G_1 + (2Y^2+2Y^4)G_2 + (1+2Y^2+2Y^4)G_3 + (2+2Y^2+2Y^4)G_4 + \ldots]$$

$$G_4 = X^2Y^8Z^4 + X^2Z^4\ [4Y^6G_1 + (3Y^4+2Y^6)G_2 + (2Y^2+2Y^4+2Y^6)G_3 + (1+2Y^2+2Y^4+2Y^6)G_4 + \ldots]$$

and similarly for $R > 4$. It follows from (22) that we have the recurrence relation

$$G_{R+2} - 2Z(1 + Y^2)G_{R+1} + Z^2(1 + 4Y^2+Y^4)G_R\ - 2Z^3(Y^2 + Y^4)G_{R-1} + Z^4Y^4G_{R-2} \tag{23}$$

$$= X^2 Z^{R+2}(1 - Y^2)^2 G_R \ .$$

Temperley pointed out that these equations are soluble in two special cases of $Y = 1$ and $Z = 1$ and he found

$$G(X\ , 1\ , Z) = X^2 Z(1 - Z)^3 \ / \ [1 - (X^2{+}4)Z + (X^2{+}6)Z^2 - (X^4{-}X^2{+}4)Z^3 + (1{-}X^2)Z^4\ ]. \quad (24)$$

In the special case of $Z = 1$, he tried $G_R = A\ \lambda^R$ and obtained the characteristic equation

$$(\lambda - 1)^2\ (\lambda - Y^2)^2 - \lambda^2 X^2 (1 - Y^2)^2 = 0\ . \quad (25)$$

The perimeter generating function is given by

$$G(X\ , Y\ , 1) = \sum_R G_R = \sum_{J=1}^{4} A_J\ \lambda_J\ /\ (1 - \lambda_J) \quad (26)$$

where λ_I are the four roots of the characteristic equation and A_I can be determined from the four equations of (22). Temperley did not calculate A_I explicitly since the algebra is too complicated to handle. Last year Brak *et al.* [19] considered the special case of $X = Y$ (square lattice) and used the computer algebra program MATHEMATICA to derive the generating function $G(Y\ , Y\ , 1)$. The general case (rectangular lattice) was solved later by Lin [20] and the result is

$$G(X\ , Y\ , 1) = (1{-}Y^2)\ [42(1 - Y^2)^2\ - 2X^2(5 - 14Y^2 + 5Y^4) - 6(1 - Y^2)S^{1/2} \quad (27)$$

$$- (1 - Y^2)(17 - X^2)T^{1/2} - (ST)^{1/2}\ ]\ /\ 8\Delta$$

where

$$\Delta = 18\ (1 - Y^2)^2 - X^2\ (2 - 5Y^2 + 2Y^4)\ ,$$

$$S = 1 - 2(X^2 + Y^2) + X^4 + Y^4 - 12X^2Y^2 - 2X^2Y^2(X^2 + Y^2) + X^4Y^4\ ,$$

$$T = 2(1 + X^2)\ (1 - Y^2)^2 + 2(1 - Y^2)\ S^{1/2}\ .$$

The idea of row-convex polygon can be extended to the honeycomb lattice and the perimeter generating function was derived by Lin and Wu. [21] . Recently we considered the row-convex polygons on a checkerboard lattice (see figure 5) and derived the recurrence relations for the perimeter generating function. The four edges of each shaded square are labelled by X , Y , Z , U . The perimeter generating function is

$$G(X,Y,Z,U) = \sum_{R,S,J,K} X^R\ Y^S\ Z^J\ U^K\ P_{R,S,J,K} = 2XYZU + \ldots \quad (28)$$

where $P_{R,S,J,K}$ is the number of row-convex polygons with R,S,J,K edges labelled respectively by X,Y,Z,U . An example of the row-convex polygon is shown in figure 5. Exact solutions have been found for the following special cases: (1) $X = Y = Z = U$ (square lattice), [19] (2) $X = U$, $Y = Z$ (rectangular lattice), [20] (3) $U = 0$ (honeycomb lattice), [21] (4) $Y = Z$ (symmetrical checkerboard lattice). [22] Very recently, Lin and Tzeng [23] obtain the general solution which is extremely complicated.

The derivation and solution of the perimeter and area generating functions for row-convex polygons on the square [10] and rectangular [8] lattices are very complicated. The original papers should be consulted for details.

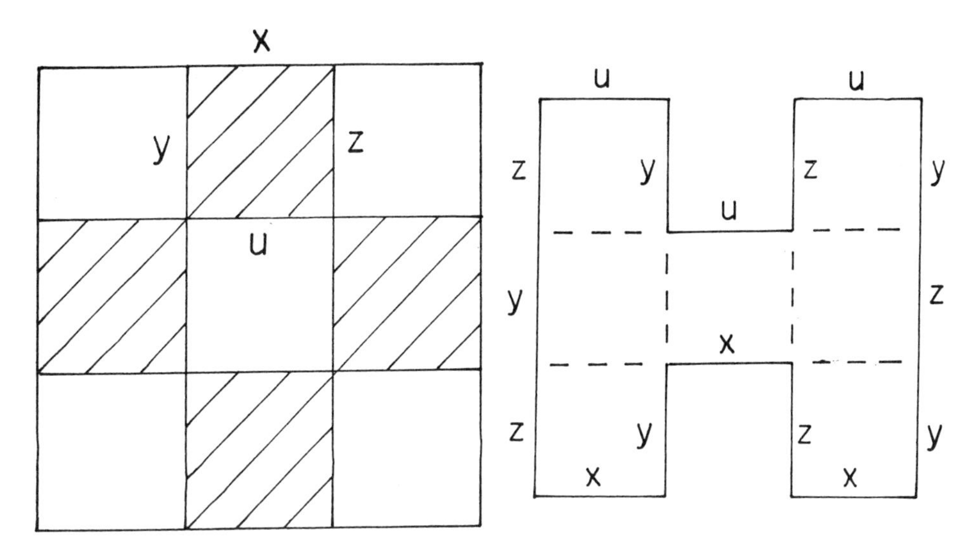

Fig. 5. The checkerboard lattice and a row-convex polygon.

REFERENCES

1. H. N. V. Temperley, Proc. Camb. Phil. Soc. 48, 683 (1952).
2. H. N. V. Temperley, Phys. Rev. 103, 1 (1956).
3. V. Privman and N. M. Švrakić, J. Stat. Phys. 51, 1091 (1988).
4. F. C. Auluck, Proc. Camb. Phil. Soc. 48, 679 (1951).
5. K. Y. Lin, Chin. J. Phys. 29, 7 (1991).
6. K. Y. Lin and S. J. Chang, J. Phys. A 21, 2653 (1988).
7. G. Pólya, J. Comb. Theor. 6, 102 (1969).
8. K. Y. Lin and W. J. Tzeng, Int. J. Mod. Phys. B 5, in press (1991).
9. K. Y. Lin, S. K. Ma, C. H. Kao and S. H. Chiu, J. Phys. A 20, 1881 (1987).
10. R. Brak and A. J. Guttmann, J. Phys. A 23, 4581 (1990).
11. M. P. Delest and G. Viennot, Theo. Comput. Sci. 34, 169 (1984).
12. D. Kim, Discrete Math. 70, 47 (1988).
13. A. J. Guttmann and I. G. Enting, J. Phys. A 21, L467 (1988).
14. I. M. Gessel, On the Number of Convex Polyominoes, Brandeis University preprint (1990).
15. I. G. Enting and A. J. Guttmann, J. Phys. A 22, 2639 (1989).
16. K. Y. Lin, J. Phys. A 22, 4263 (1989).
17. K. Y. Lin, Int. J. Mod. Phys. B 4, 1717 (1990).
18. K. Y. Lin, J. Phys. A 24, 2411 (1991).
19. R. Brak, A. J. Guttmann and I. G. Enting, J. Phys. A 23, 2319 (1990).
20. K. Y. Lin, J. Phys. A 23, 4703 (1990).
21. K. Y. Lin and F. Y. Wu, J. Phys. A 23, 5003 (1990).
22. W. J. Tzeng and K. Y. Lin, Int. J. Mod. Phys. B 5, in press (1991).
23. K. Y. Lin and W. J. Tzeng, to be published.

Transfer Matrix Spectra of Affine–D Models

D. Kim, K.-H. Kwon and J.-Y. Choi

Department of Physics and Center for Theoretical Physics
Seoul National University, Seoul 151-742, Korea

ABSTRACT

Transfer matrix spectra can be used to establish a connection between two dimensional critical lattice models and conformal field theories. Affine–D models are a class of spin models with (L+3) state per site (L =3, 4, ...) generalizing the multicritical line of the magnetic hard square model. There are two continuous parameters characterizing the models; crossing parameter λ with which critical exponents vary continuously and spectral parameter u which controls anisotropy of the interactions. Their transfer matrix spectra combined with finite-size scaling theory show that the models in the positive u regime belong to the universality class of the Ashkin–Teller model and are related to the $c = 1$ conformal field theory by $r = \{2L^2(1-\lambda/\pi)\}^{-1/2}$, where r is the orbifold radius. The models in the negative u regime are found to be related to those in the positive u regime by a certain transformation and also belong to the Ashkin–Teller universality class with the correspondence $r = \{2L^2\lambda/\pi\}^{-1/2}$.

1 Introduction

Recent developments in the conformal field theory (CFT) and its connections to critical lattice systems have provided a new paradigm for critical phenomena [1]. For example, the unitary CFTs with the central charge $c \leq 1$ have been completely classified [2, 3] and their connections to lattice models are well understood. In particular the $c = 1$ CFTs consist of three isolated cases and two sets of one–parameter family of theories. The two families describe continuum limit of two classes of lattice models with continuously varying critical exponents; the six–vertex and the Ashkin–Teller models, respectively [4]. Partition functions of lattice models on a large $N \times M$ lattice with periodic boundary conditions in both directions take the form

$$\mathcal{Z} = \exp(-NMf)\ Z(\tau) \tag{1}$$

where f is the non–universal bulk free energy per site (absorbing the usual $1/k_B T$ factor) and $Z(\tau)$ is the universal modular invariant partition function (MIPF). Here, the modular parameter τ describes geometry of torus and is determined from anisotropy of lattice models [5]. For the $c = 1$ theory in the six–vertex universality class, its MIPF is given as [6]

$$Z^{6V}(\tau; r) = \frac{1}{\eta(q)\eta(\bar{q})} \sum_{n=-\infty}^{\infty} \sum_{m=-\infty}^{\infty} q^{\frac{1}{2}(\frac{n}{2r}+rm)^2} \bar{q}^{\frac{1}{2}(\frac{n}{2r}-rm)^2} \tag{2}$$

where $\eta(q) = q^{\frac{1}{24}} \prod_{n=1}^{\infty}(1 - q^n)$ is the Dedekind eta function, $q = \exp(2\pi i \tau)$ and $\bar{q}$ is the complex conjugate of q. The continuous parameter r is called the compactification radius. The MIPF is invariant under $r \to \frac{1}{2r}$. Using a short notation $Z_n = Z^{6V}(\tau; r = n/\sqrt{2})$, the MIPF of models in the Ashkin–Teller universality class is written as [7, 8, 9]

$$Z^{AT}(\tau; r) = \frac{1}{2}[Z^{6V}(\tau; r) + 2Z_2 - Z_1]\,. \tag{3}$$

r in this case is called the orbifold radius since Z^{AT} can be obtained by the orbifold procedure from Z^{6V} [3]. Remaining $c = 1$ MIPFs are

$$\begin{aligned} \hat{E}_6 &= \frac{1}{2}(2Z_3 + Z_2 - Z_1) \\ \hat{E}_7 &= \frac{1}{2}(Z_4 + Z_3 + Z_2 - Z_1) \\ \hat{E}_8 &= \frac{1}{2}(Z_5 + Z_3 + Z_2 - Z_1)\,. \end{aligned} \tag{4}$$

Two–dimensional lattice models are said to be integrable when their Boltzmann weights satisfy the Yang–Baxter equation. There are now a large class of integrable models and their connection to CFTs has been established in many cases [10]. However, since CFTs are constructed in the boot–strap fashion, one needs to identify relationships of integrable models and CFTs in an independent way. A powerful method in this respect is the transfer matrix method since its spectra combined with the predictions of general conformal invariance principle give direct information on the scaling dimensions of operators involved [11].

One of the model for which this method is successfully applied is the magnetic hard square model [12]. It is a lattice gas on square lattice, each particle carrying an Ising spin. The particles interact with infinite nearest neighbor repulsions and finite next nearest neighbor (diagonal) attractions. They also interact ferromagnetically along the diagonals. General face weight (the Boltzmann weight of a unit square) can be written as

$$w(a,b,c,d) = \begin{cases} \exp\{La^2c^2 + Mb^2d^2 + Jac + Kbd + \frac{1}{4}(a^2+b^2+c^2+d^2)\ln z\} \\ \qquad\qquad\qquad \text{if} \quad ab = bc = cd = da = 0\,, \\ 0 \qquad\qquad\qquad \text{otherwise}\,. \end{cases} \tag{5}$$

Here z is the activity, L and M are diagonal lattice gas interactions and J and K are diagonal magnetic interactions. a, b, c and d denote the spin state of each corner of a unit square and take the value 0 (±1) when the site is vacant (occupied with spin $\pm$). The model is exactly solvable on several separate manifolds in the five–dimensional parameter space. In the so–called T–II manifold, interactions are parametrized by the two continuous parameters λ and u as

$$\begin{aligned} \tanh J &= s_- \\ \tanh K &= s \\ \exp(L) &= \frac{\sqrt{1-s_-^2}}{s^2} \\ \exp(M) &= \frac{\sqrt{1-s^2}}{s_-^2} \\ z &= 2s_-^2 s^2 \end{aligned} \tag{6}$$

where s and s_- are defined by

$$s = \frac{\sin u}{\sin\lambda} \tag{7}$$

$$s_- = \frac{\sin(\lambda - u)}{\sin\lambda}\,. \tag{8}$$

The crossing parameter λ is a continuous variable in the range

$$0 < \lambda < \frac{2\pi}{3} \tag{9}$$

and the spectral parameter u describes anisotropy of interactions. This solution manifold coincides with the multicritical line where the ferromagnetic solid, paramagnetic solid and fluid phases become critical simultaneously, and along which

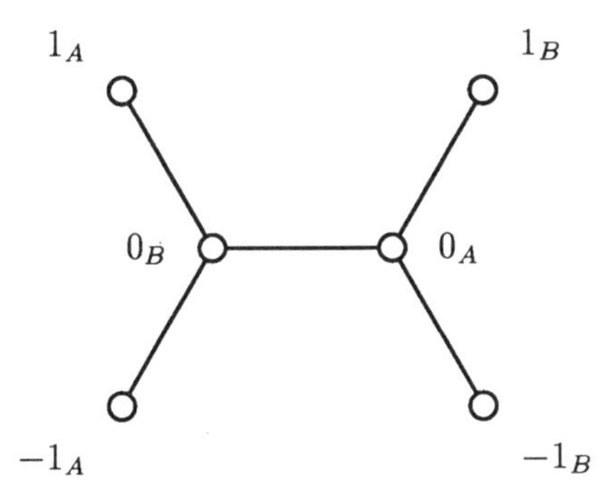

Figure 1: Adjacency condition of the magnetic hard square model in extended scheme.

the critical exponents vary continuously. This manifold is known to belong to the Ashkin–Teller universality class [13].

When one uses the extended scheme of [14] in which the internal states of a site are labelled as (s, α) with $s = 0, \pm 1$ denoting vacant and spin up (down) occupied states, respectively, and $\alpha = A, B$ denoting the sublattice label, the adjacency condition can be represented graphically by the Dynkin diagram of the affine algebra, $D_5^{(1)}$. This is shown in Fig. 1. Each node of the diagram corresponds to a local spin state while each bond specifies an allowed nearest neighbor pair of states.

Affine–D models, also denoted as $D_{L+2}^{(1)}(\lambda)$, with L=3, 4, ..., are a generalization of the multicritical manifold of the magnetic hard square model [15, 16]. These are a class of models with $(L+3)$ states per site whose adjacency condition is expressed by the Dynkin diagram of affine algebra $D_{L+3}^{(1)}$. Fig. 2 shows the diagram and labelling of the states. Boltzmann weights of the models in integrable manifold are again parametrized by the crossing parameter λ and the spectral parameter u. λ takes the range as in $L = 3$ case but u can be both positive and negative. For u positive, it ranges over $0 < u < \lambda$. For each L, this positive u region is called the regime III/IV. $-(\pi - \lambda) < u < 0$ is another region where the Yang–Baxter equation is satisfied. This regime is called the regime I/II after Baxter's solution of the hard hexagons. In this regime, positivity of Boltzmann weights is violated and hence a question arises as to whether the corresponding CFT, if there is any, is indeed unitary.

The purpose of this work is to present numerical results of the transfer matrix spectra in both regimes and to interprete the operator content in terms of

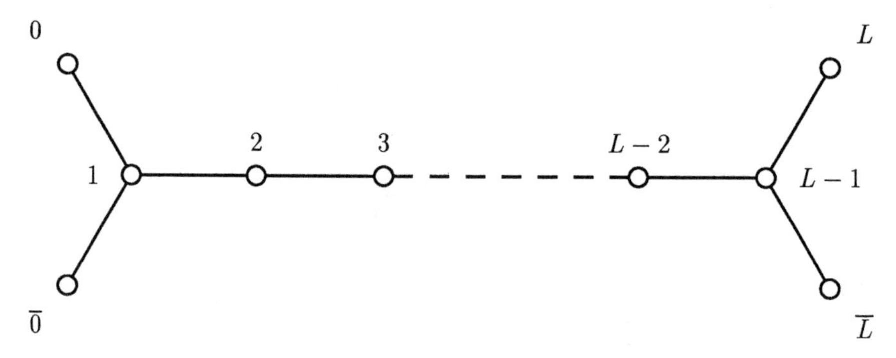

Figure 2: Diagram for adjacency conditions of affine-D models. Each node corresponds to a local spin state while each bond specifies an allowed nearest neighbor pair of states.

the $c = 1$ CFT. It turns out that the operator content in both regimes is given by that of the $c = 1$ orbifold theory. A distinguishing feature for regime I/II as compared to regime III/IV is that the strip width N should be a multiple of 4 to realize the periodic boundary condition in the continuum limit. When N is not a multiple of 4 but even, corresponding continuum theory is that with antiperiodic boundary condition and exhibits fermion operator content. In the following section, we list the face weights of the affine–D models and discuss their relation to other works. General structure of the row–to–row transfer matrix spectra is also discussed. Section 3 contains discussions on numerical results. The anisotropy factor and the central charge are determined from the largest eigenvalue. λ–dependence of excited levels are accurately determined and are shown to be consistent with the Ashkin–Teller operator content. Finally in section 4, we discuss a transformation property of the row–to–row transfer matrix between the two regimes.

2 Model and Symmetry

We label the internal states by $0,\bar{0},1,\cdots,L$ and $\bar{L}$ as shown in Fig. 2. Each model is an interaction–round–a–face model and its Boltzmann weights are product of the face weights $w(a,b,c,d|u)$, where a, b, c, and d denotes the state of the SW, SE, NE, and NW corner, respectively, of a face. We show u dependences explicitly.

With this notation, the face weights are;

$$\begin{aligned}
&w(a+1,a,a-1,a|u) = s, && a = 2,3,\cdots,L-2 \\
&w(a,a+1,a,a-1|u) = s_-, && a = 1,2,\cdots,L-1 \\
&w(a+1,a,a+1,a|u) = 1, && a = 1,2,\cdots,L-2 \\
&w(1,0,1,0|u) = 1+s && \\
&w(1,0,1,\bar{0}|u) = 1-s && \\
&w(0,1,0,1|u) = (1+s_-)/2 && \\
&w(0,1,2,1|u) = s/\sqrt{2} && \\
&w(0,1,\bar{0},1|u) = (1-s_-)/2 &&
\end{aligned} \tag{10}$$

with s and s_- given by Eqs. (7) and (8). Remaining non–vanishing weights are obtained from the reflection symmetry with respect to diagonals of a face

$$w(a,b,c,d|u) = w(c,b,a,d|u) = w(a,d,c,b|u) \tag{11}$$

and the two reflection symmetry of the Dynkin diagram;

$$a \longleftrightarrow L-a \qquad (a = 0,1,\ldots,L), \qquad \bar{0} \longleftrightarrow \bar{L} \tag{12}$$

and

$$0 \longleftrightarrow \bar{0}\,, \qquad L \longleftrightarrow \bar{L}\,. \tag{13}$$

These weights satisfy the Yang–Baxter equation. Each model is divided into two different regimes according to the sign of u. Regime I/II (III/IV) is when $-(\pi-\lambda) < u < 0$ $(0 < u < \lambda)$. Boltzmann weights in regime I/II become negative for several states of the face. It canot be removed by simple gauge transformations. The ground states of the off–critical models for $\lambda = \pi/L$ are of the form $[\cdots\ 0\,1\,\bar{0}\,1\,0\,1\,\bar{0}\,1\cdots]$ in regime II while that of regime IV is $[\cdots\ 0\,1\,0\,1\cdots]$ [17].

Next we discuss their relation to other works. Various critical lattice models based on the Temperly–Lieb algebra are constructed by Pasquier [18]. Each of Pasquier's model corresponds to a Dynkin diagram of classical algebras (A D E models) and their affine extensions ($\hat{A}$ $\hat{D}$ $\hat{E}$ models). Those associated with the Dynkin diagrams of $D_n^{(1)}$ in [18] ($\hat{D}$ models) correspond to the $\lambda = 0$ limit of the affind–D model with $u = \lambda/2$. One parameter family of weights are also discussed in Appendix of [18] and they correspond to above weights with $u = \lambda/2$ provided, using the notation of [18], $W(1,0|0,1)$ is assigned a value $w+1$ instead of w. The difference is presumably due to a misprint in [18]. In another direction, off–critical, elliptic function parametrizations for models associated with Dynkin diagrams of algebras A_n, D_n, $A_n^{(1)}$ and $D_n^{(1)}$ have been found and are summarized in [17]. Here, there are four regimes of physical interest. Critical lines are where the nome of the elliptic functions vanishes and bound regimes I and II (III and IV) when the spectral parameter is negative (positive). Thus the critical manifolds are denoted as regime I/II (III/IV) for u negative (positive). (Definition of the spectral parameter in [17] differs from ours by a factor $-\lambda$.) When specialized to the $D_{L+2}^{(1)}$ models, the critical points of general off–critical models correspond to the affine–D models with $\lambda = \pi/L$. It should be noted that critical points for A– and D–series in regime III/IV are the A and D models of Pasquier but the same is not true for other cases. Their MIPFs form a part of $c < 1$ menu. A–series in regime I/II are related to the $Z(N)$ parafermionic CFT of [19], affine–A series in both regimes are mapped to the six–vertex model ($c = 1$) [20]. Connection of the D–series in regime I/II to CFTs is not known to the authors at the present time. These models should not be confused with wider, but not inclusive class of models in which the spin states are associated with dominant integral weights of affine Lie algebras with fixed level [21]. In that point of view, above A–series (=restricted solid on solid models) correspond to the level–k $\widehat{\mathrm{SU}(2)}$ model and its critical points in regime III/IV are described by the coset CFT $\widehat{\mathrm{SU}(2)}_k \times \widehat{\mathrm{SU}(2)}_1/\widehat{\mathrm{SU}(2)}_{k+1}$.

We consider here the row–to–row transfer matrix under the periodic boundary condition and denote it by $\mathbf{T}$. Its eigenvalues are denoted by $\Lambda_r(N)$ where N is the strip width. Periodic boundary condition restricts N to be even. A consequence of the CFT is that $\Lambda_r(N)$'s give an estimation of the scaling dimensions x_r and the spin S_r of corresponding operators by the formula

$$- \mathit{Re} \ln \Lambda_r(N) = N f(u) + \frac{2\pi}{N} \frac{a_t}{a_x} \sin\theta \, (x_r - \frac{c}{12}) \tag{14}$$

$$- \mathit{Im} \ln \Lambda_r(N) = \frac{2\pi}{N} \frac{a_t}{a_x} S_r \cos\theta + \frac{2\pi}{p} q \tag{15}$$

(q=0,1,...,$p-1$) where f is the bulk free energy, c is the central charge, a_t and a_x are the lattice constants in time and space direction, respectively, and θ is the anisotropy angle. The second term in Eq. (15) is the extra phase which may present when the ground state is periodic with periodicity p [5]. In our case the reflection symmetry with respect to diagonals of a face requires $a_t/a_x = 1$.

From the periodic boundary condition, $\mathbf{T}$ commutes with the shift operator $\mathbf{S}$ defined by

$$\mathbf{S}_{\vec{a}\vec{b}} = \delta_{a_1 b_2} \, \delta_{a_2 b_3} \cdots \delta_{a_N b_1} \tag{16}$$

where $\vec{a}$ ($\vec{b}$) is a state of a row whose site values are $a_1, a_2, \cdots, a_N$ ($b_1, b_2, \cdots, b_N$). $\mathbf{T}$ also commutes with the 'top–bottom' reflection operator $\mathbf{R}$ defined by

$$\mathbf{R}_{\vec{a}\vec{b}} = \delta_{a_1\bar{b}_1}\,\delta_{a_2\bar{b}_2}\,\cdots\,\delta_{a_N\bar{b}_N} \tag{17}$$

where $\bar{a}_i = \bar{0}\,(\bar{L})$ if $a_i = 0\,(L)$, and $\bar{a}_i = a_i$ if $a_i = 1, 2, \cdots, L-1$. Alternatively, we can write

$$\mathbf{S}|abcd\cdots> = |bcd\cdots a> \tag{18}$$

$$\mathbf{R}|abcd\cdots> = |\bar{a}\bar{b}\bar{c}\bar{d}\cdots> \tag{19}$$

where $|abcd\cdots>$ stands for a basis of the transfer matrix. Using the commuting properties, we can block-diagonalize $\mathbf{T}$ into sectors identified by eigenvalues $\tilde{s}$ and $\tilde{r}$ of operators $\mathbf{S}$ and $\mathbf{R}$, respectively. $\tilde{s}$ takes the values $\exp(\frac{2\pi i}{N}k)$, with k=0,1, $\ldots$, $N-1$ and $\tilde{r}$ has the values ± 1. Spin S of a level is related to its momentum by

$$k = \frac{N}{p}q + S\ , \tag{20}$$

where q=0, ± 1, 2 when p=4 (regime I/II) and q=0, 1 when p=2 (regime III/IV). Thus only integer spin operators appear in regime III/IV. However when N is not a multiple of 4 but even in regime I/II half–integer spin operators (fermions) appear in the sector $k = \pm(N \pm 2)/4$ (q=± 1). We will use the notation $(S, \tilde{r})$ to denote a sector.

The bulk free energy can be obtained using the inversion relation and certain analyticity assumptions. The affine–D models satisfy the same inversion relation as in the six–vertex model [22] and share same bulk free energy (independent of L). It is given by

$$f(u) = -\int_{-\infty}^{\infty} dt\, \frac{\cosh(\pi - 2\lambda)t\, \sinh ut\, \sinh(\lambda - u)t}{t \sinh \pi t\, \cosh\lambda t}\ , \qquad (0 < u < \lambda)\ , \tag{21}$$

in regime III/IV and

$$f(u) = \int_{-\infty}^{\infty} dt\, \frac{\cosh(\pi - 2\lambda)t\, \sinh ut\, \sinh(\pi - \lambda + u)t}{t \sinh \pi t\, \cosh(\pi - \lambda)t}\ , \qquad (\,-(\pi - \lambda) < u < 0)\ , \tag{22}$$

in regime I/II. Note that free energies of the two regimes are related by a transformation $u \to -u$ and $\lambda \to \pi - \lambda$. .

3 Transfer Matrix Spectra

In case of the magnetic hard squares, the nonlinear functional equations for eigenvalues of the row transfer matrix had been solved for a high resolution spectroscopy [13]. For the purpose of this work, direct diagonalization of the transfer matrices which we call the low resolution spectroscopy was sufficient to draw unambigous conclusions. We have calculated all eigenvalues for the strip width N upto 12 for models with L=3, 4 and 5. When the strip width is commensurate with the ground state periodicity, the first level (the largest eigenvalue Λ_0) determines the central charge and anisotropy angle through the relation

$$c \sin\theta = \lim_{N\to\infty} \frac{6N}{\pi}(\log|\Lambda_0| + Nf(u)) \tag{23}$$

Our numerical results show that L, u and λ dependences of $\frac{6N}{\pi}(\log|\Lambda_0| + Nf(u))$ can be fitted quite accurately to the conjectured forms $\sin\{|u|/(1-\lambda/\pi)\}$ and $\sin\{\pi u/\lambda\}$ in regime I/II and III/IV respectively. This means that the central charge and the anisotropy angle θ are given by

$$c = 1 \tag{24}$$

and

$$\theta = \begin{cases} \pi|u|/(\pi-\lambda) & \text{regime I/II} \\ \pi u/\lambda & \text{regime III/IV}\ , \end{cases} \tag{25}$$

respectively, independent of L. The value $c = 1$ is in accord with Kuniba and Yajima's c [17] obtained indirectly from the local height probability solution.

Having determined c and θ, we form the N–th estimators of scaling dimensions for all levels by

$$x_r(N) = \frac{N}{2\pi\sin\theta}(-\log|\Lambda_r| - Nf(u)) + \frac{1}{12}\ . \tag{26}$$

Whenever possible, the sequences of estimators are fitted to the form $x_r(N) = x_r + a/N^b$ using data for N=4, 8, 12 (regime I/II) or 8,10,12 (regime III/IV) to get extrapolated values. Low lying levels thus identified are consistent with the operator content of the Ashkin–Teller universality class as implied by Eq. (3). First, the gaussian model operators with dimension

$$x_r = X_{n,m} = n^2 X_T/4 + m^2/X_T \tag{27}$$

for integer n and m appear in the sector $(nm,+1)$ where the thermal scaling exponent X_T is found to be

$$X_T = \begin{cases} 2\pi/L^2\lambda & \text{regime I/II} \\ 2\pi/L^2(\pi-\lambda) & \text{regime III/IV}\,. \end{cases} \tag{28}$$

All λ– and L–dependences of the scaling dimensions are embodied in this expression of X_T. The magnetization operators with dimensions (1/16, 1/16) and (9/16, 9/16) appear in the sector $(0,-1)$. Typical raw data are displayed in the following figures. Estimators of scaling dimensions for L=3, N=12, are shown as a function of λ in Figs. 3(a) and (b) for the case of u=$-\lambda/2$ and u=$\lambda/2$, respectively. The circle, triangle, square and diamond shapes denote scaling dimensions with momentum k=0,1,2 and 3, respectively. The filled (empty) points of each shape represent the sector $\tilde{r}$=+1 (-1). The solid and dashed lines are the expected exact levels obtained from Eqs. (27) and (28) for S=0 and 1, respectively. Note that the label of the x–axis in Fig. 3(b) has been inverted to emphasize the symmetry $\lambda \leftrightarrow \pi - \lambda$ between the two regimes. Fig. 4 shows the same for L=4 and u=$-\lambda/2$. The quality of numerical convergence varies. For the lowest level, we get $10^{-3} \sim 10^{-4}$ for the estimator $x_0(12)$ and its extrapolated values are of the order of 10^{-5}. Thus the uncertainty of the conjectured values of $c = 1$ is of the order of 10^{-4}. For levels around $x_r \sim 2$, the uncertainty increases to about 1%.

For N not a multiple of 4 but even in regime I/II, the (0,0) level corresponding to the identity operator and its descendants disappear. Instead, the (1,0) level is present as in the six-vertex model. In the sectors for $k = (N \pm 2)/4$, we find doublets corresponding to the fermion operators with dimension (1/16, 9/16), (9/16, 1/16), (1/16, 25/16) and (25/16, 1/16). Other levels continuously varying with λ show the same behaviors as in the case of $N/4$ = integer. This is exactly the operator content of the Ashkin–Teller model under the antiperiodic boundary condition [8, 7]. The lower levels with u=$-\lambda/2$, $N = 10$ and $L = 3$ are shown in Fig. 5.

4 Discussion

Numerical analysis shows that the operator content of the model $D^{(1)}_{L+2}(\lambda)$ in regime I/II (III/IV) is identical to that of the Ashkin–Teller model with the thermal scaling dimension given by $X_T = 2\pi/L^2\lambda$ $(2\pi/L^2(\pi-\lambda))$. Therefore the

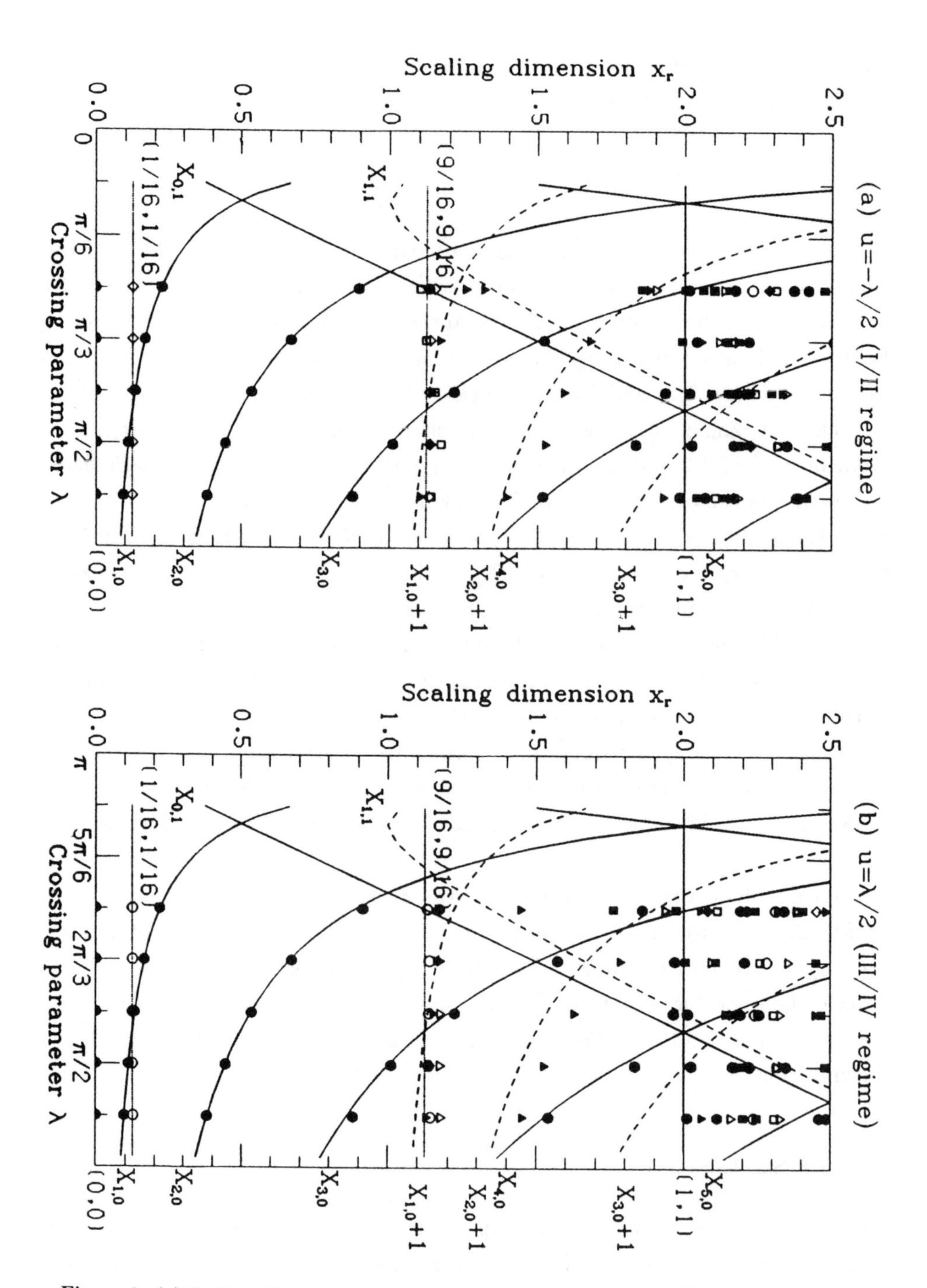

Figure 3: (a) Scaling dimension estimators $x_r(N)$ of the model $D_5^{(1)}$ $(L = 3)$ for $N = 12$, $u = -\lambda/2$ and several values of λ. See text for notations. (b)The scaling dimension estimators of $D_5^{(1)}$ in regime III/IV for $N = 12$ and $u = \lambda/2$.

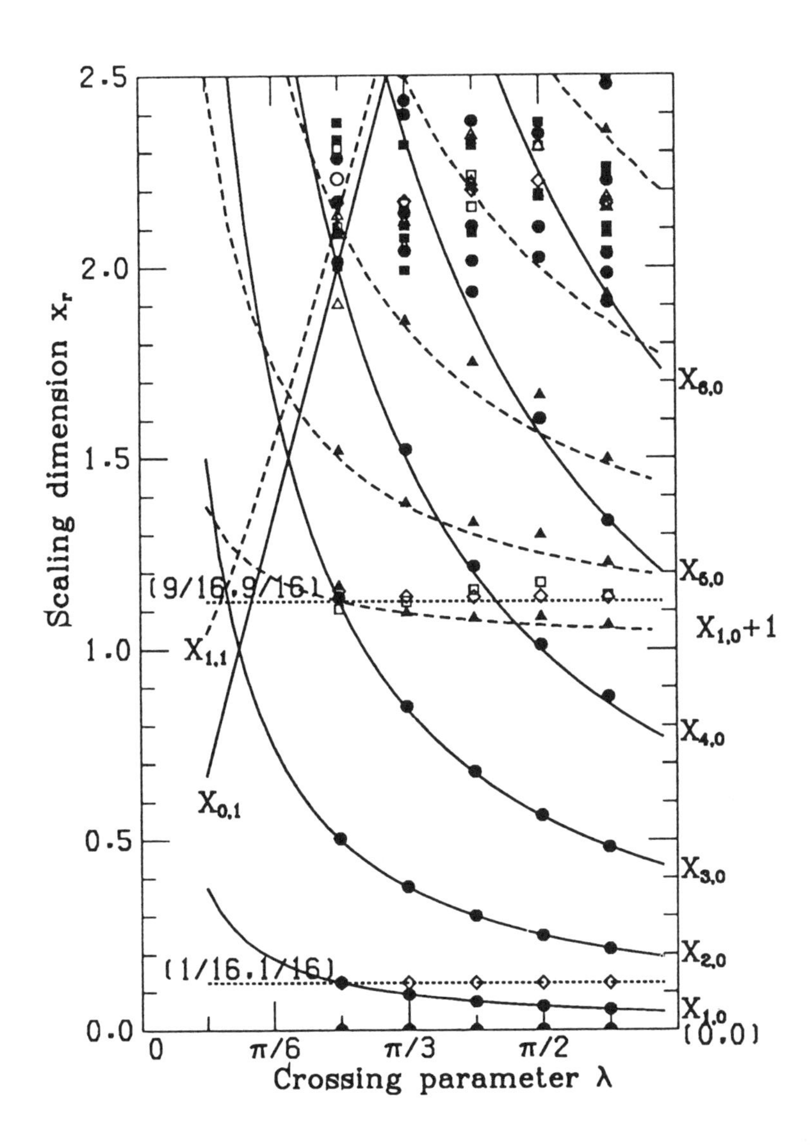

Figure 4: The same as in Fig. 3(a) for $L = 4$.

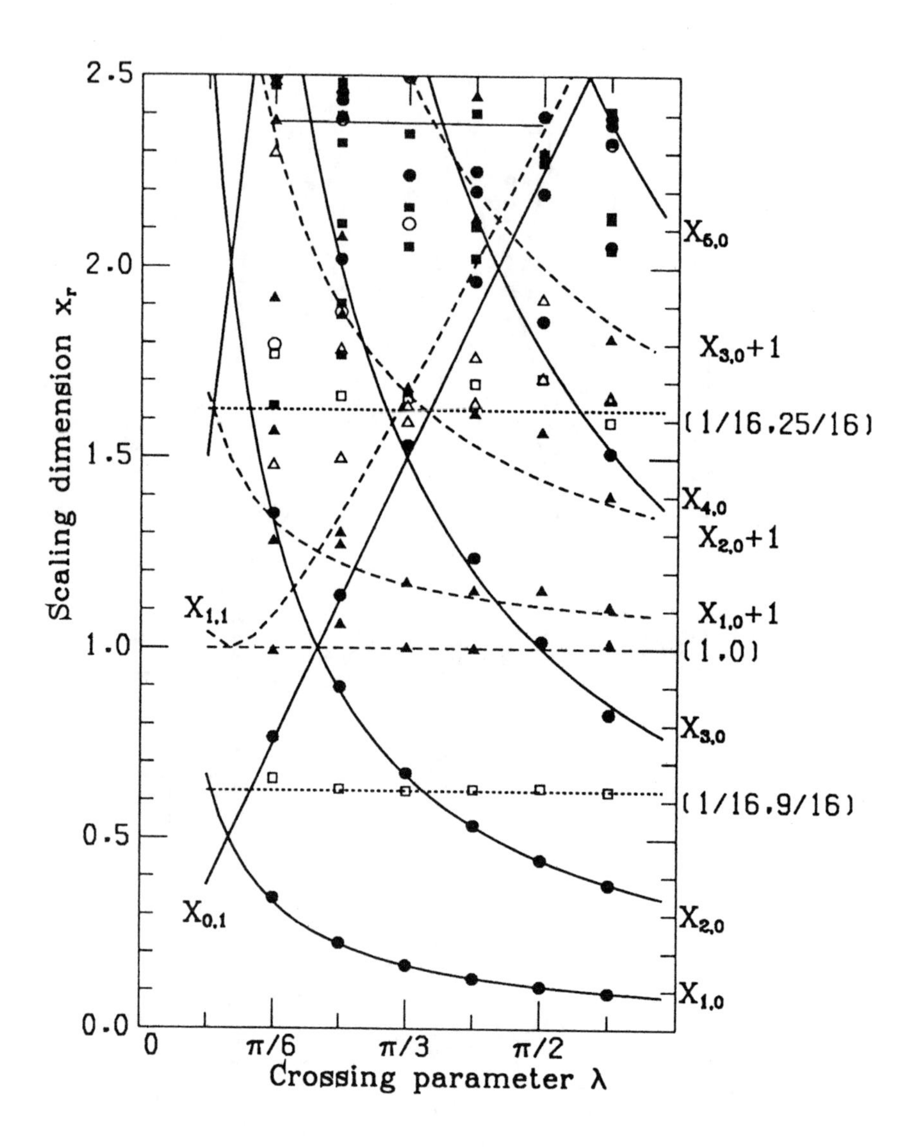

Figure 5: Scaling dimension estimators of $L = 3$ with $N = 10$, $u = -\lambda/2$ as a typical example of the spectrum for incommensurate N.

corresponding CFT is the $c = 1$ orbifold theory with the orbifold radius

$$r = \begin{cases} \{2L^2\lambda/\pi\}^{-\frac{1}{2}} & \text{regime I/II} \\ \{2L^2(1-\lambda/\pi)\}^{-\frac{1}{2}} & \text{regime III/IV}\,. \end{cases} \tag{29}$$

This result also suggests there might be a certain symmetry of the model under the transformation

$$u \to -u \qquad \text{and} \qquad \lambda \to \pi - \lambda\,. \tag{30}$$

This transformation leaves s_- invariant but changes sign of s and its effect cannot be removed by a simple gauge transformation. Therefore there appears no symmetry at the level of individual face weights. However, we do find a symmetry when two faces are considered simultaneously. It takes the form, with obvious notations,

$$\begin{array}{ccc} a & b & c \\ & (-u,\lambda) \quad (-u,\lambda) & \\ d & e & f \end{array} \;=\; \begin{array}{ccc} a & \bar{b} & \bar{c} \\ & (u,\pi-\lambda) \quad (u,\pi-\lambda) & \\ \bar{d} & \bar{e} & f \end{array} \times(-1)^{\mu}$$

where $\mu = (b-d+c-e)/2$ and the states $\bar{0}$ and $\bar{L}$ are regarded as a number 0 and L respectively in the expression for μ. This *two-face identity* has been checked for all possible configurations and for general L. The sign factor disappears when N=4n (n=integer) faces are put together under the periodic boundary condition. Let us introduce, when N=4n, four operators which reflect two adjacent spins out of every 4-spin blocks;

$$\begin{aligned} \mathbf{R}_1|abcdefgh\cdots> &= |\bar{a}\bar{b}cd\bar{e}\bar{f}gh\cdots> \\ \mathbf{R}_2|abcdefgh\cdots> &= |a\bar{b}\bar{c}de\bar{f}\bar{g}h\cdots> \\ \mathbf{R}_3|abcdefgh\cdots> &= |ab\bar{c}\bar{d}ef\bar{g}\bar{h}\cdots> \\ \mathbf{R}_4|abcdefgh\cdots> &= |\bar{a}bc\bar{d}\bar{e}fg\bar{h}\cdots> \end{aligned} \tag{31}$$

Note that

$$\mathbf{R}_{i+2} = \mathbf{R}\mathbf{R}_i \tag{32}$$

where the indices of $\mathbf{R}_i$ are defined modulo 4. We can then write, for N=4n and periodic boundary condition,

$$\mathbf{T}(-u,\lambda) = \mathbf{R}_i \mathbf{T}(u, \pi - \lambda)\mathbf{R}_{i+1} \tag{33}$$

(i=1, 2, 3, 4) where we have shown u and λ dependences of $\mathbf{T}$ explicitly. Therefore spectra in the two regimes are closely related. In fact the toroidal partition functions are simply related. Consider the model on a torus with N columns and M rows with N and M both a multiple of four; $N = 4n$, and $M = 4m$. Using Eq. (33), one can easily show that the toroidal partition function in regime I/II is thc same as that in regime III/IV after the transformation Eq. (30). Also when the torus is such that $N = 4n$ and $M = 4m + 2$, the partition function in regime I/II denoted as $Z_{PA}(-u, \lambda)$ can be written as

$$\begin{aligned} Z_{PA}(-u,\lambda) &\equiv \operatorname{tr} \mathbf{T}(-u,\lambda)^M \\ &= \operatorname{tr} \mathbf{R}_1 \mathbf{T}(u,\pi-\lambda)^M \mathbf{R}_3 \\ &= \operatorname{tr} \mathbf{T}(u,\pi-\lambda)^M \mathbf{R}\,. \end{aligned} \tag{34}$$

Therefore it is related to that for regime III/IV with antiperiodic boundary condition along the time direction. The partition function for $N = 4n+2$ and $M = 4m$, which can be obtained from $Z_{PA}(-u, \lambda)$ by the modular transformation is then given by the Ashkin–Teller partition function with antiperiodic boundary condition along the space direction. These observations fit together with the numerical result.

Acknowledgement

We thank P. A. Pearce for helpful discussions. This work is supported by the KOSEF grant 901–0203–021–2 and SNU–Daewoo Research Fund.

References

[1] Cardy J.L., in *Phase Transitions and Critical Phenomena*, ed. C. Domb and J.L. Lebowitz, Vol **11** (Academic Press, New York, N.Y.) 1987.

[2] Cappelli A., Itzykson C. and Zuber J.B., Nucl. Phys. B, **280** (1987) 445; Comm. Math. Phys. **113** (1987) 1.

[3] Ginsparg P., Nucl. Phys. B **295** (1988) 153; Kiritsis E.B., Phys. Lett. B **217** (1989) 427.

[4] Baxter R.J., *Exactly Solved Models in Statistical Mechanics*, Academic Press (1982)

[5] Kim D. and Pearce P.A., J. Phys. A **20** (1987) L451.

[6] Di Francesco P., Saleur H. and Zuber J.-B., Nucl. Phys. **B285 FS19**] (1987) 454; **B300[FS22**] (1988) 393.

[7] Yang S.-K., Nucl. Phys. B **285** (1987) 183, 639; Yang S.-K. and Zheng H.B., Nucl Phys B **285** (1987) 410.

[8] Baake M., von Gehlen G. and Rittenberg V., J. Phys. A **20** (1987) L487.

[9] Saluer H., J. Phys. A **20** (1987) L1127.

[10] Di Francesco P., *Integrable Lattice Models, Graphs and Modular Invariant Conformal Field Theories*, Princeton University Preprint #PUPT–1243 and references therein.

[11] Cardy J.L., Nucl. Phys. B **270** (1986) 186.

[12] Pearce P.A., J. Phys. A **18** (1985) 3217; Akutsu Y., Kuniba A. and Wadati M., J. Phys. Soc. Jpn. **55** (1986) 1880; Kuniba A. and Yajima T., J. Phys. A **21** (1988) 519.

[13] Pearce P.A. and Kim D., J. Phys. A **20** (1987) 6471; Kim D., Choi J.-Y. and Kwon K.-H., J. Phys. A **21** (1988) 2661.

[14] Choi J.-Y.,Kim D. and Pearce P.A., J. Phys. A **22** (1989) 1661.

[15] Choi J.-Y., Kwon K.-H. and Kim D., Europhys. Lett. **10** (1989) 703.

[16] Kwon K.-H. and Kim D., *Transfer Matrix Spectra of Affine–D Models in Regime I/II*, Seoul National University Preprint #SNUTP 91–23.

[17] Kuniba A. and Yajima T., J. Stat. Phys. **52** (1988) 829.

[18] Pasquire V., J. Phys. A **20** (1987) L1229.

[19] Zamolodchikov A.B. and Fadeev V.A., Sov. Phys. JETP **82** (1985) 215.

[20] Pearce P.A. and Seaton K.A., Phys. Rev. Lett. **60** (1988) 1347; Ann. Phys. **193** (1989) 326; Kim D. and Pearce P.A., J. Phys. A **22** (1989) 1439.

[21] Jimbo M., Miwa T. and Okado M., Lett. Math. Phys. **14** (1987) 123; Comm. Math. Phys. **116** (1988) 507; Date E., Jimbo M., Kuniba A., Miwa T. and Okado M., Lett. Math. Phys. **17** (1989) 69.

[22] Pearce P., *private communication*; Choi J.-Y., *unpublished.*

PHASE TRANSITIONS: STATICS AND DYNAMICS

PERCOLATION AND PHASE TRANSITIONS: NUMERICAL STUDIES*

CHIN-KUN HU

*Institute of Physics, Academia Sinica,
Nankang, Taipei, Taiwan 11529, R.O.C.*

ABSTRACT

From subgraph expansions of Ising-type models in external fields, phase transitions of Ising-type models have been related to percolation transitions of corresponding correlated percolation models. From Monte Carlo simulations of hard-core particles on lattices, phase transitions of such systems have been related to percolation transitions of site-correlated percolation problems. In this talk, we review recent numerical studies of lattice interacting systems based on such connections. The topics under discussion include percolation renormalization group approach to lattice systems, cluster Monte Carlo studies of the Potts model on hypercubic lattices, Monte Carlo calculations of the free energy and geometrical factors for the Potts model, etc.

I. INTRODUCTION

About two thousand years ago, a famous Chinese philosopher Chuang Chou in the Warring Period (376-221 B.C.) of Chinese historry said:" My life is limited, but the knowledge is unlimited; with limited life to pursue unlimited knowledge, it is dangerous !" Today there is much more knowledge than the time of Chuang Chou. How to solve the problem raised by Chuang Chou ?

Confucius (551-479 B.C.) once said:" In the Way I follow, there is one unifying connected principle". Similarly, in the development of physics, physicists are trying to describe many natural phenomena by one unified connected principle. If many natural phenomena may be described by one unified connected principle, the effort one should use to pursue knowledge may be greatly reduced and the problem raised by Chuang Chou may be partially solved.

In 1666 Issac Newton (1642-1727 A.D.) used the concepts of gravitational force, momentum, and Newton's equation of motion to describe the falling of an apple to the earth, the orbiting of the moon around the earth, the orbiting of planets around the sun, etc. In 1865 J. C. Maxwell (1831-1879 A.D.) used the concept of displacement current and Maxwell's electromagnetic equations to describe electric phenomena, magnetic phenomena and the generation and propagation of light. In 1967-1968 Weinberg[1] and Salam[2] used the concepts of Yang-Mills gauge fields[3] and spontaneous symmetry breaking[4] to formulate a unified theory of electromagnetic and weak interactions. The current major objective of theoretical elementary particle physics is to formulate a unified theory of all basic interactions based on some geometrical concepts.

⋆ Talk presented at the 1991 Taipei International Symposium on Statistical Physics: Computer Aided Studies in Statistical Physics, Taipei, Taiwan, 20-25 June 1991.

Besides basic interactions, there are other interesting problems in physics, such as phase transitions and critical phenomena in interacting many particle systems. Various kinds of phase transitions and critical phenomena have been observed or identified,[5-15] including gas to liquid transitions,[5] paramagnet to ferromagnet transitions,[6] superconductivity,[7] condensation of ideal Bose gas,[8] superfluid,[9,10] disorder state to order state transitions in alloy,[11] liquid crystal,[12] surface phase transitions,[13-15] etc.

In order to characterize the singular behavior of critical phenomena near the phase transition point, various critical exponents have been defined.[16] In order to understand the behavior of phase transitions and critical phenomena in interacting many particle systems, scientists have proposed various models of phase transitions. The famous models of phase transitions defined on lattices include the Ising model[17-21] and the hard-core particle models.[22-26] The Ising model has been extended to many Ising-type spin model, e.g. the Potts model,[27-28] the BEG model,[29] the Baxter model,[30] the sublattice-dilute Potts model,[31] etc. Some of these models have been used to represent the ferromagnet,[17] the binary alloy,[32] liquid-gas systems,[33] liquid crystals,[34] atoms or molecoles on surfacese,[35] the mixture[29] of He^3 and He^4, etc, while the hard-core particle models have been used to represent atoms or molecoles on surfaces.[36]

In a seminal paper, Onsager[19] has calculated the exact zero field free energy for the two-dimensional Ising model on the square lattice and found that the specific heat has logarithmic divergence with critical exponents $\alpha = \alpha' = 0$, Yang[20] has later calculated the exact spontaneous magnetization and found that $\beta = 0.125$, Fisher[21] has obtained $\gamma = 1.75$. For the hard-core particles on the triangular lattice with nearest-neighbor exclusion, i.e. hard hexagon model, Baxter[23,24] has calculated the exact partition function and order parameter as a function of chemical potential Δ and found $\alpha = \frac{1}{3}$, $\beta = \frac{1}{9}$, and the critical point $\Delta_c = \ell n \frac{1}{2}(11 + 5\sqrt{5})$. Baxter[24-25] has also calculated the exact partition function of the hard square model with the next-nearest-neighbor attractive coupling K along a line in the (Δ, K) plane which includes a tricritical point[37] $(K, \Delta_c) = (1.65557\cdots, -3.2538\cdots)$. It is of interest to know that the critical behavior of the two dimensional Ising model and the hard-core particle models have been realized in experiments .[13-15,35-36] However, only in exceptional case one can find exact solutions for phase transition models and the traditional general method, such as mean-field methods,[38] usually do not give reliable results.

Studies from experiments and model calculations indicate that the critical exponents of a given system satisfy certain equalities, called scaling relations,[16] e.g. $\alpha = \alpha'$, $\gamma = \gamma'$, $\alpha' + 2\beta + \gamma' = 2$, $\beta\delta = \beta + \gamma$, $(2 - \eta)\nu = \gamma$, $d\nu = 2 - \alpha$ and that phase transitions of various systems may be classified into universal classes such that the systems in the same class have the same set of critical exponents.[16]

In order to understand these scaling relations of critical exponents, it has been proposed[39,40] that in the critical region the singular part of the Gibbs free energy, $G(\epsilon, h)$, and the pair correlation function $\Gamma(r, \epsilon)$ for two particles with a separation r are generalized homogeneous functions of their arguments, so that

$$G(\lambda^{y_T}\epsilon, \lambda^{y_h}h) = \lambda^d G(\epsilon, h), \tag{1.1}$$

for any value of the number λ and $\Gamma(r,\epsilon)$ may be written as:

$$\Gamma(r,\epsilon) = \epsilon^{2(d-y_h)/y_T} \; f(r\epsilon^{1/y_T}). \tag{1.2}$$

The critical exponents may be written in terms of the magnetic scaling power y_h and the thermal scaling power y_T, e.g.

$$\beta = \frac{d-y_h}{y_T}, \quad \delta = \frac{y_h}{d-y_h}, \quad \gamma = \gamma' = \frac{2y_h - d}{y_T}, \tag{1.3}$$

$$\alpha = \alpha' = 2 - \frac{d}{y_T}, \quad \nu = \frac{1}{y_T}. \tag{1.4}$$

Scaling relations follow from such equations. However, studies show that y_T and y_h are usually nonintegers.

In this talk, I will point out that the idea of "Percolation" is useful for formulating the unified theory and general numerical calculation method for phase transitions of interacting systems.

The random percolation model was first proposed by Flory[41] in 1941 to explain the sol-gel phase transitions. Since then percolation has emerged as an important branch of sciences in recent decades.[42–43] In percolation problems one may also define critical exponents and relate such exponents to the "magnetic scaling power" and the "thermal scaling power".[42,43] The random percolation problem has been well understood and developed in the sense that the singularity at p_c is due to the onset of the appearance of percolating clusters at p_c, the magnetic scaling power and the thermal scaling power may be related to the fractal dimensions of incipient percolating clusters[44] and the fractal dimensions of cutting bonds,[45] D_c, the scaling relations of critical exponents may be related to the scaling form of the cluster-size distribution[42] for large clusters near p_c, the percolation renormalization group methods[46,47] have been formulated to calculate the critical point, the critical exponents, and the percolation probability.

In the following sections we will point out that many aspects of the above understanding and developments for the random percolation problems may be extended to the Ising-type spin models and hard-core particle models. In such interacting systems, the phase transitions may be related to percolation transitions of correlated percolation problems.[48–66]

II. PERCOLATION FOR ISING-TYPE SPIN MODELS

To begin with, we give some historical remarks. The singular behavior of the percolation transition in the percolation problem is clearly related to the onset of the appearance of the percolating cluster in the system. It is of interest to know whether the phase transitions in interacting systems, e.g. the simple Ising model, have a mechanism similar to that of the percolation transition. If so, then one could have a geometric picture of the former as clear as that of the latter. To solve this problem one must unambiguously define clusters for the Ising system such that at high temperatures the system has only small nonpercolating clusters and at low temperatures the system has at least one percolating cluster. When the temperature is decreased from high temperatures to low temperatures, there is a percolation transition at a temperature T_p. As in the case of the random percolation problems,[41-47] one may also define the percolation probability P, the mean cluster size S, pair connectedness function and the corresponding correlation length ξ, etc. for the clusters of the Ising model. If the phase transition of the Ising model is a percolation transition, then $T_p = T_c$ and the critical exponents $(\beta_p, \gamma_p, \gamma'_p, \nu_p, \nu'_p, \eta_p, etc)$ of the clusters are the same as the corresponding Ising critical exponents $(\beta, \gamma, \gamma', \nu, \nu', \eta, etc)$.

In 1967 Fisher[67] published a droplet model in which he considered the lattice sites with an Ising spin $\sigma = -1$ (opposite to the external magnetic field) occupied and the lattice sites with an Ising spin $\sigma = +1$ unoccupied; the nearest neighbor (NN) occupied sites were considered to be in the same cluster. The ferromagnetic interactions between Ising spins make the occupation of lattice sites correlated and the Ising model has been mapped into a site-correlated percolation model (SCPM). For the mean number of clusters (per site), he wrote down a semiphenomenological equation with two parameters, say σ and τ. The critical exponents of the Ising model may be expressed in terms of σ and τ. From such relations, one may derive scalings laws of critical exponents, e.g. $\alpha' + 2\beta + \gamma' = 2$. However, in 1974 Fisher's semiphenomenological equation was found to be inconsistent with Monte Carlo data.[68] Despite such failure, since 1974 Fisher's idea of Ising clusters has been used in studying the connection between the phase transition of the Ising model and the percolation transition. It was found[69] that for the two dimensional lattice $T_p = T_c$ but $\gamma_p = 1.91 > \gamma = 1.75$ and for the three dimensional lattice $T_p \neq T_c$.

In 1969-1972 Kasteleyn and Fortuin[70,71] showed that the spontaneous magnetization and the magnetic susceptibility of the q-state Potts model for $q \to 1$ correspond to the percolation probability and the mean cluster size for the bond random percolation model. They also wrote down following subgraph expansion for the Potts model in zero external field for general q:

$$Z(G,p,q) = \sum_{G'} p^{b(G')} \, (1-p)^{E-b(G')} \, q^{n(G')} \, . \tag{2.1}$$

Here the summation is over all subgraphs of G, $b(G')$ is the number of occupied bonds in G', $n(G')$ is the number of clusters in G', and p is related to the nearest-neighbor coupling constant K by

$$p = 1 - e^{-K} . \tag{2.2}$$

It seems that they considered the generation of subgraphs using the probability weight

$$\pi(G', p) = p^{b(G')} \, (1-p)^{E-b(G')} \tag{2.3}$$

for the bond random percolation model and then calculate the average values of geometrical quantities in such subgraph distributions. Therfore they wrote the partition function and the pair correlation function in the following way.

$$Z(G,p,q) = < q^{n(G')} \; ; G, p >, \tag{2.4a}$$

$$< \sigma_\nu \quad \sigma_{\nu'} > = < \gamma_{\nu\nu'} \, q^{n(G')} \; ; G, p > / < q^{n(G')} \; ; G, p > . \tag{2.4b}$$

Namely, Kasteleyn and Fortuin only had the idea of "random percolation", but did not have the idea of "correlated percolation". The partition function $Z(G,p,q)$ is the average value of $q^{n(G')}$ in the random bond percolation. Such interpretation of the partition function was also taken by Essam in his review paper on percolation.[72]

In 1981, I and Kleban[73] studied low-temperature limits of Syozi model model[74] on the semidilute honeycomb lattice and the bond decorated square (sq), plane triangular (pt), and honeycomb (hc) lattices. The order-disorder phase boundary for $T \to 0$ gives the critical concentration P_c. We found that:

$$P_c(decorated \;\; sq) = 0.5, \tag{2.5a}$$
$$P_c(decorated \;\; pt) = 1 - P_c(decorated \;\; hc) = \frac{5}{18}(3-\sqrt{3}) = 0.3522..., \tag{2.5b}$$
$$P_c(semidilate \;\; hc) = 0.5. \tag{2.5c}$$

On the other hand, it has been known that for the random bond or site percolation problems, the critical probabilities are given by

$$p_c(bond, sq) = 0.5, \tag{2.6a}$$
$$p_c(bond, pt) = 1 - p_c(bond, hc) = 2sin\frac{11}{18} = 0.349..., \tag{2.6b}$$
$$p_c(sublattice \;\; site, hc) = 0.5. \tag{2.6c}$$

Therefore the values of Eqs(2.5a) and (2.5c) are respectively equal to the values of Eqs(2.6a) and (2.6c), but the value of Eq.(2.5b) is not equal the value of Eq(2.6b), i.e.

$$P_c(decorated\ \ pt) \neq p_c(bond, pt). \tag{2.7}$$

In order to understand Eq.(2.7), in the summer of 1981 I used the method of Wu[75] to study the subqraph expansion of the Syozi model in the external field. I found that in the low-temperature limit the spontaneons magnetization of the Syozi model is corresponding to the percolation probability of a bond-correlated percolation model (BCPM). In early 1982, I extended the Syozi model to a sublattice dilate q-state potts model (SDQPM) and found that the SDQPM for $T \to 0$ is corresponding to a q-state bond-correlated prcolation model(QBCPM).[31]

In the summer of 1982, I found that the 2 state BCPM of Ref 31 is also corresponding to the simple Ising model at finite temperatures,[55] such that the probability weight of a subgraph G' is given by

$$\pi(G', p) = p^{b(G')}\ (1-p)^{E-b(G')}\ \ 2^{n(G')}\ , \tag{2.8}$$

where $p = 1 - e^{-K}$ with K being the nearest neighbor coupling constant of the Ising model. Soon after finding this result, I found that the connection between the Ising model and 2 state BCPM may be obtained directly from the subgraph expansion of the Ising model in an external field. I also realized that such a connection may be extended easily to Ising-type spin model.[48,76]

It should be noted that in Ref.48, I considered subgraph expansion of the Ising model and the q-state potts model in external fields, therefore I am able to relate the spontaneons magnetization and the magnetic susceptibility of the spin model to the percolation probability and the mean cluster-size of the corresponding correlated percolation models. Based on such a connection, the phase transition of the spin model is the percolation transition of the corresponding percolation problems.

In Ref.48a, I pointed out: "hence, it is desirable to generate directly the distribution of subgraphs G' according to the probability factor $\pi(G, p)$ (i.e. Eq(2.8) of the present paper) by MC (i.e. Monte Carlo) or SE (i.e. series expansion) techniques and study the behavior of clusters in such subgraphs. This could be a useful way to study the thermodynamic propertices of the Ising model." However, at that time I did not know how to construct an efficient MC algorithm to simulate such clusters. Near the date I submitted Refs.48a and 55, Sweeny[77] submitted a paper, in which he used a method to simulate clusters for two dimensional q-state Potts model. In 1987, Swendsen and Wang[78] proposed a fast algorithm to simulate clusters for the QPM. Based on such algorithm, we have written a very general computer program to study the QPM on hypercubic lattices.[79,80] The program gives accurate critical properties of the QPM on hypercubic lattices.[79,81] The results will be reviewed in Sec.V. In the next subsection, we will take the QPM as an example to show how to relate an Ising-type spin model to a correlated percolation model using subgraph expansions of the partition function for the spin model.

In the simple QPM, each site of a lattice G of N sites and E nearest-neighbor (NN) bonds is occupied by a spin s_i with spin components $-j$, $-j+1$, ..., $j-1$, and j, where $1 \leq i \leq N$, $2j+1=q$, and q is an integer. The partition function of the QPM with normalized NN ferromagnetic coupling constant $K = J/k_B T$ and the normalized magnetic field B may be written as

$$
\begin{aligned}
Z_N &= \sum_{s_i=-j}^{j} \prod_{<i,k>} exp[K\delta(s_i, s_k)] \prod_i exp(Bs_i) \\
&= \sum_{s_i=-j}^{j} \prod_{<i,k>} [1+(e^K-1)\delta(s_i, s_k)] \prod_i exp(Bs_i).
\end{aligned}
\tag{2.9}
$$

The first product is over all NN pairs $< i, k >$ and the second product is over all lattice sites. Now we expand the first product in (2.9) and use the subgraphs $G' \subseteq G$ to represent the terms in the expansion. After summing over all spin states, we have[48,57,59]

$$
\begin{aligned}
Z_N &= \sum_{G' \subseteq G} (e^K - 1)^{b(G')} \prod_c \{e^{(Bn_c j)} + e^{[Bn_c(j-1)]} + ... + e^{(-Bn_c j)}\} \\
&= e^{KE} \sum_{G' \subseteq G} p^{b(G')} (1-p)^{E-b(G')} \prod_c [e^{(Bn_c j)} + ... + e^{(-Bn_c j)}],
\end{aligned}
\tag{2.10}
$$

where $p = 1 - e^K$, the sum is over all subgraphs G' of G, $b(G')$ is the number of occupied bonds in G', the product extends over all clusters c in a given G', and $n_c = n_c(G')$ is the number of sites in the cluster c.

Let $N^*(G')$ denote the total number of sites in the percolating clusters of G'. Using Z_N of (2.10), we obtain following expressions for the spontaneous magnetization M, the zero-field magnetic susceptibility χ, the internal energy U, and the specific heat C_h of the q-state Potts model in $N \to \infty$:

$$
\begin{aligned}
M &= \lim_{B \to 0^+} \lim_{N \to \infty} \frac{\partial}{\partial B} lnZ_N/N \\
&= \lim_{N \to \infty} W^{-1} \sum_{G' \subseteq G} \pi(G', p, q)[N^*(G')/N]\, j \\
&\equiv j < N^*(G') >_0 \equiv j\, P,
\end{aligned}
\tag{2.11}
$$

$$
\chi = \lim_{B \to 0^+} \lim_{N \to \infty} \frac{\partial^2}{\partial B^2} lnZ_N/N = \frac{1}{12}(q^2-1)S + \chi_p, \tag{2.12}
$$

$$U = - \lim_{B \to 0^+} \lim_{N \to \infty} \frac{\partial}{\partial \beta} lnZ_N/N = -\frac{zJ}{2p}\bar{p}, \tag{2.13}$$

$$C_h = \frac{\partial}{\partial T} U = \frac{k_B K^2}{p^2}[-(1-p)\frac{z}{2}\bar{p} + F]. \tag{2.14}$$

Here χ_p is the contribution from fluctuations of percolating clusters, which does not change the critical point and exponent of first term,[82] $\pi(G', p, q)$ is the probability weight of G', W is the sum of $\pi(G', p, q)$ over all G', z is the coordinate number of G and P, S, $\bar{p}$, and F in (2.11)-(2.14) are respectively the percolation probability, the mean cluster size, the average number of occupied bonds, and the fluctuations of occupied bonds of the following q-state bond-correlated percolation model (QBCPM):

a. All sites of G' are occupied and each bond of G is attached with the bond probability p (being $1 - e^{-K}$). This process generates subgraphs $G' \subseteq G$.

b. The overall probability weight of a $G' \subseteq G$ is enhanced by a factor q for each finite cluster in G'.

Using a similar subgraph expansion, we may show that the correlation function of the QPM is related to the pair connectedness function of the QBCPM. Therefore, the clusters in the QBCPM have the same critical point and order of phase transition as the QPM. When $q \leq q_c$, their critical exponents $\beta_p, \gamma_p, \gamma'_p, \alpha_p, \alpha'_p, \nu_p, \nu'_p$ and η_p are the same as the corresponding values for the QPM. Thus the phase transition of the QPM is a percolation transition.

To establish the connection between an Ising-type model and a correlated percolation or cluster model, the most important step is to use the subgraph expansion to express the partition function of the Ising-type model in an external field as the generating function of the correlated percolation or cluster model. Besides the example of the q-state Potts model considered in Eqs.(2.9)-(2.14), the Ising-type models which have been mapped into the correlated percolation models include a sublattice-dilute q-state Potts model (SDQPM),[31,48b] a dilute q-state Potts model,[60] the Ising model with multi-spin interactions (e.g. the Baxter model),[57c] a lattice model for the hydrogen bonding in water molecules,[56] a model for reversible sol-gel phase transitions in solvent,[58] etc.

Based on the connection between the QPM and the QBCPM, the scaling laws of the critical exponents may be related to the scaling behavior of the cluster-size distribution,[50,59] a percolation renormalization group method (PRGM) has been proposed to calculate the critical point, the critical exponent, the free energy, and the spontaneous magnetization for the QPM.[61] The essential idea of the PRGM will be reviewed in Sec. IV of the present paper.

III. PERCOLATION FOR HARD-CORE PARTICLE MODELS

The densities of hard-core particles (HCP's) on lattices may be controlled by normalized chemical potential Δ. When Δ is increased from small values to large values, the density fluctuations F for a given system diverge at a phase transition point Δ_c.

We consider a system of "hard-core" particles on a lattice G with N sites: the pair potential $V(r) = \infty$ for sites separated by a distance $r \leq R$, and $V(r) = -J$ when two particles have the shortest possible separation in the system, while $V(r) = 0$ otherwise. The grand partition function for such a system can be written as

$$\Omega(G, R, p, K) = (1 + e^{\Delta})^N \sum_{G' \subseteq G}{}' p^{\nu(G')} (1-p)^{N-\nu(G')} \; exp(u(G')K), \tag{3.1}$$

where $p = exp(\Delta)/(1+exp(\Delta))$ and $K = \beta J$. Each possible particle configuration corresponds to a section graph G' of G, such that G' contains only occupied sites and there is no occupation of sites separated by less than the exclusion distance R. The summation includes all the section graphs G' of G , $\nu(G')$ is the number of occupied sites in the G', and $u(G')$ is the number of pairs of particles in G' which have the shortest possible separation in the system. Examples of G' are shown in Figs. 1-2 of Ref. 64(a).

The free energy per unit volume, f, is given by $(ln\Omega(G, R, p, K))/V$ in the limit $V \to \infty$, where $V = N_m \nu_0$ with N_m being the total number of occupied particles and ν_0 being the volume per particle in the closest packed configuration. Here we take ν_0=1. The first and the second derivatives of f with respect to Δ give the density of particles (the average number of particles per unit volume) ρ and the density fluctuations F, respectively. The first and the second derivatives of f with respect to K give the average energy U and the energy fluctuations F_u, respectively. At the phase transition point Δ_c or p_c (being $exp(\Delta_c)/(1+exp(\Delta_c))$, F and F_u diverge.

To give an answer to the question about whether the singularity of F and F_u at p_c is due to the percolation transition, we must define '*cluster*' for hard-core particles. We define the following cluster. Two hard-core particles are in the same cluster when they have the shortest possible separation in the system. With such a definition of clusters for hard-core particles, the mean cluster size $S(G, R, p, K)$ for nonpercolating clusters and the percolation ratio $P(G, R, p, K)$ for percolating clusters may be easily defined.

We use the standard Monte Carlo simulation technique to generate a Markov chain of configurations of particles on a finite lattice of linear length L, subject to the constraint of excluded volume interactions. The typical values of ρ, F, U, F_u, and S may be estimated from the maximum of F, $p_c(F)$, the maximum of F_u, $p_c(F_u)$, and the maximum of S, $p_c(S)$, respectively. We have studied the hard-core particles on the plane triangular (pt), the square (sq), and the simple cubic (sc) lattices with $K = 0$ or $K \neq 0$. For the pt and sc lattices, we also consider the next-nearest-neighbor exclusion interactions. For the hard square model, we have also considered the application of an external driving field so that the particles prefer to move in one direction. We have found that in every studied case the percolation transition point $p_c(S)$ is consistent with the phase transition point $p_c(F)$ and p_c within numerical uncertainty.[64] Our results indicate that the phase transitions of the studied system are percolation transitions of the site correlated percolation models (SCPM) defined above. Each SCPM is expected to be in the same universal class as the corresponding hard-core particle model. The connection between fractal dimensions and cluster-size distributions of the clusters near p_c and critical exponents of HCP's is expected to be similar to the case of Ising-type spin models. We have also used the PRGM to study the hard square model. The result suggest that it is in the universality class of the Ising model.[65] The method and the result of the present section may be extended to other hard-core particle models.

IV. PERCOLATION RENORMALIZATION-GROUP METHOD

The renormalizaion group theory for critical phenomena[46,47,83–85] is an important development in physics. Based on the connection between the QPM and the QBCPM, Hu[49,51] proposed a percolation renormalization group method (PRGM) that may be used to calculate the free energy, the critical point, and the critical exponents for the QPM. Hu and Chen[61b] have considered several versions of the PRGM. For the sake of simplicity, here we present the method in one version of the PRGM which is exact for one-dimensional system.[61a]

For $B = 0$, Z_N of Eq.(2.10) may be written as:

$$\begin{aligned} Z_N &= \sum_{G' \subseteq G} (e^K - 1)^{b(G')} \; q^{n(G')} \\ &= e^{KE} \sum_{G' \subseteq G} x^{b(G')} \; y^{E-b(G')} \; q^{n(G')} \; , \end{aligned} \tag{4.1}$$

where $b(G')$ and $n(G')$ are the numbers of occupied bonds and clusters in G', respectively, and

$$x = p = 1 - e^{-K} \quad , \quad y = 1 - x = e^{-K} \; . \tag{4.2}$$

To construct a percolation renormalization group (PRG) transformation for the system of Eq.(4.1), we divide the lattice G into cells of linear length λ. We generate all possible configurations of b occupied bonds and $\bar{E} - b$ empty bonds on the cell and give each configuration a weight according to Eq.(4.1). Here $\bar{E}$ is the total number of links which connect NN pairs of sites in the cell D and the sites neighboring one side of the cell.

We may classify the bond configurations on a cell D into percolating configurations D'_p and nonpercolating configurations D'_f. In the former case, there is a path of occupied bonds which connect a site in one side of the cell to a site in the opposite side of the cell. In the later case, there is no such a path. Based on such classification, we may define the following partial sums:

$$R(D,x,y,q) = \sum_{D'_p \subseteq D} x^{b(D'_p)} \; y^{\bar{E}-b(D'_p)} \; q^{n(D'_p)} , \tag{4.3a}$$

$$Q(D,x,y,q) = \sum_{D'_f \subseteq D} x^{b(D'_f)} \; y^{\bar{E}-b(D'_f)} \; q^{n(D'_f)} , \tag{4.3b}$$

where the sum in (4.3a) is over all D'_p on D and the sum in (4.3b) is over all D'_f on D; $b(D'_p)$ and $n(D'_p)$ are respectively the number of occupied bonds and the number of clusters in D'_p, similar definitions are given for $b(D'_f)$ and $n(D'_f)$.

Now consider a PRG transformation from a cell of linear dimension λ_1, denoted by D_1, to a cell of linear dimension λ_2, denoted by D_2, where $\lambda_1 > \lambda_2$. Such RG transformation will be denoted as λ_1/λ_2 transformation. We propose that the PRG transformation equation may be writtten as:

$$e^{K'_0} \; R(D_2, x', y', q) = R(D_1, x, y, q), \tag{4.4a}$$

$$e^{K'_0} \; Q(D_2, x', y', q) = Q(D_1, x, y, q), \tag{4.4b}$$

where

$$x' = 1 - e^{-K'} \quad , \quad y' = 1 - x' = e^{-K'} , \tag{4.5}$$

and K'_0 is a constant which arises from the background energy of the PRG transformation.[86] In Eq.(4.5), K' is the coupling constant for the QPM after the PRG transformation.

As in the usual real space RG transformation, from the fixed point of Eq.(4.4), x_c, we have the phase transition point K_c (being $J/k_B T_c$) which is related to x_c by

$$x_c = 1 - e^{-K_c} \equiv p_c . \tag{4.6}$$

From the linearized RG transformations near x_c, we may calculate the thermal scaling power y_T and the critical exponent ν.

$$\frac{1}{\nu} = y_T = \frac{(\ln \frac{\partial x'}{\partial x})_{x_c}}{\ln \frac{\lambda_1}{\lambda_2}} . \tag{4.7}$$

Let $\langle S(D'_{1p}) \rangle$ and $\langle S(D'_{2p}) \rangle$ denote respectively the average numbers of sites in the percolating clusters of D'_{1p} and D'_{2p} at p_c, then the fractal dimension

$D\ (=y_h)$ may be calculateed from the equation[86]

$$\left(\frac{\lambda_1}{\lambda_2}\right)^D = \frac{\langle S(D'_{1p})\rangle}{\langle S(D'_{2p})\rangle}. \tag{4.8}$$

The zero field free energy f may be calculated from a series of background energies arising from step by step PRG transformations.[49,51,61] Our RG transformations give exact f for one dimensional system.[61a] The results of our calculations of K_c, free energies, y_T and D for the two dimensional QPM are shown in Ref.61. The exact free energy for the Ising model and conjectured exact results[28] for y_T and D are shown for comparison. We have found that the calculated results approach exact results as λ_1 and λ_2 increase.

Another method has also been used to calculate y_T for the QPM. The result is similar to that obtained by Eq.(4.7).[62]

Following the case of the random percolation problem,[47] we associate with each site of the lattice an adimensional "mass" m_0 and consider the renormalization of m_0 under the PRG transformation

$$m'_0\ \langle S(D'_{2p})\rangle = m_0\ \langle S(D'_{1p})\rangle. \tag{4.9}$$

After a series of PRG transformations, we have a series of renormalized masses $m_0, m_0^{(1)}\ (=m'_0), m_0^{(2)}, \ldots, m_0^{(n)}$. The percolation probability of the QBCPM, P, and the spontaneous magnetizations of the QPM, M, may be obtained from the equation[61c]:

$$P = M/j = \lim_{n\to\infty} \frac{m_0^{(n)}}{\lambda^{nd}\ m_0}, \tag{4.10}$$

for $p > p_c$ with $\lambda = \lambda_1/\lambda_2$ and $2j+1 = q$.

Let $\langle S_B(D'_{1p})\rangle$ and $\langle S_B(D'_{2p})\rangle$ denote respectively the averge numbers of sites in the backbones of the percolating clusters of D'_{1p} and D'_{2p}, then the renormalization of the backbone "mass" may be written as

$$m'_B\ \langle S_B(D'_{2p})\rangle = m_B\ \langle S_B(D'_{1p})\rangle, \tag{4.11}$$

where m_B is the "mass" at a backbone site of D'_{1p} and m'_B is the "mass" at a backbone site of D'_{2p}. After a series of PRG transformations, we have a series of renormalized backbone masses: $m_B, m_B^{(1)}\ (=m'_B),\ m_B^{(2)}, \ldots, m_B^{(n)}$. The backbone mass density P_B of the QBCPM and the QPM may be obtained from the equation

$$P_B = \lim_{n\to\infty} \frac{{m_B}^{(n)}}{\lambda^{nd}\ m_B}, \tag{4.12}$$

for $p > p_c$.

The fractal dimension of the backbone at p_c, D_B , may be calculated from the equation

$$(\frac{\lambda_1}{\lambda_2})^{D_B} = \frac{\langle S_B(D'_{1p})\rangle}{\langle S_B(D'_{2p})\rangle}, \tag{4.13}$$

which is similar to Eq.(4.8) for D. The backbone critical exponent β_B is defined by the equation

$$P_B(p) = A_B(p-p_c)^{\beta_B}, \tag{4.14}$$

for p near and larger than p_c. It is easy to show that D_B and β_B are related by the equation

$$D_B = d - \frac{\beta_B}{\nu}. \tag{4.15}$$

The percolation probabilities $P(p)$ of the QBCPM for $q = 2$, which corresponds the simple Ising model,[17-19] calculated from 4/3, 3/2, 3/1, and 2/1 PRG transformation are shown in Fig.1 of Ref.61c, where Yang's exact solution[20] is represented by dotted line. The Figure shows that when λ_1 and/or λ_2 is increased, our calculated results become closer to the exact solution. $P(p)$ of the QBCPM for q being 0.001, 1, 2, 3, and 4, which are calculated from 4/3 PRG transformations, are shown in Fig.2 of Ref.61c. For $q \neq 2$, there is no exact solution for comparison.

The backbone mass densities $P_B(p)$ calculated from 3/2 PRG transformation are shown in Fig.5 of Ref.61c for q being 0.001, 1, 2, 3, and 4. The backbone fractal dimension, D_B , calculated from 2/1, 3/1, and 3/2 PRG transiformation are shown in Fig.6 of Ref.61c. It should be noted that D_B increases with q while D decreases with q for small values of q.

It seems that P_B, D_B , and β_B for the QBCPM and QPM with $q \neq 1$ have not been calculated before. Such calculated results help us to understand the effect of correlation on values of P_B, D_B , and β_B .

The percolation renormalization group method (PRGM) presented in this section may be extended to other correlated percolation problems corresponding to Ising-type spin models.

V. MONTE CARLO STUDIES OF THE POTTS MODEL

Based on the subgraph expansion for the Potts model, Swendsen and Wang proposed a cluster Monte Carlo simulation method (CMCSM) to study distribution of Potts spins and occcupied bonds in the lattice.[78] From the energy-energy correlation function at the critical point, they found that the system simulated by the PMCSM reaches equilibrium much quickly than the traditional spin flip method. However, they did not use this method to calculate equilibrium physical quantities for the QPM.

Using the CMCSM and the labeling technique of Hoshen and Kopelman,[87] we have written down a general computer program to simulate clusters of the QBCPM corresponding to the QPM.[79,80] In this program, the space dimensions d, the linear length of the lattice L, the number of spin components q, and p of Eq. (2.10) are input parameters and the percolation probability P, the mean cluster size S, the mean number of occupied bonds $\bar{p}$, and the fluctuations of occupied bonds F related to the spontaneous magnetization M, the magnetic susceptibility χ, the internal energy U, and the specific heat C_h of the QPM may be calculated, see Eqs. (2.11)-(2.14). The periodic boundary condition is chosen to reduce the boundary effect. We use our program to calculate P, S, $\bar{p}$ and F as a function of p for the QPM on d-dimensional hypercubic lattices with $2 \leq d \leq 6$. For $d = 2$, q is chosen to be 2-6; for $3 \leq d \leq 6$, q is chosen to be 2, 3, 4, 8, 16, 32, 64, 128, 256, 512, 1024, and 2048.[79,81] For each choice of d, L, and q, p is increased from high temperature region ($p < p_c$) to low temperature region ($p > p_c$) with the increment Δp which is taken to be a smaller value near the critical region than the value far away from the critical region. At each p, we let the system evolves N_s cycles, abandon the first N_e cycles and take data from the remaining $(N_s - N_e)$ cycles among which a bond configuration G' is taken from every N_I cycles. The multiple labeling technique[87] is applied to such bond configurations to calculate $N^*(G')$ and $n_c(G')$ for Eqs. (2.11) and the mean cluster size S, respectively. From $\bar{p}$ and F, we may calculate U/J and C_h/k_B using Eqs. (2.13) and (2.14).

Typical P, S, $\bar{p}$, F, U/J and C_h/k_B diagrams are shown in Ref.79 for the square (sq) and simple cubic (sc) lattices. The curves of U/J in a given space dimensions d indicate the changeover from second-order phase transitions to first-order phase transitions when q is increased.[79] From the peaks of S and F, we obtain our $p_c(S)$ and $p_c(F)$. It is well known[28] that the exact critical point K_c of the sq lattice QPM satisfies the equation: $exp(K_c)=\sqrt{q}+1$. Therefore the exact p_c for the sq lattice QPM is given by:

$$p_c = 1 - e^{-K_c} = \sqrt{q}/\ (1 + \sqrt{q}\ ). \tag{5.1}$$

We find that the calculated values of $p_c(F)$ and $p_c(S)$ for the sq lattice are consistent with the exact p_c.[79]

For the QPM on hypercubic lattices with $d \geq 3$, there is no exact solution of the critical point. But the high temperature series analysis method (HTSA), the low temperature series analysis method (LTSA), and the other Monte Carlo simulation method (MCSM) give approximate values of K_c and p_c.[28] For $q = 2$, $d = 3$, and $L = 16$, the deviation of our $p_c(F)$ and $p_c(S)$ from the series analysis result, which is very reliable, is about $1.4 \sim 2.2$ %. Such large deviation might be due to the fact that the L used is not large enough in such second-order phase transition system. In the next section, we will point out that a simple percolation Monte Carlo renormalization group method may be used to calculate much more accurate critical point. For $q = 3$ and $q = 4$, our $p_c(F)$ and $p_c(S)$ are consistent with the Monte Carlo and Monte Carlo renormalization group results[79] but deviate from the low temperature series analysis result and high temperature series analysis results. However, it has been believed that the phase transitions of the QPM with $d \geq 3$ and $q \geq 3$ are first order. It is generally believed that the first-order phase transition temperature $(\frac{1}{K_c})$ determined by the LTSA is higher than the exact value and the first-order phase transition temperature $(\frac{1}{K_c})$ determined by the HTSA is lower than the exact value. The results of Ref.79 with $d = 3$ and q being 3 and 4 suggest that our and other Monte Carlo data[28] are reliable and the deviations of the results of HTSA and LTSA from Monte Carlo data are due to the fact that phase transitions are first order.

Hu and Hsiao[81] have found that in space dimensions d=3, 4, 5, and 6, the critical points increase slowly with q. As q increases, the critical points for different d approach to each other such that at q=2048, the absolute values of the differences between such critical points are less than 1%.

In the traditional spin flip canonical Monte Carlo simulations, if the system is simulated for very long time then every spin has equal probabilities to be in its q different components and the long time average of magnetizations in a given direction is zero even for $T < T_c$. Special technique, such as average over absolute values of magnetizations, must be used to obtain the spontaneous magnetization for $T < T_c$. A related difficultly also appears for the magnetic suceptibility χ, where χ for a finite or semi-infinite system does not decrease when T is decreased from T_c. Therefore such χ may not be used to approximate χ for thermodynamic system for $T < T_c$. In our Monte Carlo simulations, the spontaneous magnetization M is calculated from the percolation probability P of Eq.(2.11) and the magnetic susceptibility χ is calculated from of the mean cluster size (of the nonpercolating clusters) S, therefore we do not have the difficulties in the traditional spin flip calculation of M and χ and transfer matrix calculation of χ.

The method used in this section and the discussion given above for the QPM may be extended to other Ising-type spin models considered in Sec. II, which also have corresponding correlated percolation problems.

VI. MONTE CARLO RENORMALIZATION GROUP METHOD

Based on the method of Sec.IV and Sec.V, we may formulate a very simple percolation Monte Carlo renormalization group method (PMCRGM) to study the critical property of the QPM. From $R(D,x,y,q)$ and $Q(D,x,y,q)$ of Eq.(4.3), we may define

$$E_p(D,p,q) = \frac{R(D,x,y,q)}{R(D,x,y,q)+Q(D,x,y,q)}. \tag{6.1}$$

$E_p(D,p,q)$ is the probability that a randomly chosen subgraph is a percolating subgraph and may be called "the existence probability". We use the CMCSM of Sec. V to calculate $E_p(D,p,q)$ as a function of p for two different D's, say D_1 and D_2, with linear dimensions L_1 and L_2 respectiely. Based on the method of Sec.IV, the intersection of $E_p(D_1, p, q)$ and $E_p(D_2, p, q)$ gives the critical point p_c of Eq.(4.6). From the slopes of $E_p(D_1, p, q)$ and $E_p(D_2, p, q)$ at p_c, we may obtain the critical exponent ν of Eq.(4.7). For $d = 3$ and $q = 2$, we obtain $p_c(S)$=0.351 for L=12 and $p_c(S)$=0.352 for L=16 using the method of Sec.V. From the intersection of $E_p(D, p, q)$ for L=12 and 16, we obtain p_c=0.357±0.001, which is consistent with the value p_c=0.3580913 obtaied by Pawley et. al..[88] However, this PMCRGM may only give critical behavior of the QPM. In the following we will present a method that may be used to calculate global behavior of the QPM.

Based on the connection between the QPM and the QBCPM, Hu has written the partition function for the QPM as a sum over products of an interacting factor $I(b, n)$ and the geometrical factors: $G_f(b, n)$ and $G_p(b, n)$, where $G_f(b, n)$ and $G_p(b, n)$ are respectively the numbers of nonpercolating subgraphs and percolating subgraphs with b occupied bonds and n clusters.[50,57,59] Similar geometrical factors may also be defined for hard-core particle models. From geometrical factors, Hu and Chen have formulated percolation renormalization group methods (PRGM) to study critical properties of the QPM[61] and the hard square model.[65] In this section we extend the static Monte Carlo simulation method of Huang[89] to calculate $G_f(b, n)$, $G_p(b, n)$ and the free energy for the QPM.[63,90] The calculated results are quite accurate when compared with exact results obtained by a fast algorithm proposed by Chen and Hu.[66] To calculate the free energy of the QPM as a function of the temperature, our method is simpler than the method considered by Ferrenberg and Swendsen,[91] because in the latter one should solve coupled nonlinear equations.

$G_f(b, n)$ and $G_p(b, n)$ obtained by the Monte Carlo method may be combined with the PRGM[61] to formulate a Monte Carlo renormalization group method for the QPM. Similar studies may be extended to other Ising-type models[56−60] and the lattice hard-core particle models[64] whose phase transitions are percolation transitions.

Equation(4.1) may be rewritten as[50,57,59]

$$Z_N = \sum_{b=0}^{E} \sum_{n=0}^{N} G(b, n) I(b, n), \tag{6.2}$$

$$I(b, n) = (e^K - 1)^b q^n. \tag{6.3}$$

In Eq.(6.2), $G(b, n)$ is the number of subgraphs with b occupied bonds and n clusters. The calculation of $G(b, n)$ by Monte Carlo simulations is similar to the calculation of the number of energy states $\Gamma(E)$ in Ref.89.

In the random sampling of subgraphs, the probability of the appearance of any subgraph is $p_R = 2^{-E}$. Let $g(b, n)$ be the number of subgraphs contained in a set S of N_s states generated in a random sampling of subgraphs. When N_s is very large, we have

$$\frac{G(b,n)}{2^E} \approx \frac{g(b,n)}{N_s}, \tag{6.4}$$

i.e.

$$G(b,n) \approx \frac{1}{N_s} \sum_{G'}{}' \frac{1}{p_R}, \tag{6.5}$$

where the summation is over all G' in S with occupied bonds and n clusters. In the random sampling, subgraphs with certain b and n values, say $\bar{b}$ and $\bar{n}$, dominate. It is difficult to simulate G' with other b and n values and to calculate the corresponding $G(b,n)$. To overcome this problem we may use an important sampling so that subgraphs with b bonds and n clusters will appear with probabilities $p_I(G') > p_R$. The summand in Eq.(6.5) must be reduced by the factor $p_R/p_I(G')$. Therefore in the important sampling we have:

$$G(b,n) \approx \frac{1}{N_s} \sum_{G'}{}' \frac{1}{p_I(G')}. \tag{6.6}$$

For given K and q, we use the following procedure to generate G' and to calculate $p_I(G')$. We begin with the subgraph with no occupied bond and N clusters and attempt to put bonds on the lattice in sequence. If an occupied bond reduce the number of cluster by 1, the probability to put the bond is $p_1 = (e^K - 1)/(e^K + q - 1)$. If an occupied bond does not reduce the number of clusters, the probability to put the bond is $p_2 = 1 - e^{-K}$. The product of the factors $p_1, (1-p_1), p_2$, or $(1-p_2)$ over E bonds of G gives $p_I(G')$ of the generated G'. We may vary K and q to obtain $G(b,n)$ in different region of the (b,n) plane. We may classify the generated subgraphs into nonpercolating subgraphs G'_f and percolating subgraphs G'_p. It is obvious that $G_f(b,n)$ and $G_p(b,n)$ are respectively given by:

$$G_f(b,n) \approx \frac{1}{N_s} \sum_{G'_f}{}' \frac{1}{p_I(G'_f)}, \tag{6.7}$$

and

$$G_p(b,n) \approx \frac{1}{N_s} \sum_{G'_p}{}' \frac{1}{p_I(G'_p)}, \tag{6.8}$$

The sums in Eqs.(6.7) and (6.8) are over G'_f and G'_p, respectively. Please note that $G(b,n)$ is the sum of $G_f(b,n)$ and $G_p(b,n)$.

From the calculated $G(b,n)$, the free energy f, the internal energy U and the specific heat C_h may be calculated from the equations:

$$f = \frac{1}{N}\ell n Z_N \ , \tag{6.9}$$

$$U/J = -\frac{z}{2p}\bar{p} = -\frac{z}{2pZ_N}\sum_{b=0}^{E}\sum_{n=0}^{N} G(b,n)I(b,n)b/E, \tag{6.10}$$

$$C_h = \frac{k_B K^2}{p^2}[-(1-p)\frac{z}{2}\bar{p} + F], \tag{6.11}$$

$$F = \frac{1}{Z_N}\sum_{b=0}^{E}\sum_{n=0}^{N} G(b,n)I(b,n)(b-\bar{p}E)^2/N, \tag{6.12}$$

where z is the coordination number of the lattice. We may also write down equations for contributions come from nonpercolating subgraphs or percolating subgraphs only.

In Ref.63, we simulate the QPM on a 6×6 square lattice with the periodic boundary condition. K is varied from 0.1 to 1 with the increment $\Delta K = 0.1$ and q is varied from 1 to 10 with the increment $\Delta q = 1$. The total number of the combinations (K,q) is 100. For each (K,q) we generate 10^5 different subgraphs which may be used to calculate $G(b,n)$. For given b and n, different (K,q) combinations usually give different $G(b,n)$ values. We choose the result from a (K,q) combination which gives the maximum number of terms in the summation of Eq.(6.6).

The calculated f, U/J and F for q being 2, 2.5, 3, 3.5, 4, and 5 are plotted in Fig.1 of Ref.63 which shows that our calculations give quite reliable results.

In Ref.90, we have simulated the QPM on a 5×5 square lattice with the free boundary condition. We have chosen q=2 and do simulations at K=0.2, 0.4, 0.6, 0.8, 0.88, 1.0, and 1.2 with 2×10^6 MC sweeps at each K, except at K=0.88, where they have used 6×10^6 sweeps. Equations (6.7) and (6.8) are used to calculate $G_f(b,n)$ and $G_p(b,n)$. For given b and n, different K values usually give different $G_f(b,n)$ and $G_p(b,n)$ values; we choose results from K values which give maximum numbers of terms in the summations of Eqs.(6.7) and (6.8), respectively. Such $G_f(b,n)$ and $G_p(b,n)$ are then used to calculate *partial bond fluctuations* and *partial free energies*. In Figure 1 of Ref.90, the calculated *partial bond fluctuations* have been compared with the exact results obtained from a fast algorithm to calculate exact geometrical factors.[66] It is obvious from Fig. 1 of Ref.90 that our simulations give quite accurate results. The difference between the Monte Carlo and exact values of the *partial free energies* is only about 0.01 per cent and therefore we do not plot them.

The method of this section may be easily extended to study other spin models[50,53,56–60] and hard-core particle models[64] whose partition functions have similar subgraph expansions.

From $G_p(b,n)$ and $G_f(b,n)$ we may use the percolation renormalization group method[61,65] to calculate the critical point, the critical exponent, and the thermodynamic free energy. Our method to calculate the free energy for the QPM is simpler than the method of Ferrenberg and Swendsen,[91] in which one should solve a set of coupled nonlinear equations in order to obtain absolute free energy over wide range of K values.

VII. PERCOLATION AND OTHER PHASE TRANSITIONS

Until now, we consider only lattice systems. In 1988 Kratky[92] considered the case where two hard spheres of radius R in the contiuum space are in the same cluster if their separation r is smaller than $4R$. The hard spheres are assumed to have a uniform negative background potential depending on the density of the system, so that they have a gas-liquid phase transition. Kratky has found that such gas-liquid phase transition corresponds to the percolation transition of the cluster defined above.

However, this is still a classical system. It is well known that for temperatures smaller than a critical temperature T_c a finite fraction of particles in the Bose gas will be in the zero momentum state,[8,38] which may be considered as a percolating state. Therefore, the Bose-Einstein condensation of quantum Bose gas may be considered as a percolation transition.

VIII. SUMMARY REMARKS

In the development of theoretical physics, geometrical concepts have often been found to be useful for the formulation of physical theories, e.g. geometrical optics, classical mechanics, general relativity, and gauge field theory of basic interactions. In previous sections, we have found that percolation is a useful geometrical concept for understanding the singular behavior of phase transitions and critical phenomena in Ising-type models, hard-core particle models, hard spheres, and Bose gases. On the other hand, in the random bond (site) percolation problem, the fluctuation of the number of occupied bond (site) is an analytic function of bond (site) occupation probability. Such results suggest that the percolation transition might be a general necessary, but not sufficient, condition for the singular behavior in the interacting many particle systems.

Based on the connection between phase transitions and percolation transitions, we have new methods to calculate the global and singular behavior of interacting many particle systems.

It is of interest to extend above results to other phase transition systems, e.g. hard-core particles in continuum space with finite range attractive interactions, quantum spin models , superconductors, etc., and to develop new calculation methods.

From such studies, we might be able to have a unified connected principle and calculational method for phase transitions and critical phenomena in interacting many particle systems and the problem raised by Chung Chou may be partially solved.

ACKNOWLEDGMENTS

The author thanks Mr. Chi-Ning Chen, Mr. Jau-Ann Chen, Dr. Li-Jen Chen, Dr. S.-S. Hsiao, and Mr. Kit-Sing Mak for their cooperation on the work reviewed in Secs. III-VI and Professor Bambi Hu for critical reading of the manuscript and giving useful comments. We thank Computing Center of Academia Sinica (Taiwan) for providing research facilities. This work was supported by National Science Council of Republic of China (Taiwan) under contract number NSC81-0208-M001-55.

REFERENCES

1. S. Weinberg, *Phys. Rev. Lett.* **19** (1967) 1964.
2. A. Salam, in *Proc. 8th. Nobel Symposium* (Amqvist and Wiksell, Stockholm, 1968).
3. C. N. Yang and R. L. Mills, *Phys. Rev.* **96** (1954) 191.
4. P. W. Higgs, *Phys. Rev.* **145** (1966) 1156.
5. T. Andrews, *Phil. Trans. R. Soc.* **159** (1869) 575.
6. P. Curie, *Journal de Physique* **4** (1895) 263.
7. H. Kamerlingh Onnes, *Akad. Van Wetenschappen (Amsterdam)* **14** (1911) 113, 818.
8. A. Einstein, *Preussische Akad. Wiss., Phy.-math. Klasse* (1924) 261 and (1925) 3.
9. W. H. Keesom and K. Clusius, *Proc. Roy. Acad. Ams.* **35** (1932) 307.
10. F. London, *Nature* **141** (1938) 643; *Phys.Rev.* **54** (1938) 947.
11. H. Moser, *Physik Zeits.* **37** (1936) 737.
12. G. W. Gray, *Mole. Struc. and Prop. of Liq. Crys.*, (Academic Press, London,1962).
13. P. Nordblad, *et. al.*, *Phys. Rev. B* **28** (1983) 278.
14. H. K. Kim and M. H. W. Chan, *Phys. Rev. Lett.* **53**(1984) 170.
15. J. C. Campuzano, *et.al.*, *Phys. Rev. Lett.* **54** (1985)2684.
16. H. E. Stanley, *Introduction to Phase Transitions and Critical Phenomena*, (Oxford Univ. Press, New York,1971).
17. W. Lenz, *Physik Z.* **21** (1920) 613.
18. E. Ising, *Z. Physik* **31**(1925) 253.
19. L. Onsager, *Phys. Rev.* **65** (1944) 117.

20. C. N. Yang, *Phys. Rev.* **85** (1952) 809.
21. M. E. Fisher, *J. Math. Phys.* **5** (1964) 944.
22. L. K. Runnels, in *Phase Transitions and Critical Phenomena vol.2,* ed. C. Domb and M.S. Green (London, Academic, 1972).
23. R. J. Baxter, *J. Phys. A* **13** (1980) L61-L70.
24. R. J. Baxter, *J. Stat. Phys.* **22** (1980) 465.
25. R. J. Baxter and P. A. Perace, *J. Phys.* **A16** (1983) 2239.
26. N. C. Bartelt and T. L. Einstein, *Phys. Rev.* **B30** (1984) 5339.
27. R. B. Potts, *Proc. Cambridge Phil. Soc.* **48** (1952) 106.
28. F. Y. Wu, *Rev. Mod. Phys.* **54** (1982) 235.
29. M. Blume, V. J. Emery and R. B. Griffiths, *Phys. Rev.* **A4** (1971) 1071.
30. R. J. Baxter, *Phys. Rev. Lett.* **26** (1971) 832.
31. C.-K. Hu, *Physica* **116A** (1982) 265-271, see also Ref.48b and references therein.
32. R. Peierls, *Proc. Cambridge Phil. Soc.* **32** (1936) 417.
33. C. N. Yang and T.D. Lee, *Phys. Rev.* **87** (1952) 404.
34. A. Caille, *et.al.*, *Can. J. Phys.* **58** (1980) 581.
35. M. N. Barber, *Phys. Rep.* **59** (1980)375-409.
36. J. W. Evans, *Surf. Sci.* **215** (1989) 319-331.
37. D. A. Huse, *Phys. Rev. Lett.* **49** (1982) 1121.
38. K. Huang, *Statistical Mechanics, 2nd. ed.*, (Addison-Wiley, New York, 1987).
39. L. P. Kadanoff, *Physics* **2** (1966) 263.
40. B. Widom, *J. Chem. Phys.* **43** (1965) 3892 and 3898.
41. P. J. Flory, *J. Am. Chem Soc.* **63** (1941) 3083,3091,3096.
42. D. Stauffer, *Phys. Rep.* **54** (1979) 1; *Introduction to Percolation Theory,* (Taylor and Francis, London, 1985).
43. G. Deutscher, R. Zallen and J. Adler ed., *Percolation Structures and Processes*, (AIP, New York, 1983).
44. H. E. Stanley, *J. Phys.* **A10** (1977) L211.
45. A. Coniglio, *J. Phys.* **A15** (1982) 3829.
46. P. J. Reynolds, H. E. Stanley and W. Klein, *J. Phys.* **A10** (1977) L203;, *J. Phys.* **A11** (1978) L199;, *Phys. Rev.* **B21** (1980) 1223.
47. C. Tsallis, A. Coniglio and G. Schwachheim, *Phys. Rev.* **B32** (1985) 3322.
48. C.-K. Hu, *Phys. Rev.* **B29** (1984) (a) 5103-5108 and (b) 5109-5116.

49. C.-K. Hu, *History of Science Newsletter (Taipei) Suppl.* Vol. **5**(1986)80-101.

50. C.-K. Hu, in *Progress in Statistical Mechanics*, ed. C.-K. Hu, (World Scientific, Singapore,1988) pp.91-117.

51. C.-K. Hu, *National Science Council Monthly* **16** (1988) 1287-1314.

52. C.-K. Hu, K.-S. Mak, and C.-N. Chen, in *3rd. Asia Pacific Phys. Conf.*, ed. Y. W. Chan, A. F. Leung, C. N. Yang, and K. Young, (World Scientific, Singapore, 1988), p.668.

53. C.-K. Hu, *Proc. Natl. Sci. Counc. R.O.C. (A)* **14**(1990)73.

54. C.-K. Hu, in *Proceedings of the Fourth Asia Pacific Physics Conf.*, ed. S.H. Ahn, S.H. Choh, I1-T. Cheon, and C. Lee, (World Scientific, Singapore, 1991), p.21-30.

55. C.-K. Hu, *Physica* **119A** (1983) 609-614, see the sentence below Eq.(16) on page 613.

56. C.-K. Hu, *J. Phys.* **A16** (1983) L321-326.

57. C.-K. Hu, a. *Chin. J. Phys. (Taipei)* **22(1)**(1984)1-12 ; b. *Chin. J. Phys. (Taipei)* **22(4)**(1984)1-20 ; c. *Chin. J. Phys. (Taipei)* **23** (1985) 47-63.

58. C.-K. Hu, *Ann. Rep. Inst. Phys. Acad. Sin (Taiwan)* **14** (1984)7-12.

59. C.-K. Hu, *J. Phys.* **A19** (1986) 3067;, *Phys. Rev.* **B34** (1986) 6280; *Chin. J. Phys. (Taipei)* **24** (1986)183-187;, *J. Phys.* **A20** (1987) 6617.

60. C.-K. Hu, *Phys. Rev.* **B44** (1991) 170-177.

61. C.-K. Hu and C.-N. Chen, a. *Phys. Lett. A* **130** (1988) 436; b. *Phys. Rev. B* **38** (1988) 2766; c. *ibid.* **B 39** (1989) 4449.

62. C.-K. Hu and C.-N. Chen, *Phys. Rev. B* **40** (1989) 854.

63. C.-K. Hu and J.-A. Chen, in *Proceedings of the Fourth Asia Pacific Phys. Conf.*, ed. S.H. Ahn, S.H. Choh, I1-T. Cheon, and C. Lee, (World Scientific, Singapore, 1991), p.730-733.

64. C.-K. Hu and S.-K. Mak, a. *Phys. Rev. B* **39** (1989) 2948; b. *ibid.* **42** (1990) 965; unpublished.

65. C.-K. Hu and C.-N. Chen, *Phys. Rev.* **B43** (1991) 6184-6185.

66. C.-N. Chen and C.-K. Hu, *Phys. Rev.* **B43** (1991) 11519-11522.

67. M. E. Fisher, *J. Appl. Phys.* **38** (1967) 981; *Physics* **3** (1967) 255.

68. H. Muller-Krumbhaar, *Physs. Lett.* **48A** (1974) 459; **50A** (1974) 27.

69. See Sec.I of Ref. 48a and related references for brief review of such results and other developments, e.g. the work of Coniglio and Klein in *J. Phys.* **A13** (1980) 2775; see also Refs.49-54.

70. P. W. Kasteleyn and C. M. Fortuin, *J. Phys. Soc. Jpn. (Suppl)* **26** (1969) 11.

71. C. M. Fortuin and P. W. Kasteleyn, *Physica* **57** (1972) 536-564.

72. J. W. Essam, in *Phase Transitions and Critical Phenomena*, edited by C. Domb and M. S. Green (Academic, New York, 1973), Vol.2, pages 266-267; after presenting an equation similar to Eq.(2.4a) of the present paper, Essam said: "where the mean value is for the random bond percolation problem."

73. C.-K. Hu and P. Kleban, *Phys. Rev.* **B25** (1982) 6760-6764.

74. I. Syozi, *Prog. Theor. Phys. Jpn.* **34** (1965) 189; I. Syozi and S. Miyazima, *ibid* **36** (1966) 1083.

75. F. Y. Wu, *J. Stat. Phys.* **18** (1978) 115.

76. See last sections of Ref.48a and 48b.

77. M. Sweeny, *Phys. Rev.* **B27** (1983) 4445. This paper did not formally prove the connection between the QPM and the QBCPM, but emphasize the numerical verification of such connction.

78. R. H. Swendsen and J. S. Wang, *Phys. Rev. Lett.* **58** (1987) 86.

79. C.-K. Hu and S.-K. Mak, *Phys. Rev. B* **40** (1989) 5007.

80. L.-J. Chen, C.-K. Hu, and S.-K. Mak, *Comp. Phys. Comm.* **66** (1991) 377.

81. C.-K. Hu and S.-S. Hsiao, unpublished.

82. M. D. D. Meo, D. W. Heermann, and K. Binder, *J. Stat. Phys.* (1990) 585-617; see also Eq.(4) of Ref.48a.

83. K. G. Wilson, *Phys. Rev.* **B4** (1971) 3174 and 3184.

84. K. G. Wilson, *Sci. Am.* **241(2)** (1979) 158. This paper used the directions of Ising spins to define clusters, however such clusters will not give correct critical behavior of Ising models.

85. S. K. Ma, *Modern Theory of Critical Phenomena* (Benjamin, Reading, Mass., 1976).

86. T. A. Larsson, *J. Phys.* **A19** (1986) 2383, this paper first considered to costruct percolation renormalization group equations for the QPM, but did not include K_0' and hence was incomplete.

87. J. Hoshen and R. Kopleman, *Phys. Rev.* **B14** (1976) 3438.

88. G. S. Pawley, R. H. Swendsen, D. J. Wallace, and K. G. Wilson, *Phys. Rev.* **B29** (1984) 4030.

89. H.-M. Huang, *Int. J. Mod. Phys.* **B3** (1988) 473.

90. C.-K. Hu and J.-A. Chen, in *Spontaneous Formation of Space-time Structures and Criticality*, edited by T. Riste and D. Sherrington, NATO ASI Series (Plenum, New York and London, 1991).

91. A. M. Ferrenberg and R. H. Swendsen, *Phys. Rev. Lett.* **63** (1989) 1195.

92. K. W. Kratky, *J. Stat. Phys.* **52** (1988)1413.

Acceleration and Analysis Methods in Statistical Physics

Robert H. Swendsen
Department of Physics, Carnegie-Mellon University,
Pittsburgh, Pennsylvania 15213

July 8, 1991

Abstract

This review describes some of the recent progress in simulation techniques and data analysis for Monte Carlo computer simulations. Cluster methods for dealing with problems of slow relaxation at critical points and in spin glasses are introduced. Recent developments in histogram methods for improving the efficiency of the data analysis are also discussed.

1 Introduction

Although Monte Carlo (MC) computer simulations have now been well established as a reliable method for investigating the properties of thermodynamic systems,[1] the computational requirements of many problems of current interest are so great that major improvements in both hardware and algorithmns will be necessary to deal with them. This paper reviews some recent progress that has been made in both the simulational methods themselves and in the analysis of data obtained from those simulations.

The standard Monte Carlo method, as introduced by Metropolis *et al.* [2] thirty-five years ago, is remarkably efficient for many applications. However, it runs into problems in at least two regimes of great interest. The first is in the neighborhood of a critical point, where the large clusters of correlated spins require very long relaxation times. Single-spin-flip algorithms are at a fundamental disadvantage because they only have an important effect on the boundaries of such clusters - the larger the cluster, the less efficient the simulations. One method for dealing with this problem involves the use of a mapping between thermodynamic systems and a percolation problem, which was introduced by Fortuin and Kasteleyn.[3, 4] This type of algorithm, known as a cluster algorithm, is extremely effective in increasing the efficiency of computer simulations at phase transitions.[5]-[14]

The second area concerns systems with frustrated interactions, and especially spin glasses. These systems have many low-energy states that all contribute to the equilibrium properties. However, these low-energy states are separated by high energy barriers, making it extremely difficult to bring them into equilibrium. Characteristic relaxation times in excess of 10^6 sweeps must be overcome if standard MC methods are used. Unfortunately, the cluster methods mentioned in the previous paragraph lose their efficiency in the presence of frustration. However, other methods have been developed that extend the idea of using clusters of spins to frustrated systems. I shall discuss such a method, known as Replica Monte Carlo, which is especially effective in two dimensions.

An additional issue is the question of how to best analyze the information contained in the configurations produced by the MC simulation. For this problem, I will discuss the recent progress in applying histogram methods to new problems. I will also discuss the new multiple histogram equations that greatly improve efficiency in combining the results of several MC studies of a given system at different temperatures and magnetic fields. These methods turn out to be extremely useful for determining the locations and heights of peaks, as well as giving the entropy and free energy over a wide parameter range.

2 Critical Slowing Down

It is useful to have a particular example in mind for the discussion of critical slowing down. Consider the Ising model of a ferromagnet. The Hamiltonian is given by

$$H = K_I \sum_{\langle i,j \rangle} \sigma_i, \sigma_j \tag{1}$$

where σ_i is the spin on lattice site i, which can take on the values +1 or −1, and K_I $(= J/k_B T)$ is the coupling strength,. Factors of $-1/k_B T$ have been absorbed into the constant K.

At high temperatures the spins are weakly correlated, but as the temperature is lowered clusters of positive and negative spins are formed. These clusters are irregular, and can consist of one spin or many. At any temperature, all cluster sizes will be found in the system, up to a maximum size given by the correlation length ξ. The magnitude of the correlation length can be determined from the spin-spin correlation function

$$f_\sigma(\vec{r} - \vec{r'}) = \frac{\langle \sigma_{\vec{r}} \sigma_{\vec{r'}} \rangle - \langle \sigma_{\vec{r}} \rangle^2}{\langle \sigma_{\vec{r}}^2 \rangle - \langle \sigma_{\vec{r}} \rangle^2} \tag{2}$$

which, away from the critical point, decays exponentially as $exp(-t/\xi)$.

At the critical temperature T_c, $\xi \to \infty$ and a phase transition occurs. The properties of the system at and near the phase transition are related to the divergence of the correlation length. For example, the divergence of the magnetic susceptibility can be understood from the large effect on the magnetization when the spins in a large cluster are reversed.

Near T_c, the size of the clusters affects the relaxation of the system, which can be measured in various ways. The most important method uses the time-dependent correlation function

$$f_E(t - t') = \frac{\langle E(t)E(t') \rangle - \langle E \rangle^2}{\langle E^2 \rangle - \langle E \rangle^2} \tag{3}$$

where E is the total energy (or some other property) of the system.

Several different correlation (or relaxation) times can be extracted, but for our purposes, the most useful is the integrated correlation time.

$$\tau_{int} = \sum_{t=1} f(t) \tag{4}$$

where time is discrete and measured in units of lattice sweeps. The upper limit of the sum would be infinity if $f(t)$ could be computed exactly, but for practical calculations it is usually cut off at the first negative value of $f(t)$. Müller-Krumbhaar and Binder have shown that the error in the MC determination of the average energy is determined by the total number of configurations divided by $g = 1 + 2\tau_{int}$.[15]

There is generally a power-law relationship between the correlation length ξ and the characteristic correlation time τ.

$$\tau \propto \xi^z \tag{5}$$

where z is known as the dynamical critical exponent. The value of z is usually about 2, which means that the correlation time increases rapidly as the critical temperature is approached. This increase in correlation time is responsible for inefficiency in standard MC simulations near second-order phase transitions. To obtain independent configurations from the MC simulation, it is necessary to run the program for a time much longer than τ, so that the necessary computer time grows as L^{d+z}. The factor L^d is not really a problem, since the larger system contains proportionally more information. However, the factor of L^z is a real disadvantage, leading to inefficient simulations of large systems.

3 Fortuin-Kasteleyn Transformation

In 1969, Fortuin and Kasteleyn introduced an exteremly interesting transformation, linking a large class of thermal problems to a class of percolation problems.[3] This transformation is the basis for a cluster algorithm introduced by Jian-Sheng Wang and myself in 1987.[4]

The Fortuin-Kasteleyn transformation applies to a generalized Ising model known as the Potts model. It is described by the Hamiltonian

$$H = K \sum_{\langle i,j \rangle} (\delta_{\sigma_i,\sigma_j} - 1) \tag{6}$$

where the spin on lattice site i, σ_i, can take on the values $1, 2, \ldots, s$ where s is the number of states. If $s = 2$, this is equivalent to the Ising model described in Eq. (1).

The Potts partition function Z is given by

$$Z = \sum_{\sigma} exp(H). \tag{7}$$

To derive the transformation, we will first treat the interaction between sites l and m. Define a restricted Hamiltonian with the interaction between l and m removed

$$H_{lm} = K \sum_{\langle i,j \rangle \neq \langle l,m \rangle} (\delta_{\sigma_i,\sigma_j} - 1) \tag{8}$$

and two restricted partition functions

$$Z_{lm}^{same} = \sum_{\sigma} exp(H_{lm})\delta_{\sigma_l,\sigma_m} \tag{9}$$

and

$$Z_{lm}^{diff} = \sum_{\sigma} exp(H_{lm})(\delta_{\sigma_l,\sigma_m} - 1). \tag{10}$$

These three partition functions are clearly related, since the sum over all configurations for which $\sigma_l = \sigma_m$ and all those for which $\sigma_l \neq \sigma_m$, must include all possible configurations.

$$Z = Z_{lm}^{same} + e^{-K} Z_{lm}^{diff} \tag{11}$$

The extra factor of $exp(-K)$ is due to the term that was dropped in H_{lm}.

The requirement that $\sigma_l = \sigma_m$ is very similar to the concept of a percolation bond. However, the requirement that $\sigma_l \neq \sigma_m$ is inconvenient. It is much more convenient to remove all restrictions on σ_l and σ_m.

$$Z_{lm}^{ind} = \sum_{\sigma} exp(H_{lm}) \tag{12}$$

These partition functions are also related, since

$$Z_{lm}^{ind} = Z_{lm}^{same} + Z_{lm}^{diff} \tag{13}$$

Combining Eqs. 11 and 13, we find

$$Z = (1 - e^{-K})Z_{lm}^{same} + e^{-K} Z_{lm}^{ind} \tag{14}$$

The coefficients can be interpreted as probabilities of the presence or absence of a bond. Defining

$$p = 1 - exp(-K) \tag{15}$$

as the probability of a bond, and $q = 1 - p$ as the probability for no bond, Eq.14 becomes

$$Z = pZ_{lm}^{same} + (1-p)Z_{lm}^{ind} \tag{16}$$

The next step is to repeat this for all interactions in the system. This eliminates all interactions between spins and replaces them with configurations of bonds. All spins on a given cluster must have the same value, but that value is independent of the spins on any other cluster. This problem, with both spins and bonds present, is known as "site-bond percolation", and was introduced by Coniglio and Klein.[16]

The spin values are then integrated out, leaving a factor of s for each cluster. Denoting the total number of clusters by N_c, the number of bonds by b, and the number of missing bonds by m, the Potts partition function is expressed as

$$Z = \sum_{bonds} p^b (1-p)^m s^{N_c} \tag{17}$$

which is the central result of Fortuin and Kasteleyn.[3]

4 Swendsen-Wang Algorithm

Applying the FK transformation directly to the spin configurations leads to a new MC algorithm.[4] The algorithm first replaces the spins by a configuration of bonds, and then uses the bond configuration to construct a new configuration of spins.

To be explicit, begin with some configuration of Potts spins on the lattice of interest. The assignment can be uniform, random, or taken from some previous simulation.

Bonds between nearest-neighbor sites with the same value of the spin are assigned with the probability $p = 1 - exp(-K)$. No bonds are placed between sites with different spin values. Each cluster in the resulting configuration of bonds will contain only spins with the same value.

Now use the random number generator to assign new spin values to each cluster (and the same value to every site within a given cluster). The bonds are then erased, and we have a new spin configuration.

This new algorithm is clearly ergodic, since there is always a non-zero probability to go from any configuration to any other configuration in a single sweep.

The only aspect of this algorithm that can cause difficulties is the identification of the clusters. For scalar computers, there is a very efficient algorithm due to Hoshen and Kopelman,[17] which does this calculation with an amount of computer time proportional to the total number of spins. The total time to update the entire lattice is about twice that for one standard MC sweep, independent of the size of the lattice. However, on a vector or parallel machine, the problem of efficiently identifying clusters is still being worked on, and although progress has been made, this is still the weakest link in the algorithm.

Even with the difficulties in vectorization, the efficiency of the SW algorithm is very high, and for large lattices two or three orders of magnitude improvement have been achieved. The original calculations by Jian-Sheng Wang and myself indicated that the value of the dynamical critical exponent for the $d = 2$ Ising model was reduced from $z \approx 2.1$ to $s_{SW} \approx 0.35$. Recent work with larger lattices and better statistics suggests that the algorithm is even more efficient, and $\tau_{SW} \propto ln(L)$, which would imply that $z_{SW} = 0$.[18]

For the $d = 3$ Ising model, the improvement is also very good, although not as striking as in two dimensions. The dynamical critical exponent is reduced from a vlaue near 2 to about .45[19, 20] In four dimensions, $z_{SW} \approx 1.0$[21].

The extension of the SW algorithm to include magnetic fields can be carried out in two ways, which are both quite simple. The first is to include an extra "ghost" spin that interacts with every spin in the system with a coupling constant equal to the magnetic field.[3] The second method is to treat each cluster as a single spin in a magnetic field equal to the true magnetic field times the number of sites in the cluster. The heat bath algorithm will then provide an efficient simulation.

It is also easy to include negative (antiferromagnetic) interactions for the case of $s = 2$ (Ising model).[4] This is actually an extension of the original FK transformation, noting that for $s = 2$, the condition that the spins are different is a unique specification. Z_{lm}^{same} can therefore be eliminated in favor of Z_{lm}^{diff}. The result can be interpreted in terms of anti-bonds with a probability

$$p = 1 - exp(- \mid K \mid) \tag{18}$$

In principle, this allows us to treat spin glasses. However, although the generalization of the FK transformation is valid for spin glasses, the algorithm is inefficient. The problem is that a theorem due to Fortuin and Kasteleyn, that the correlation functions, and therefore the correlation lengths, for the thermal and percolation problems are identical,[3, 22, 23] is no longer valid. With frustration, the correlation length in the percolation representation is always larger, and actually diverges before the phase transition is reached. This means that most of the lattice sites in the system belong to the same cluster, so that no important changes can be made.

The bond configurations generated by this method have a probability distribution proportional to $p^b(1-p)^m s^{N_c}$, as in Eq. 17. This means that a side effect of the SW algorithm is to provide equilibrium configurations of the bond configurations. This turns out to be extremely useful, since several properties of the thermal system can actually be evaluated more accurately in the percolation representation. This has been exploited by a number of people, beginning with Sweeny[22], who had developed a method for simulating two-dimensional clusters directly.[5, 13],[23]-[26]

A simple example of the use of the bond configurations would be in the calculation of the Ising model magnetization, which is given by

$$M = \lim_{h \to 0^+} \lim_{L \to \infty} L^{-d} \langle \sum_i \sigma_i \rangle \tag{19}$$

where h is the magnetic field.[3, 23] Transforming to "site-bond correlated percolation,"[16] and denoting the number of sites in the n-th cluster by C_n, we find that

$$M = \lim_{h \to 0^+} \lim_{L \to \infty} L^{-d} \langle \sum_n C_n \sigma_n \rangle \tag{20}$$

If we let the field go to zero, all finite clusters will take on positive and negative values with equal probability, so that their contribution to the magnetization will vanish. Only the largest cluster will contribute, and its contribution will be non-zero only if $p > p_c$. This gives us an equation for the magnetization evaluated in the percolation representation.[3, 22, 23]

$$M = \lim_{L \to \infty} L^{-d} \langle \sum_n C_{largest} \rangle \tag{21}$$

5 Wolff's Extensions

Although I do not have space to discuss everthing that has been done in developing these methods, I feel that it is essential to include two very important extensions due to Ulli Wolff.[10]

The first modification made by Wolff was to introduce the single-cluster algorithm.[10] Instead on transforming the entire lattice, a single site is picked, and bonds are created with the probability given in Eq. 15 with those neighbors with the same spin. The process is repeated until the cluster can grow no further, at which point all spins in it are flipped. At first sight it might seem that this reduces to the SW algorithm, but actually the single-cluster algorithm preferentially flips larger clusters with a probability proportional to their size.

In two dimensions, the single-cluster algorithm seems to have the same dynamical critical exponent as the orginal SW algorithm, but the correlation time is reduced by a factor of about 2, since clusters are flipped with probability one instead of one-half.

In three dimensions, the single-cluster algorithm has $z_{sc} \approx 0.4$, and in four dimensions, the single-cluster algorithm seems to give $z_{sc} \approx 0$,[21] as opposed to $z_{SW} \approx 1.0$.[27] This is a very dramatic improvement, but one which has still not been completely understood.

Wolff has also made an even more important extension of the cluster approach by applying it to general O(n) models. He introduced an embedding of an Ising variable into the O(n) model, and simulated the resulting Ising model with a cluster algorithm. For example, in the xy model, a two-dimensional unit vector is chosen at random, and each spin is reflected through the plane described by this vector. For each spin, an Ising value of $+1$ can be assigned to the original direction and -1 to the reflected direction. This gives an effective Ising Hamiltonian without frustration, which can be easily simulated. This approach completely eliminates critical slowing down in the low-temperature phase,[11, 28] which is an even better result than has been achieved for the simpler Ising model.

It should be noted that similar embedding of Ising variables was developed independently by Brower and Tamayo to simulate the ϕ^4 model.[12]

Wang, Kotecký, and I have applied the idea of a two-dimensional embedding to solve the problem of how to do efficient MC simulations of antiferromagnetic Potts models with more than three states.[29, 30] By picking two spin values at random and then freezing all spins that do not have one of these two values, we obtain a simple Ising antiferromagnet on a diluted lattice. For large s, we can arrange the spin values in pairs and treat the system as being composed of several non-interacting Ising antiferromagnets. It turns out that this approach is extremely efficient in both two and three dimensions.

6 Replica Monte Carlo

As mentioned in the previous section, the presence of frustration in the Hamiltonian prevents the efficient application of a cluster MC algorithm based on the Fortuin-Kasteleyn transformation. An extreme example of such a Hamiltonian is the Ising spin glass.

$$H = K \sum_{\langle i,j \rangle} B_{i,j} \sigma_i \sigma_j \tag{22}$$

where K is the dimensionless inverse temperature as before, and the $B_{i,j}$ are quenched random variables that take on the values $+1$ and -1. At low temperatures, the generalized bonds from the Fortuin-Kasteleyn transformation will percolate around the frustrated plaquettes and almost all spins will belong to a single cluster.

To treat such systems, Wang and I introduced a different kind of cluster method, which we called Replica Monte Carlo.[31]-[34]

The first feature of the method is that instead of simulating different temperatures sequentially, n replicas of the system are simulated simultaneously. Each replica contains the same bond distribution, $B_{i,j}$, but a different inverse temperature K^a. The entire system of replicas can be described by a single Hamiltonian

$$H_{rep} = \sum_{a=1}^{n} K^a \sum_{\langle i,j \rangle} B_{i,j} \sigma_i^a \sigma_j^a \tag{23}$$

The advantage of setting up the simulation in this manner is that it enables us to pass information between the replicas. Consider two replicas, a and b. The Hamiltonian for the pair of replicas is

$$H_{pair} = \sum_{\langle i,j \rangle} B_{i,j} (K^a \sigma_i^a \sigma_j^a + K^b \sigma_i^b, \sigma_j^b) \tag{24}$$

Introduce new variables for the simulation. Instead of using σ_i^a and σ_i^b, use σ_i^a and $\tau_i^{ab} = \sigma_i^a \sigma_i^b$. The pair Hamiltonian then becomes

$$H_{pair} = \sum_{\langle i,j \rangle} B_{i,j}(K^a + K^b \tau_i^{ab} \tau_j^{ab}) \sigma_i^a \sigma_j^a \tag{25}$$

If we now hold the τ_i^{ab}'s fixed, we see that the couplings between σ_i^a's with the same value of τ_i^{ab} become $K_a + K_b$, so that they are bound strongly. On the other hand, the couplings between σ_i^a's with different values of τ_i^{ab} become $K_a - K_b$, so that they are only weakly coupled when the temperature difference between the replicas is small, even when both replicas are at very low temperatures. By using the distribution of τ_i^{ab} to define clusters of sites with the same value of τ's, we form clusters of spins which are tightly coupled to each other, but only weakly coupled to spins in a different τ cluster. This allows us to write a cluster Hamiltonian

$$H_{cluster} = \sum_{\alpha,\beta} k_{\alpha,\beta} \eta_\alpha \eta_\beta \tag{26}$$

where the interactions between the clusters are given by

$$k_{\alpha,\beta} = \sum_{\alpha,\beta boundary} B_{i,j} \sigma_i^a \sigma_j^a (K^a - K^b) \tag{27}$$

This Hamiltonian can now be simulated with standard MC methods, and the orginal spins then become $\sigma_i^a \eta_\alpha$ and $\sigma_i^b \eta_\alpha$ for each spin in the α cluster.

In practice, replicas are arranged at successive temperatures, usually separated by a constant $\Delta K = K^a - K^{a+1}$. The parameter ΔK and the number of replicas n can be varied to maximize efficiency.

Although this procedure clearly satisfies detailed balance, it does not by itself satisfy ergodicity. Therefore, it is necessary to perform these MC cluster moves as part of a full simulation that includes standard MC single spin flips, which are efficient at high temperatures.

The application of this method to the two-dimensional Ising spin glass has been extremely succcessful, reducing correlation times by three or four orders of magnitude at low temperatures.[31] In three dimensions, Replica Monte Carlo also provides considerable improvement,[32, 34] but since it is only about a factor of 100, it has not eliminated the difficulties in three-dimensional simulations.

7 The single-histogram method

The single-histogram equations actually go back over thirty years, and have been discussed by many authors.[35]-[39] To derive the single-histogram equations, consider a general form for the Hamiltonian

$$H(\sigma) = \sum_{\alpha} K_\alpha \hat{\mathcal{S}}_\alpha(\sigma) \tag{28}$$

where σ represent a spin (or particle) configuration, $\beta = 1/k_B T$ and K_α is the dimensionless coupling constant associated with the operator $\hat{\mathcal{S}}_\alpha(\sigma)$.

To simplify the notation, we will consider a Hamiltonian which contains only one term, $K\hat{\mathcal{S}}$. Everything is still valid if any term H_o that does not depend on the parameter K is added to the Hamiltonian. Extensions to more than one parameter are straightforward.

In the limit of infinite run length, the probability of observing the configuration σ is simply

$$P_{K_0}(\sigma) = \frac{\exp[K_0 \hat{\mathcal{S}}(\sigma)]}{\sum_{\{\sigma\}} \exp[K_0 \hat{\mathcal{S}}(\sigma)]} \tag{29}$$

From this distribution, the thermal average of any operator on σ can be determined.

$$< f(\sigma) > (K_0) = \sum_{\{\sigma\}} f(\sigma) P_{K_0}(\sigma) \tag{30}$$

It is convenient to work with a related probability distribution, $P_{K_0}(S)$, where S is one of the values in the spectrum of the operator $\hat{\mathcal{S}}(\sigma)$. For the simple one-term Hamiltonian, S is one of the possible values which the energy can assume. $P_{K_0}(S)$ can be written in the form

$$P_{K_0}(S) = \frac{W(S) \exp[K_0 S]}{\sum_{\{S\}} W(S) \exp[K_0 S]} \tag{31}$$

where

$$W(S) = \sum_{\{\sigma\}} \delta(\hat{\mathcal{S}}(\sigma), S)) \tag{32}$$

is the number of configurations with energy S, and the sum in the denominator goes over the entire spectrum of $\hat{\mathcal{S}}(\sigma)$. The information in this distribution can be used to calculate the average values of functions of S.

$$< f(S) > (K_0) = \sum_{\{S\}} f(S) P_{K_0}(S) \tag{33}$$

Note that $W(S)$ is independent of K_0 so that the distribution with a different value of K is

$$P_K(S) = \frac{W(S)\exp[KS]}{\sum_{\{S\}} W(S)\exp[KS]} \tag{34}$$

From the form of Eqs. (31) and (34), it is clear that knowledge of the distribution for some value of K is sufficient to determine it for all values of K. Eq. (31) can be re-written to give $W(S)$:

$$W(S) = Z(K_0) P_{K_0}(S)\exp[-K_0 S] \tag{35}$$

($Z(K_0) = \sum_{\{S\}} W(S)\exp[K_0 S]$ is an estimate for the partition function of the system.) If this expression is inserted into Eq. (34), the distribution with coupling K can be expressed as

$$P_K(S) = \frac{P_{K_0}(S)\exp[(K - K_0)S]}{\sum_{\{S\}} P_{K_0}(S)\exp[(K - K_0)S]} \tag{36}$$

This is the single-histogram equation. When applying Eq. (36) to actual Monte Carlo data, the effects of a finite run length must be taken into account. An MC simulation of length N generates a histogram of S values $H(S, K_0, N)$ which gives the exact distribution $P_{K_0}(S)$ in the limit of an infinite run length

$$P_{K_0}(S) = \lim_{N\to\infty} \frac{H(S, K_0, N)}{N} \tag{37}$$

However, for a finite run, the histogram will be subject to statistical errors

$$\delta^2 H(S, K_0, N) = gH(S, K_0, N) \tag{38}$$

with $g = 1+2\tau_{int}$ [15] where τ is the integrated correlation time. In particular, the relative uncertainty will become large in the wings of the distribution where the number of histogram entries is small. This is important because of the shift in the peak of the distribution when Eq. (36) is used. When $| K - K_0 |$ becomes too large, the peak in the new distribution will occur for an S value in the wings of the measured distribution resulting in unacceptable errors.

The relationship between the width of the distribution and the range of reliable K values can be worked out rather easily. For a single-peaked distribution, the width of the peak is related to the specific heat by

$$\delta S \propto \left(C(K_0) L^d \right)^{1/2} \tag{39}$$

where L^d is the volume of the system. (L is a characteristic linear dimension of the system.) The shift of the peak position is related to the specific heat and ΔK by

$$\Delta S \propto \left(L^d C(K_0) \right) \Delta K \tag{40}$$

The maximum ΔK which gives reliable results occurs when $\Delta S \propto \delta S$ which gives

$$\Delta K_{\mathrm{MAX}} \propto \left(C(K_0) L^d \right)^{-1/2} \tag{41}$$

which supports the observation of Chesnut and Salsburg [36] that the range of reliable results for this method decreases as the system size increases. From this result, it was concluded (prematurely) that this method would work only up to a certain system size.

If one is interested in the properties at a critical point, another effect must be taken into account. The finite-size scaling region, which usually coincides with the region of the specific heat peak, also decreases as the system size increases. The rate of decrease, for a thermal operator, is given by L^{-y_T}.

At a critical point, the specific heat at the peak increases as $L^{\alpha/\nu}$ so that ΔK_{MAX} decreases as $L^{-\frac{1}{2}(\alpha/\nu + d)}$. Using the scaling result $\alpha = 2 - d\nu$, ΔK_{MAX} is seen to decrease as $L^{-\nu}$. Since $y_T = 1/\nu$, ΔK_{MAX} decreases at the same rate as the finite-size scaling region. This means that if a single simulation provides information over the entire finite-size scaling region for some value of L, similar simulations on larger lattices will also provide information over the entire finite-size scaling region.

Away from the critical point, the specific heat is approximately independent of system size so that ΔK_{MAX} decreases as $L^{-d/2}$. (The width of the distribution increases as $L^{d/2}$.) Because most operators of interest are extensive, the number of simulations needed to provide information over the entire range of S increases as $L^{d/2}$. Similar arguments hold for each of the operators in the Hamiltonian.

For first-order phase transitions, a probability distribution will typically have two or more peaks. The y value corresponding to a first-order transition is $y = d$. This is due to latent heat of the transition, or the distance between peaks of the distribution, which is extensive. The quantity which governs the shift in the peak, however, is not the specific heat itself, but rather the "specific heat" associated with the individual peaks in the distribution.

A demonstration of the single-histogram method for the d=2 Ising model has been given by Ferrenberg and Swendsen,[39] comparing the results to the exact finite-lattice solution of Fisher and Ferdinand [40]. For a 16 by 16 lattice, they were able to determine the position of the specific heat maximum to within 0.04% and the value of the maximum to within 0.2%. Within the peak region, the agreement between the MC results and the exact solution is excellent, and for all temperatures in the range $T_c \pm 20\%$ the error is less than 0.5%.

8 The Multiple-Histogram Method

To investigate the behavior of the system over a wider range of parameter values, it is necessary to perform simulations at more than one temperature (or magnetic field) to get sufficient statistical accuracy). In this section, I will describe an optimized method for combining the data from an arbitrary number of simulations to obtain information over a wide range of parameter values. [41]-[43] The multiple-histogram method described here goes beyond earlier methods[44]-[52] in that it provides an optimized combination of data from different sources, produces results in the form of continuous functions for all values of interest, and can be applied to an arbitrary number of simulations. Errors can be calculated and provide a clear and simple guide to optimizing the length and location of additional simulations to provide maximum accuracy.

For simplicity, we will again present the derivation for a one-parameter Hamiltonian $H = KS$, but the generalization to an arbitrary number of parameters is just as simple and straightforward as it is for the single-histogram equations.

Consider R MC simulations. The nth simulation, with N_n MC updates, is performed at K_n and provides a histogram $H_n(S)$. As shown in in the previous section, the histogram provides an estimate for the equilibrium probability distribution

$$P_{eq}(S) = \frac{H_n(S)}{N_n} = W(S)\exp[K_n S - f_n] \tag{42}$$

where $W(S)$ is the density of states and f_n is a parameter equal to the free energy at K_n. This equation can be solved to get an estimate for the density of states W(S).

$$W_n(S) = \frac{H_n(S)}{N_n}\exp[f_n - K_n S] \tag{43}$$

Of course, due to statistical errors in the histogram, this estimate from one simulation will be reliable only over some range of S values (as discussed in the previous section). However, each of the R simulations performed will give a different estimate for the density of states. An improved estimate for $W(S)$ can be determined as a weighted sum over each individual estimate for the density of states:

$$W(S) = \sum_{n=1}^{R} p_n(S) W_n(S) \tag{44}$$

subject to the condition

$$\sum_{n=1}^{R} p_n(S) = 1 \tag{45}$$

This estimate is then optimized, *for each value of* S, by choosing the $p_n(S)$ so as to minimize the error in the estimate for $W(S)$.

$$\delta^2 W(S) = \sum_{n=1}^{R} p_n^2(S)\delta^2 W_n(S) \tag{46}$$

The important idea of optimizing separately for each value of S was first suggested by Bennett in connection with the related problem of calculating the difference in free energy between two temperatures when MC data at each of these temperatures is available.[53, 54]

The only errors in $W_n(S)$ come from errors in the histogram $H_n(S)$.

$$\delta^2 W_n(S) = N_n^{-2} \exp[2f_n - 2K_n S] \delta^2 H_n(S) \tag{47}$$

These errors are given by

$$\delta^2 H_n(S) = g_n \overline{H_n(S)} \tag{48}$$

where a bar over an expression indicates the expectation value with respect to all MC simulations of length N_n. If successive MC configurations are independent, then $g_n = 1$, otherwise

$$g_n = 1 + 2\tau_n \tag{49}$$

where τ_n is the correlation time for the nth simulation. [15]

Inserting this into the expression for $\delta^2 W_n(S)$ gives

$$\delta^2 W_n(S) = N_n^{-2} \exp[2f_n - 2K_n S] g_n \overline{H_n(S)}. \tag{50}$$

At this point we must make an estimate of $\overline{H_n(S)}$. Since $\overline{H_n(S)}$ is directly related to the density of states (see Eq. (43) by

$$\overline{H}_n(S) = N_n W(S) \exp[K_n(S) - f_n]. \tag{51}$$

Using the approximation in Eq. (51), we find an estimate of the error in the density of states to be

$$\delta^2 W_n(S) = N_n^{-1} \exp[f_n - K_n S] g_n W(S). \tag{52}$$

Inserting this into Eq. (46) gives

$$\delta^2 W(S) = \sum_{n=1}^{R} p_n^2(S) N_n^{-1} \exp[f_n - K_n S] g_n W(S) \tag{53}$$

The condition in Eq. (45) can be taken into account explicitly by rewriting Eq. (53) as

$$\delta^2 W(S) = (1 - \sum_{n=2}^{R} p_n(S))^2 N_1^{-1} \exp[f_1 - K_1 S] g_1 W(S) + \sum_{n=2}^{R} p_n^2(S) N_n^{-1} \exp[f_n - K_n S] g_n W(S) \tag{54}$$

This expression can now be minimized with respect to one of the $p_n(S)$ values, resulting in

$$p_1(S)N_1^{-1}exp[f_1 - K_1S]g_1 = p_n(S)N_n^{-1}exp[f_n - K_nS]g_n \tag{55}$$

for any $n \neq 1$. Solving this for $p_n(S)$ gives

$$p_n(S) = \frac{n_n g_n^{-1} \exp[K_nS - f_n]}{\sum_{m=1}^{R} n_m g_m^{-1} \exp[K_mS - f_m]} \tag{56}$$

If we then define

$$P(S, K) = W(S)\exp[KS] \tag{57}$$

we obtain the essential multiple-histogram equations as

$$P(S, K) = \frac{\left(\sum_{n=1}^{R} g_n^{-1} H_n(S)\right) \exp[KS]}{\sum_{m=1}^{R} n_m g_m^{-1} \exp[K_mS - f_m]} \tag{58}$$

where

$$\exp[f_n] = \sum_S P(S, K_n) \tag{59}$$

Eqs. (58) and (59) must be iterated to determine the f_n values self consistently. As with other Monte Carlo techniques for calculating free energies, these equations determine the free energy to within an additive constant. The convergence can be accelerated by making use of derivatives of f values on one iteration with respect to those of the previous iteration.

If the optimized $p_n(S)$ in Eq. (56) is inserted into the expression for the error in the density of states Eq. (46), the final expression for the statistical errors in $W(S)$ is

$$\delta W(S) = \left[\sum_n g_n^{-1} H_n(S)\right]^{-1/2} W(S) \tag{60}$$

from which it is clear that this method always reduces the statistical errors when additional MC simulations are added to the analysis. This expression also provides a clear guide for planning a series of simulations. The locations and heights of peaks in the relative error, plotted as a function of S, give direct quantitative indications of the optimum locations and lengths of additional MC simulations.

Once the f_n values are determined, Eq. (58) can be used to calculate the average value of any operator on S as a function of K.

$$< A > (K) = \frac{\sum_{\{S\}} A(S)P(S,K)}{\sum_{\{S\}} P(S,K)} \tag{61}$$

In particular, one can calculate the specific heat

$$C(K) = \frac{K^2}{V}\left(< S^2 > - < S >^2\right) \tag{62}$$

and accurately determine the location and height of its peak.

As a practical matter, it is useful to handle most of the calculations in terms of the logarithms of the various quantities in these equations. Numerical problems involved in implementing the multiple-histogram method will be discussed in the Appendix.

If the method is restricted to two MC simulations, the calculated difference in the free energies between the simulated temperatures is identical to that obtained by Bennett's method.[53, 54]

The free energy $F(K)$ is calculated using

$$F(K) = \ln Z(K) \tag{63}$$

where

$$Z(K) = \sum_{\{S\}} P(S,K) \tag{64}$$

and has a factor of $-\beta$ absorbed in it so that the actual free energy $\mathcal{F}$ is

$$\mathcal{F} = -F(K)/\beta \tag{65}$$

Information about the free energy can also be used to calculate the entropy $\mathcal{S}$ as a function of K.

$$\mathcal{S}(K) = \ln Z(K) - KZ^{-1}\sum_{\{S\}} SP(S,K) \tag{66}$$

Like F, the entropy $\mathcal{S}$ is determined to within an additive constant.

9 Conclusions

In this review, I have given a general introduction to cluster Monte Carlo algorithms for critical slowing down at second order phase transitions and frustrated systems, such as spin glasses.

I have also discussed some of the new applications of histograms for the analysis of second order transitions, as well as giving the basic equations for the multiple histogram method.

A great deal of progress has been in using these new techniques, but much work need to be done. We still have only a rudimentary understanding of why the cluster methods are so good, and we have no way of predicting the value of the effective dynamical critical exponents.

The situation for simulations of spin glasses in three and four dimensions is far from satisfactory, and more work is needed to devise a truly effective simulation method for these cases.

A number of directions are open for future developments. One of the most interesting is the combination of cluster methods with multi-grid techniques. Progress has already been made which shows that significant reductions in correlation times can be achieved by such a combination.[6, 7] This approach involves a considerably greater challenge to implement the program, but it also introduces considerable flexibility in designing further improvements.

In summary, the cluster approach has greatly expanded our ability to carry out high accuracy computer simulations for a large class of models. Prospects for future development look very promising.

10 Acknowledgements

I would like to acknowledge support by the National Science Foundation Grant No. DMR-9009475.

References

[1] *Monte Carlo Methods in Statistical Physics*, 2nd ed., edited by K. Binder (Springer, Berlin, 1986); *Applications of the Monte Carlo Method in Statistical Physics*, 2nd ed., edited by K. Binder (Springer, Berlin, 1987).

[2] N. Metropolis, A.W. Rosenbluth, A.H. Teller, and E. Teller, J. Chem. Phys. **21**, 1087 (1953).

[3] P.W. Kasteleyn and C.M. Fortuin, J. Phys. Soc. Jpn. Suppl. **26**s, 11 (1969); C.M. Fortuin and P.W. Kasteleyn, Physica (Utrecht) **57**, 536(1972).

[4] R. H. Swendsen and J.-S. Wang, Phys. Rev. Lett. **58**, 86 (1987).

[5] U. Wolff, Phys. Rev. Lett. **60** 1461 (1988); Nucl. Phys. B**300** [FS22] 501 (1988).

[6] D. Kandel, E. Domany, D. Ron, A. Brandt, and E. Loh, Phys. Rev. Lett. **60**, 1591 (1988).

[7] D. Kandel, E. Domany, A. Brandt, Phys. Rev. B **40**, 330 (1988).

[8] F. Niedermayer, Phys. Rev. Lett. **61**, 2026 (1988).

[9] R. G. Edwards and A. D. Sokal, Phys. Rev. D **38**, 2009 (1988).

[10] U. Wolff, Phys. Rev. Lett. **62**, 361 (1989).

[11] U. Wolff, Nucl. Phys. B**322**, 759 (1989).

[12] R. C. Brower and P. Tamayo, Phys. Rev. Lett. **62**, 1087 (1989).

[13] M. Hasenbusch, Nucl. Phys. B**333**, 581 (1990).

[14] R. Ben-Av, D. Kandel, E. Katznelson, P. G. Lauwers, and S. Solomon, J. Stat. Phys. **58**, 125 (1990).

[15] H. Müller-Krumbhaar and K. Binder, J. Stat. Phys. **8**, 1 (1973).

[16] A. Coniglio and W. Klein, J. Phys. A **13**, 2775 (1980).

[17] J. Hoshen and R. Kopelman, Phys. Rev. B**14**, 3438 (1976).

[18] D.W. Heermann and A.N. Burkitt, Physica A, **162**, 210 (1990).

[19] X.-J. Li and A. D. Sokal, Phys. Rev. Lett. **63**, 827 (1989).

[20] J.-S. Wang, Physica A **164**, 240 (1990).

[21] P. Tamayo, R. C. Brower, and W. Klein, J. Stat. Phys. **58** 1083, (1990).

[22] M. Sweeny, Phys. Rev. B**27**, 4445 (1983).

[23] C.-K. Hu, Phys. Rev. B**29**, 5103, 5109 (1984).

[24] U. Wolff, Phys. Lett. B**228**, 379 (1989).

[25] C.-K. Hu and K.-S. Mak, Phys. Rev. B **40**, 5007 (1989).

[26] F. Niedermayer, Phys. Lett. B**237**, 473 (1990).

[27] W. Klein, T. Ray, and P. Tamayo, Phys. Rev. Lett. **62**, 163 (1989).

[28] R. G. Edwards and A. D. Sokal, Phys. Rev. D **40**, 1374 (1989).

[29] J.-S. Wang, R.H. Swendsen, and R. Kotecký, Phys. Rev. Lett., **63**, 109 (1989).

[30] J.-S. Wang, R.H. Swendsen, and R. Kotecký, Phys. Rev. B, **42**, 2465 (1990).

[31] R. H. Swendsen and J.-S. Wang, Phys. Rev. Lett. **57**, 2607 (1986).

[32] J.-S. Wang and R. H. Swendsen, Phys. Rev. B **37**, 7745 (1988).

[33] J.-S. Wang and R. H. Swendsen, Phys. Rev. B **38**, 4840 (1988).

[34] J.-S. Wang and R. H. Swendsen, Phys. Rev. B **38**, 9086 (1988).

[35] Z.W. Salsburg, J.D. Jackson, W.Fickett, and W.W. Wood, J. Chem. Phys. **30**, 65 (1959);

[36] D.A. Chesnut and Z.W. Salsburg, J. Chem. Phys. **38**, 2861 (1963);

[37] I.R. McDonald and K. Singer, Disc. Far. Soc. **43**, 40 (1967);

[38] G. Torrie and J.P. Valleau, Chem. Phys. Lett. **28**, 578 (1974); J. Comp. Phys. **23**, 187 (1977).

[39] A.M. Ferrenberg and R.H. Swendsen, Phys. Rev. Lett. **61**, 2635 (1988).

[40] A.E. Ferdinand and M.E. Fisher, Phys. Rev. **185**, 832 (1969)

[41] A.M. Ferrenberg and R.H. Swendsen, Phys. Rev. Lett. **63**, 1195 (1989).

[42] A.M. Ferrenberg and R.H. Swendsen, Computers in Physics Sep/Oct, 101 (1989).

[43] R.H. Swendsen and A.M. Ferrenberg, "Histogram Methods for Monte Carlo Data Analysis," in "Computer Simulation Studies in Condensed Matter Physics: Recent Developments" (Athens, Georgia, 20-24 February 1990), p. 179.

[44] J.P. Valleau and D.N. Card, J. Chem. Phys. **57**, 5457 (1972);

[45] M. Falcioni, E. Marinari, M.L. Paciello, G. Parisi and B. Taglienti, Phys. Lett. **108B**, 331 (1982).

[46] E. Marinari, Nucl. Phys. **B235[FS11]**, 123 (1984).

[47] C.N. Yang and T.D. Lee, Phys. Rev. **87**, 404 (1952).

[48] G. Bhanot, S. Black, P. Carter and R. Salvador, Phys. Lett. **B183**, 331 (1987).

[49] G. Bhanot, K.M. Bitar, S. Black, P. Carter and R. Salvador, Phys. Lett. **B187**, 381 (1987).

[50] G. Bhanot, K.M. Bitar and R. Salvador, Phys. Lett **B188**, 246 (1987).

[51] G. Bhanot, R. Salvador, S. Black and R. Toral, Phys. Rev. Lett. **59**, 803 (1987).

[52] M. Karliner, R. Sharpe and Y.F. Chang, Nucl. Phys. **B302**, 204 (1988).

[53] C.H. Bennett, J. Comp. Phys. **22**, 245 (1976).

[54] Hyungyu Park, private communication.

NEW NUMERICAL METHOD TO STUDY PHASE TRANSITIONS AND ITS APPLICATIONS

Jooyoung Lee
Material Science Division, Argonne National Laboratory,
Argonne, IL60439, USA

J. M. Kosterlitz
Department of Physics, Brown University,
Providence, RI 02912, USA

ABSTRACT

We present a powerful method of identifying the nature of transitions by numerical simulation of finite systems. By studying the finite size scaling properties of free energy barrier between competing states, we can identify unambiguously a weak first order transition even when accessible system sizes are $L/\xi < 0.05$ as in the five state Potts model in two dimensions. When studying a continuous phase transition we obtain quite accurate estimates of critical exponents by treating it as a field driven first order transition. The method has been successfully applied to various systems

INTRODUCTION

There are not many statistical models which can be solved analytically with non trivial solutions. With rapid development of computer technology, computer simulation methods appear to be a realistic substitute for theoretical solutions which may not be available at all.

Mathematically speaking there occur no phase transitions in finite systems and the success of computer simulations near phase transitions depends on the finite size scaling,[1-3] where we extrapolate to the thermodynamic limits (N, $V \to \infty$) from small system sizes. This paper will describe how one can extend the idea of finite size scaling into the task of identifying the nature of transition. As a by-product it, in case of continuous transitions, we also get a surprisingly accurate way to obtain correlation-length exponent ν and critical temperature T_c.

The philosophy of all the existing methods for identifying the nature of transition relies on the outcomes of the finite size scaling of first order transitions with system size L much larger than the correlation length ξ ,which remains finite at first order transitions contrary to the second order. Therefore it is not surprising that they fail to give unambiguous answer (even with large size simulations[4-6] up to $L = 256$) to identifying weak first order transitions like $q =$ 5 Potts model[7] in two dimensions, where $\xi > 1000$ lattice spacings.[4] For example, the hysteresis effect, which has been the most popular method, comes from the competition between observation time and equilibration time in computer simulations of a finite size system. At a very strong first order transition the equilibration time near T_c can be much larger than the order of observation time, even though the latter can be as long as allowed by CPU time. One might be able to identify the nature of transition correct from a very careful analysis of the hysteresis effect,[8-10] but as the transition becomes weaker the effect becomes more ambiguous.

Recently[11], Binder has introduced a quantity that is believed to be very sensitive to the nature of transition. For energy it is defined as

$$V_L = 1 - \frac{< E^4 >_L}{3 < E^2 >_L^2} \tag{1}$$

With a continuous transition, $V_L \to 2/3$ for all temperature as $L \to \infty$. For a first order transition it takes on the value 2/3 for high and low temperature, but in a temperature range of $O(L^{-d})$ of T_c it can take on the value less than 2/3. Finite size scaling of the first order transition predicts $V_L|_{min}$ up to $O(L^{-d})$ as[12]

$$V_L|_{min} = 2/3 - (e_1/e_2 - e_2/e_1)^2/12 + AL^{-d} \tag{2}$$

which correct some earlier work.[13] Here e_1 and e_2 are bulk energy value for two competing states and A is a rather complicated expression involving T_c, e_1, e_2 and bulk specific heat of each phase. In principle $V_L|_{min}$ can be used to distinguish between continuous and first order (temperature driven) transition, but we find[12] that the asymtotic approach of L^{-d} sets in only for $L \gg \xi$ and more over $V_L|_{min} = 0.66622..$ for the case of $q = 5$ Potts model in two dimensions which is quite close to 2/3 to be distinguished numerically.

When one is forced to a numerical investigation on a totally unknown system, the first thing he has to do is to identify its nature, which dictates the next step. With a first order transition one may wants to evaluate, critical temperature, correlation length, discontinuities in various quantities, etc., whereas with a continuous transition, various exponents, etc. Therefore, without information on the nature of the transition, it makes little sense to intend to obtain critical exponents which is relevant only to continuous transitions.

FINITE SIZE SCALING OF BULK FREE ENERGY BARRIER[14]

Distinguishing a weak first order transition from continuous transitions is one of the ultimate tasks that computer simulation methods would like to provide. To do better than what has been done in the literature, it is obvious that we need to investigate the origin of the first order transitions consequence of which makes them so much different from continuous transitions when $L \gg \xi$. The essential feature of a discontinuous transition is that there exist a set of competing bulk phases with infinite free energy barriers of $O(L^{d-1})$ between them in the thermodynamic limits. Rather than studying the outcome of this large free energy barrier (which gives rise to the results of finite size scaling of strong first order in the limit $L \gg \xi$), we want to focus on how this barrier will evolve as a function of system size depending on the nature of the transition.

The key idea is, taking Ising model in two dimensions as an example of field-driven first order transition, how the two delta-function-like distribution of magnetization at $m = \pm m_0 \neq 0$ below T_c develops to a single gaussian centered at $m = 0$ as the temperature is increased through T_c. First, we consider a case of strong (field driven) first order transition with $L \gg \xi$. For Ising model this can be realized simply with $T \ll T_c$ in zero magnetic field $h = 0$. In this limit, the bulk free energy of restricted magnetization can be expanded as power of $1/L$ as follows.[15,16]

$$F(m,T,L) = L^d f_0(m,T) + L^{d-1} f_1(m,T) + O(L^{d-2}) \qquad (3)$$

Note that the form of L^{d-1} in Eq. 3 is relevant for systems with a discrete symmetry of the ground states such as Potts models. In case of a system with continuous symmetry of the ground states such as XY or Heisenberg model it is replaced by L^{d-2}. The bulk free energy density f_0 is minimum and constant for $-m_0 \leq m \leq m_0$ and the interfacial term f_1 has maximum at $m = 0$ due to the $\pm$ symmetry of the Ising model. Then it is obvious that F has minima at $m_1(L) = -m_0 - O(L^{-1})$ and $m_2(L) = m_0 + O(L^{-1})$ with a barrier of

$$\Delta F(t,L) = F(0,t,L) - F(m_1,t,L) \sim A(t)L^{d-1} + O(L^{d-2}) \qquad (4)$$

Therefore we find the known result of symmetry breaking below T_c that, even though there are two equivalently most probable states, once the system sits on one of them it is not allowed to reach the other equivalent state by infinite barrier of $O(L^{-d})$.

Here, $t = T/T_c - 1$ drives the system along the field driven first order transition line at $t < 0$ through the critical point at $t = 0$ into the disordered phase $t > 0$. For a temperature driven first order transition the roles of $h = 0$ and m are played by T_c and internal energy E. There may be a field g which drives the system along the temperature driven first order transition line at $g < 0$ through the critical or multicritical point at $g = 0$ into the second order transition line $g > 0$. For two dimensional Potts model $g < 0$ for $q > 4$ and $g = 0$ for $q = 4$. Therefore $F(E,g,L)$ is the equivalent quantity of $F(m,t,L)$ in Eq. 3.

When $L \ll \xi$, $F(X,g,L)$ is dominated by its singular part and finite size scaling theory tells us that it may be written in terms of scaling variables $x = XL^{\lambda_x}$ and $y = gL^{\lambda_g}$ as $F(X,g,L) = B(x,y)$.[16,17] For small x and y, $B(x,y)$ has an analytic expansion in x and y. At a first order transition, $y < 0$, $B(x,y)$ must have a set of minima of equal depth corresponding equivalent states, at $x_i = x_i(y)$. Two local minima guarantee that there exist at least one maximum (or saddle point) at $x_m = x_m(y)$, which corresponds to the configuration with maximum interface between competing states. So that the barrier between equivalent local minima $\Delta F(y) \equiv B(x_m(y), y) - B(x_i(y), y)$ will grow with increasing $-y$, eventually crossing over to the strong first order behavior of L^{d-1}.

The simplest possible scenario, which holds for simple systems like ferromagnetic q-state Potts models with continuous transitions, is with $y = tL^{1/\nu}$ and $x = ML^{\beta/\nu}$. Here one can show explicitly that for small x and y

$$\Delta F(t, L) \simeq a - btL^{1/\nu} \tag{5}$$

With periodic boundary conditions and L^d cubic geometry of the system, a and b are positive constant independent of L and they depend on boundary conditions and the geometry of the system. An assumption has been made here that the system size L is sufficiently large so that all the irrelevant variables may be ignored. Given this, it follows that $\Delta F(g, L)$ increases with L in the first order regime ($g < 0$), is an $L-$independent constant (possibly zero) at the critical point ($g = 0$) and decreases with L in a continuous transition (or disordered) regime ($g > 0$). This general argument constitute a very sensitive test of the nature of a transition even with $L < \xi$. Rather than depending on the quantitative estimate of thermodynamic values as $L \to \infty$, such as Binder's cumulant or exponent d of specific heat and susceptibility, one can determine the nature of a transition from the qualitative behavior of $\Delta F(g, L)$ which is the characteristic of the type of transition. However a note of cation is needed here. One can safely argue that the transition is of first order if $\Delta F(g, L)$ is found to increase with L, but with $a = 0$, if no peak structure is seen, it may merely mean that L is too small for the peak to be noticeable above numerical noise or that irrelevant variables have not scaled out. In this case one needs to seek how to increase a without affecting the critical properties of the model

TEMPERATURE DRIVEN FIRST ORDER TRANSITIONS

To study temperature driven first order transition, one simply need to simulate a given system at near phase transition accumulating the histogram of internal energy, which is the conjugate to the temperature. For the case of pressure driven first order transition one need the histogram of volume, which is the conjugate to the pressure. In N Monte Carlo (MC) sweeps, standard probability theory implies that the number of times the energy E is realized is

$$H(E, \beta, L) = NZ^{-1}(\beta, L)\Omega(E, L)\exp(-\beta E) \equiv \exp\{-A(E, \beta, L)\} \tag{6}$$

where $\Omega(E, L)$ is the number of states with energy E and $Z(\beta)$ is the partition function. $A(E, \beta, L)$ differs from the bulk free energy $F(E, \beta, L)$ by a temperature- and N-dependent additive quantity. Therefore, at fixed β, L, N

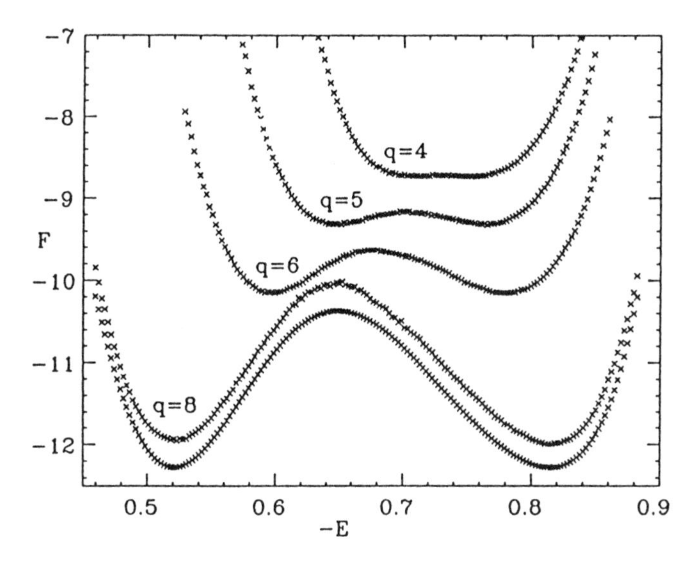

Fig. 1. $A(E, \beta_c(L), L)$ for $q = 4, 5, 6, 8$ Potts models for 32×32. Bottom curve is $q = 8$ data smoothed by polynomial fit. The energy scale is normalized to unity for complete order.

the shape of $A(E)$ will be same to that of $F(E)$ and we immediately obtain the crucial result

$$\Delta F_E(L) = \Delta A_E(L) \equiv A(E_1, \beta_c, L) - A(E_m, \beta_c, L) \tag{7}$$

where the function $A(E, \beta, L)$ have minima of equal depth at $E = E_1, E_2$ and maxima at E_m, at the pseudo-critical temperature $\beta_c(L)$ corresponding to the coexistence of ordered and disordered states. Eq. 7 is important since a measurement of ΔA from simulation gives a direct evaluation of the corresponding ΔF. The technical problem of computing $A(E, \beta, L)$ at the transition has been solved by the histogram method as used by Ferrenberg and Swendsen[18] who demonstrated how to extrapolate data from one value of β to nearby values. Since such extrapolations are accurate only for $\delta\beta \sim O(L^{-d})$ near a first order transition, one needs to locate $\beta_c(L)$ reasonably accurately by some preliminary simulations then perform one long simulation of about 5×10^6 MC steps to obtain reasonable statistics. Then one extrapolate the data to $\beta_c(L)$ to measure $\Delta F_E(L)$. In practice we also smoothed the data by an eighth-order polynomial fit.

Fairly extensive testing of temperature driven first order transitions was carried out for Potts models in two and three dimensions. Sizes were limited to $L \leq 60$ for $d = 2$ and $L \leq 14$ for $d = 3$ by the computer time available since good statistics are much more important than large system sizes. Despite $L/\xi < 0.05$ for $q = 5$, the first order nature was unambiguously shown for $q = 5, 6, 8, 10$

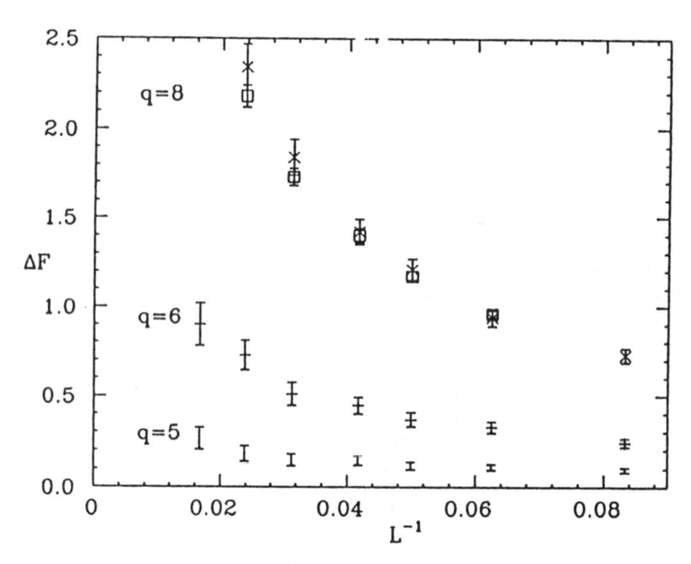

Fig. 2. Peak height $\Delta F_E(L)$ for $q = 5, 6, 8$ Potts models. Open squares for $q = 8$ were obtained by Swendsen-Wang algorithm[19]; all others by Metropolis algorithm.

[12,14] Potts models (see figures 1 and 2). From three dimensional Potts model with $L \leq 14$ we find[20], for the first time by simulation, $q_c = 2.45 \pm 0.1$. We also conclude that there is no essential singularity in correlation length or latent heat near $q = q_c$ in contrast to two dimensions.

FIELD DRIVEN FIRST ORDER TRANSITIONS

When applied to a continuous transition, the general method described in previous sections provides an easy and surprisingly accurate way of calculating the correlation length exponent ν, critical temperature T_c and somewhat less accurate order parameter exponent β/ν. Test[14] on Ising model in two dimensions up to $L \leq 60$ gives $\nu = 1.003(10)$, $2\beta/\nu = 0.247(8)$. We get $\nu = 0.634(6)$, $2\beta/\nu = 1.02(3)$ in three dimensions with $L \leq 14$. Here we describe, somewhat in details, direct application of the general method to $q = 3$ ferromagnetic Potts model in two dimensions to obtain T_c and $1/\nu$.

Let's consider a square lattice of $L \times L = N$ with periodic boundary conditions. If we denote N_i as the number of spins with $q = i$, we have a order parameter vector $\vec{N} = (N_1, N_2, N_3)$ and $N = N_1 + N_2 + N_3$ with two degrees of freedom. We want rewrite $\vec{N}$ in terms of two basis vectors $\vec{a} = 1/\sqrt{6}(2, -1, -1)$ and $\vec{b} = 1/\sqrt{2}(0, 1, -1)$ which are perpendicular to (N,N,N)/3 as follows

$$\vec{N} = (N, N, N)/3 + N_a\vec{a} + N_b\vec{b} \tag{8}$$

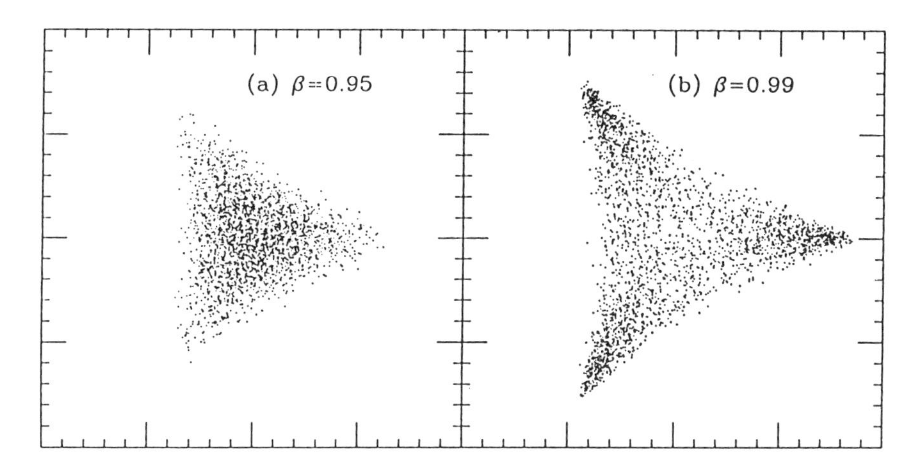

Fig. 3. Scatter plot of order parameter of $q = 3$ Potts model for 30×30.

where $N_a = (2N_1 - N_2 - N_3)/\sqrt{6}$ and $N_b = (N_2 - N_3)/\sqrt{2}$. We take the order parameter as the last two terms in Eq. 8

In this way the scatter plot of the order parameter for $N = 30 \times 30$ is shown in Fig. 3. For $\beta = 0.99$, one can clearly see the three dense areas corresponding to states where one of q is dominating the others. In the deconfinement transition[21,22] of quarks in $SU(3)$, one obtain similar distributions of order parameter from the Lattice QCD simulations. A criterion based on the certain ratio of occurrence between the states of $\theta = 0$ and $\pi/3$ is often used[22] to determine the nature of transition and critical temperature. But one has to be very careful using this kind of criterion since direct application of it to this model would lead to $\beta_c \simeq 0.97$ with first order transition. The three dense areas in Fig. 3. is only an artifact of finite size system (we know that the model undergoes a second order transition at $\beta = \ln(1 + \sqrt{3}) \simeq 1.005$), and they will disappear with larger system size just as the case of $\beta = 0.95$.

Near phase transition the free energy as a function of order parameter can be written in terms of scaling variables $x_i = (N_i - N/3)L^{\beta/\nu}$ and $y = tL^{1/\nu}$ with $x_1 + x_2 + x_3 = 0$. The symmetry of the system allows that the free energy can be expanded only by two terms of $x_1^2 + x_2^2 + x_3^2$ and $x_1 x_2 x_3$. With some algebra one can show that up to $O(r^4)$

$$F(r, \theta, y) = Ar^2 + Br^3 \cos\theta(4\cos^2\theta - 3) + Cr^4 + Dyr^2 \cdots \tag{9}$$

where (r, θ) comes in as the polar coordinates of x_is. We find ΔF defined as the difference between minima $(\theta = 0, \pm 2\pi/3)$ and saddle points $(\theta = \pi, \pm\pi/3)$ reads just same in the form of Eq. 5. A plot of $\Delta F_M(L, t)$ is shown in Fig. 4 for $10 \leq L \leq 36$. A very similar structure is seen for Ising models in two and three dimensions.[14]

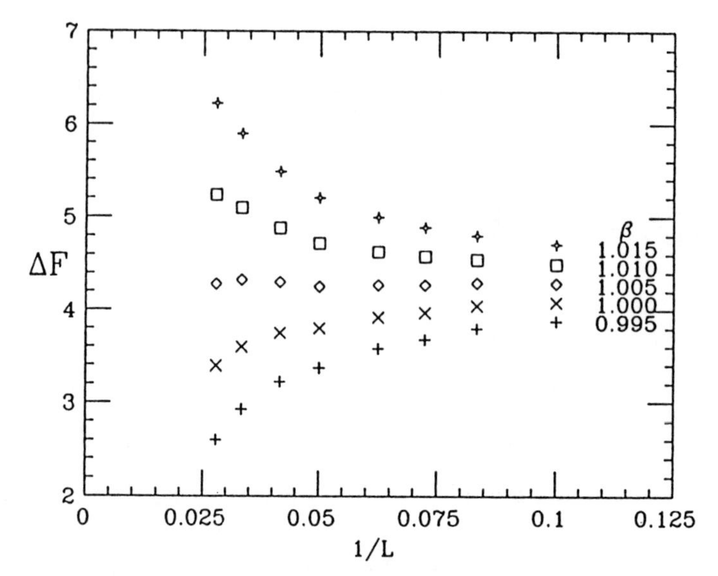

Fig. 4. $\Delta F_M(\beta, L)$ near second order transition of $q = 3$, $d = 2$ Potts model. All points at fixed L are obtained from $\beta = 1.005$ by extrapolation

With Eq. 5 the temperature derivative

$$S(L) \equiv \delta \Delta F(t, L)/\delta t \sim L^{1/\nu} \tag{10}$$

provides a numerical way of obtaining $1/\nu$ from a linear fit of $\ln S$ and $\ln L$ without exact value of T_c. This is very important since T_c and $1/\nu$ are obtained by independent two separate ways. För the case of Ising models $S(L)$ was obtained by a numerical derivative using the histogram method[18] mentioned earlier. For $q = 3$ model, observation that the maxima and saddle points occur only along the symmetry axis of $\theta = n\pi/3$ with $n = 1, \cdots, 6$ enables us to get $S(L)$ from[23]

$$S(L) = < E >_0 - < E >_{\pi/3} \tag{11}$$

where $< E >_{0,\pi/3}$ is the average internal energy for the symmetry axis of $\theta = 0, \pi/3$.

We show $S(L)$ in Fig. 5 for $5 \leq L \leq 36$ where we find $1/\nu = 1.19 \pm 0.02$ taking large sizes ($L \geq 12$). The order parameter exponent β/ν can be obtained by measuring the separation between minima (or saddle points) and the origin at T_c which scales as $L^{-\beta/\nu}$. The major source of error for β/ν is from the small uncertainty in T_c. Once T_c is found, simulations in microcanonical ensemble along the symmetry axes will be more effective to find exponents.

The method has been applied to simple models with great success but a note of caution is necessary here. First it is very important to know that the

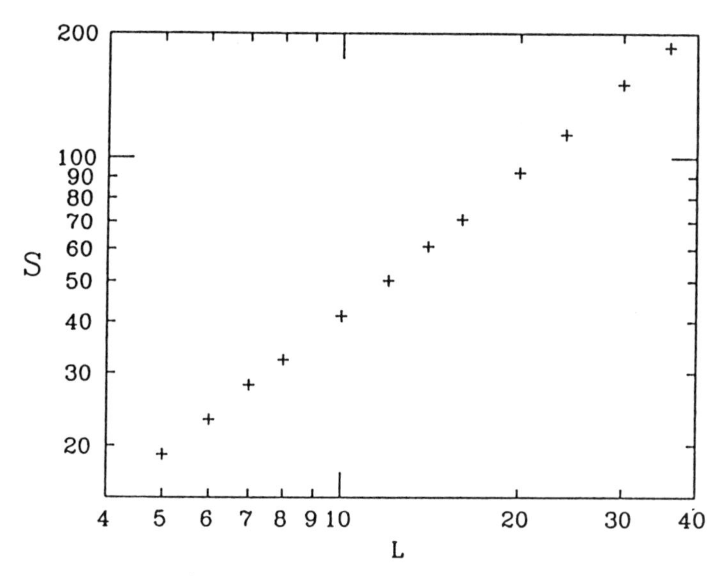

Fig. 5. $S(L) \sim L^{1/\nu}$ for $5 \leq L \leq 36$. Note curvature for $L \leq 12$ indicating an irrelevant variable.

transition is continuous so this kind of approach can give a very good estimates of exponents. Secondly, the presence of slowly decaying irrelevant variables may effect the estimates. For the case of $q = 3$ Potts model in two dimensions we observe small upward curvature in $S(L)$ for $L \leq 12$

MORE APPLICATIONS

The original motivation for this general method was to study the problem of fully frustrated Josephson junction arrays which is a two-dimensional regular periodic array of superconducting grains in a perpendicular magnetic field of half flux quantum per plaquette. This has been intensively studied[24] for many years both experimentally and theoretically with rather inconclusive results. Here we leave the results of the direct application of the method to references 25–27.

The method also has been applied to the study of Hopfield network with limited-connectivity.[28] Two-dimensional melting study of hard disk system with this method shows[29] that it is of first order transition unambiguously. Also some thermodynamic quantities such as the critical pressure and correlation length is obtained. Applications to other melting systems are underway

CONCLUSION AND DISCUSSION

We have developed a remarkably powerful method for numerically detecting first order transitions since the qualitative behavior of $\Delta F(L)$ is the characteristic of the type of transition. At a continuous transition good estimates of the critical exponents and critical temperature can be obtained as a by product. Extensive tests were carried out to known systems showing the power and the simpleness of the method. Applications to many other systems bring out new results[20,25–29] as mentioned in the text.

Numerical proof of the second order phase transition ($q = 3.9$ Potts model in two dimensions, for example) is more difficult subject especially when $a = 0$ in Eq. 5, which seems to be the most of the cases we tested. The success depends on how one can increase a without affecting the critical properties of the system. One candidate which is under investigation, is using the effect of dimensional crossover.

There are lots of rooms here one can investigate more on the method not to mention on the application side. A theory for $A(E, M, \beta, L)$ near a tricritical point would be very helpful since it is notoriously difficult to locate by numerical methods. We have not yet studied systems with a continuous symmetry such as $O(n)$ model. This kind of study will help to understand, for example, the problem of general two-dimensional melting. With growing interest in flux line lattice in HTC superconductors, this method should, in principle, give the best out of the numerical simulations

This work was supported by the NSF grant no. DMR-8918358 (JMK and JL), by US Department of Energy under contract no. W-31-109-ENG-39 (JL).

REFERENCES

1. M. N. Barber, Phase Transitions and Critical Phenomena, edited by C. Domb and J. L. Lebowitz (Academic, New York, 1983), Vol. 8, p.145.
2. V. Privman and M. E. Fisher, J. Stat. Phys. **33**, 385 (1983); Phys. Rev. B **32**, 447 (1985).
3. K. Binder, Rep. Prog. Phys. bf 50,783 (1987).
4. P. Peczak and D. P. Landau, Phys. Rev. B **39**,11932 (1989).
5. J. F. McCarthy, Phys. Rev. B **41**,9530 (1990)
6. M. Fukugita, H. Mino, M. Okawa and A. Ukawa, J. Phys. **A23**, L561 (1990).
7. R. B. Potts, Proc. Camb. Phil. Soc. **48**,106 (1952); F. Y. Wu, Rev. Mod. Phys. **54**,235 (1982).
8. K. Binder, Monte Carlo Method in Statistical Physics, edited by K. Binder, Topics in Current Physics, (Springer, Berlin, Heidelberg and New York, 1979) Vol. 7, p.1.
9. D. P. Landau and K. Binder, Phys. Rev. B **17**, 2328 (1978).
10. K. Binder and D. P. Landau, Phys. Rev. B **21**, 1941 (1980).
11. K. Binder, Phys. Rev. Lett. **47**, 693 (1981); Z. Phys. B **43**, 119 (1981).
12. J. Lee and J. M. Kosterlitz, Phys. Rev. B **43**, 3265 (1991).
13. M. S. S. Challa, D. P. Landau, K. Binder, Phys. Rev. B **34**, 184 (1986).
14. J. Lee and J. M. Kosterlitz, Phys. Rev. Lett. **65**, 137 (1990).
15. K. Binder, Rep. Prog. Phys. **50**, 783 (1987).

16. For recent reviews see Finite Size Scaling and Numerical Simulations of Statistical Systems, edited by V. Privman (World Scientific, Singapore, 1990).
17. E. Eisenriegler and R. Tomaschitz, Phys. Rev. B **35**, 4876 (1987).
18. A. M. Ferrenberg and R. H. Swendsen, Phys. Rev. Lett. **61**, 2635 (1988).
19. R. H. Swendsen and J-S. Wang, Phys. Rev. Lett. **58**, 86 (1987).
20. J. Lee and J. M. Kosterlitz, Phys. Rev. B **43**, 1268 (1991).
21. A. D. Kennedy, et al., Phys. Rev. Lett. **54**, 87 (1985); S. A. Gottlieb, et al., ibid., **55**, 1958 (1985); M. Fukugita, M. Okawa and A. Ukawa, Nucl. Phys. B **337**, 181 (1990).
22. N. H. Christ and A. E. Terrano, Phys. Rev. Lett. **56**, 111 (1986); N. H. Christ and H.-Q. Ding, ibid., **60**,1367 (1988).
23. J. Lee, Ph. D. Thesis (unpublished).
24. S. Teitel and C. Jajaprakash, Phys. Rev. Lett. **51**, 199 (1983); W. Y. Shih and D. Stroud, Phys. Rev. B **32**, 158 (1985).
25. E. Granato, J. M. Kosterlitz, J. Lee and P. M. Nightingale, Phys. Rev. Lett. **66**, 1090 (1991)
26. J. Lee, J. M. Kosterlitz and E. Granato, Phys. Rev. B **43**, 11531 (1991)
27. J. Lee, E. Granato and J. K. Kosterlitz, Phys. Rev. B (in press).
28. K. J. Strandburg, M. Peshkin and D. F. Boyd (unpublished).
29. J. Lee (unpublished).

Monte Carlo Analysis of the Three-Dimensional Ising Model

Nobuyasu Ito*

Department of Physics, University of Tokyo,

Hongo, Bunkyo-ku, Tokyo 113, Japan

The values of the spontaneous magnetization, m_S, of cubic-lattice Ising model which have been estimated using the Monte Carlo simulation and the extrapolation method are analyzed. The critical indices are estimated from its behavior using the least square fitting and Akaike's information criterion (AIC). The AIC analysis indicates that the m_s is a simple single-pole function. Based on this type of function, the values of K_c and β are estimated to be 0.2216566(17) and 0.31977(38), respectively.

I. INTRODUCTION

The Monte Carlo simulation has been used to study behavior of the model in the statistical physics for more than thirty years. At the early time, that is, in 1950s and in 1960s, binary alloy model and ferromagnetic Ising model on three-dimensional lattices were simulated, although the simulation scale was so small that only qualitative features were studied. The simulation speed at that time was at most dozens of spins update per second[1,2]. The progress of the computer keeps tremendous speed. The computing speed has been increasing exponentially. Giga FLOPS calculation for several hundreds or several thousands hours is not so difficult today and the construction of tera FLOPS machines is already realistic projects.

Today it is possible to update $1G(1 \times 10^9)$ spins per second[3]. When the available amount of simulation power becomes larger than the relaxation time of some model, the model comes into the scope of computer study. Random systems and stochastic optimizations of non-linear problems often have very long relaxation times if the conventional Monte Carlo or molecular dynamics methods are applied and many problems remain unsolved because of their long relaxation time. This is one aspect of the computing study in qualitative direction. There are another aspect in the high-speed simulation. It is the improvement of the accuracy. The accuracy of the estimated values by the stochastic method behaves as $1/T^{1/2}$ where T denotes the length of the simulation. This relation means that one more digit for the estimate requires hundred times amount of CPU. This relation between the computation time and the accuracy is, however, much better than the combinatorial algorithms which require exponentially longer CPU for every one digit of the result. But we should treat the estimated result and its error carefully. One danger is the underestimation of the error bar. It produces incorrect result. The other is the overestimation and it is especially severe when one attacks the problem which requires very large CPU time. If the error is overestimated two times, it means that three-quarters of the CPU is wasted. This is the other aspect of the computing study in quantitative direction.

The above mentioned two aspects of large-scale simulation provide several physical and technical problems. What are the most efficient simulation algorithms and codes[3,4,5,6]? Is it possible to make the simulation dynamics which have shorter relaxation time [7,8,9]? What is the most efficient analyzing method[10,11]? In this paper, an analyzing method which is useful to select the correct model is discussed. If there is some knowledge about the behavior of the relevant quantities, we can estimate the characteristic quantities based on the theory and the results. But it often happens that we do not know how to analyze the results of the simulation. In such case, the method which selects the model based on the results of the simulation is necessary.

We will analyze the spontaneous magnetization of the cubic-lattice ferromagnetic Ising model using the Akaike's Information criterion (AIC)[12]. For that model, the expectation values of the squared magnetization of finite lattices were extrapolated to the thermodynamic limit and the values of the spontaneous magnetization were obtained[13]. The Hamiltonian of that model is $H = -J\sum_{<i,j>}\sigma_i\sigma_j$ where the Ising spin σ_i takes the value of +1 or −1. In the following, K denotes βJ ($\beta = 1/k_{\mathrm{B}}T$) and it is used to indicate the temperature. These values are given in Table I.

TABLE I. The estimated values of the square of spontaneous magnetization, m_s^2, are given. The values estimated by the low temperature series expansion, $m_s \approx 1-2u^3-u^5\sum_{r=0}^{15} b_r u^r$ ($u = \exp(-4K)$)[14] are also listed. The values of LTE up to the order of u^{20} is consistent with those of Monte Carlo and extrapolation method for $K \geq 4.0$ as far as the present accuracy.

K	m_s^2 (extrapolation)	m_s^2 (LTE)
0.2219	0.03305(29)	
0.2220	0.04123(20)	
0.2240	0.138708(39)	
0.2250	0.172717(39)	
0.2260	0.202463(26)	
0.2280	0.253695(29)	
0.2300	0.297541(17)	
0.2350	0.386471(24)	
0.2400	0.456726(18)	
0.2500	0.563919(13)	1.30
0.2600	0.643006(17)	0.9472936
0.2800	0.752312(16)	0.8110804
0.3000	0.823000(12)	0.8347336
0.3200	0.8710284(79)	0.87335925
0.3400	0.9046911(64)	0.90515008
0.3600	0.9288144(45)	0.92889587
0.4000	0.9593242(46)	0.95932213
0.4500	0.9791523(28)	0.97915266
0.5000	0.9890727(23)	0.98907259
0.5500	0.9941851(14)	0.99418664
0.6000	0.99687516(95)	0.996876367

Based on those values, the critical indexes, that is, the values of critical point and critical exponent were estimated. The form of the spontaneous magnetization for cubic lattice Ising model has not been known yet but the following two kinds of functions are plausible near the critical point. One is the simple single-pole-type function,

$$m_s = \left(\sum_{k=1}^{N} a_k (K - K_c)^k\right)^{\beta}. \tag{1}$$

The least square fitting of this type of functions produced the estimates, $K_c = 0.221657(3)$ and $\beta = 0.3205(6)$. The other is the functions of the form of,

$$m_s = \sum_{k=0}^{N} b_k (K - K_c)^{k+\beta} + \sum_{k=1}^{M} c_k (K - K_c)^k \tag{2}$$

and these functions produced the estimates, $K_c = 0.221651(4)$ and $\beta = 0.3265(9)$. There is small but definite discrepancy between these two estimates for the value of β. But there was no reason to select one of these two kinds of functions. Therefore the conclusions were $K_c = 0.221654(6)$ and $\beta = 0.324(4)$ which covered the estimates using the two kinds of functions[13].

In the present paper, the AIC idea is applied to select the appropriate function form for the spontaneous magnetization of cubic lattice Ising model. The AIC is given in the next section and it is applied to analyzing the m_s in the third section. The last section is summary and discussion.

II. MODEL SELECTION AND AIC

When a set of statistical values are given by experiments, observations or simulations, these data should be analyzed to obtain the essential properties underlying them. If the model behavior is known a priori, we can describe the data by the most likelihood method, for example, least square fitting. If it is not known, some model selection method is necessary. The Akaike's information criterion, which is abbreviated to AIC, is a very useful method for this purpose. The basic idea of the AIC is that the model which has the minimum Kullback-Leibler mean information is considered to be the best model. Akaike's theory treats general situation[12], but in this section the AIC is derived for the present purpose.

The Monte Carlo estimates of the square of spontaneous magnetization and their errors are denoted by y_i and $\sigma_i (i = 1, \cdots, n)$, respectively. The y_i is the estimated value at the inverse temperature K_i. The distribution function of $\{y_i \ (i = 1, \cdots, n)\}$ is the Gaussian form and it is denoted by $P(\{y_i\})$, which is:

$$P(\{y_i\}) = \left(\prod_{i=1}^{n} \frac{1}{\sqrt{2\pi}\sigma_i}\right) \exp\left[-\sum_{i=1}^{n} \frac{(y_i - y^0(K_i))^2}{2\sigma_i^2}\right], \tag{3}$$

where $y^0(K)$ is the exact solution of spontaneous magnetization, which is not known yet. In other word, $y^0(K)$ is true model for $\{y_i\}$. Now a candidate $\hat{y}(K)$ is considered. It may hold free parameters and they are determined so that the $\hat{y}(K)$ fits the data $\{y_i\}$ best. When the model $\hat{y}(K)$ is fitted to the data, the distribution function $P(\{y_i\})$ is estimated to be

$$\hat{P}(\{y_i\}; \hat{y}(K)) = \left(\prod_{i=1}^{n} \frac{1}{\sqrt{2\pi}\sigma_i}\right) \exp\left[-\sum_{i=1}^{n} \frac{(y_i - \hat{y}(K_i))^2}{2\sigma_i^2}\right]. \tag{4}$$

The Kullback-Leibler mean information of this $\hat{P}$ in terms of P is,

$$I(P, \hat{P}) = -\int_{-\infty}^{\infty} \cdots \int_{-\infty}^{\infty} P(\{y_i\}) \log \frac{P(\{y_i\})}{\hat{P}(\{y_i\})} dy_1 \cdots dy_n \tag{5}$$

$$= \frac{1}{2}S + r - \frac{n}{2}, \tag{6}$$

where S and r denote $\sum_{i=1}^{n} (y_i - \hat{y}(K_i))^2/\sigma_i^2$ and the number of fitting parameters in $\hat{y}(K)$, respectively. The higher-order nonlinear terms are neglected in the derivation of the equation (6). The model which has the minimum value of this $(S + 2r - n)/2$ is considered to be the best model for the data $\{y_i\}$.

In the next section we will use $2I$ as AIC, that is,

$$\mathrm{AIC}(\{y_i\}, \{\sigma_i\}, \hat{y}(K)) = S + 2r - n. \tag{7}$$

III. AIC ANALYSIS OF SPONTANEOUS MAGNETIZATION

In this section, the estimated values of m_s given in Table I are analyzed based on AIC. All the 21 data in Table I are not used. The data near the critical point are used for the analysis. The used sets of data are $\{m_s^2(K); 0.2219 \leq K \leq K_{\max}\}$ which is denoted by $D(K_{\max})$. Several functions of the form of eqs. (1) and (2) are fitted for each region $D(K_{\max})$.

The results of the fittings are given in Table II. It is observed that the function of the form of eq. (1) is the better model than the function of the form of eq. (2). There are two sets of data in which the eq. (2) has the smallest value of AIC. But the eq. (1) is in the next position in such cases, that is, it has the secondly smallest AIC. In other sets, the (1) has the smallest AIC.

This result shows that the spontaneous magnetization of cubic lattice Ising model has the form of (1). The results of fittings which have the smallest AIC in the (1)-type function for each set of data are named A, B, C, F, K, Q, M and N which are given in the last column of Table II. (These discontinuous names originate in the notations in the previous paper[13].) These eight fittings are assumed to be independent and they are used to estimate the values of K_c, β and a_1. The results are

$$K_c = 0.2216566(17), \quad \beta = 0.31977(38) \quad \text{and} \quad a_1 = 20.065(65). \tag{8}$$

The errors show the regions of one standard deviation. These results are shown in Fig. 1 with the adopted results in Table II. These figures indicate that the results in (8) are correct estimates from the data, A, B, C, F, K, Q, M and N.

IV. SUMMARY AND DISCUSSION

The spontaneous magnetization of the three-dimensional Ising model is analyzed based on its values estimated by the Monte Carlo simulation. The AIC idea is applied to select the correct fitting function. The present results indicate that the simple pole function (1) is better than (2). Of course it may be possible that the true behavior is not the form of neither (1) nor (2). If any other candidate is find, it can be tested by the present method. The estimated values of the critical point, critical exponent and critical amplitude are accompanied by smaller errors compared with the previous estimates which did not use any method to distinguish the correct functions. Physically and naively speaking, the eq. (1) is natural and the present analysis also supports this possibility. It is also shown that the present model selection method based on the AIC idea will be very useful for the analysis of the results of Monte Carlo simulation.

ACKNOWLEDGMENTS

The present author thank Prof. M. Suzuki (University of Tokyo) and Prof. S. Miyashita (Kyoto University) for fruitful discussion.

TABLE II. The results of fittings using two forms of functions are listed. The value of K_{max} specifies the set of data, $D(K_{max})$, fitted by functions. The number of data in $D(K_{max})$ is denoted by N. N and M denote the order of the fitting functions and they are defined by equation (1) and (2). The column named "a_1 or b_1"holds a_1 or b_1 if the fitting function is (1) or (2), respectively. Both a_1 and b_1 correspond to the critical amplitude although their values do not the same because of their definitions.

K_{max}	N_s	form	N	M	AIC	K_c	β	a_1 or b_1	
0.225	4	(1)	2	-	4.00	0.221660(26)	0.318(16)	19.6(36)	A
0.226	5	(1)	2	-	3.06	0.2216543(76)	0.3218(39)	20.46(83)	B
		(2)	0	1	3.21	0.2216461(95)	0.3312(63)	2.84(12)	
		(1)	3	-	5.00	0.221661(30)	0.316(23)	19.1(54)	
		(2)	1	1	5.00	0.221664(38)	0.310(44)	2.41(83)	
0.228	6	(2)	0	1	2.31	0.2216489(45)	0.3293(19)	2.805(34)	
		(1)	2	-	2.61	0.2216590(40)	0.3190(12)	19.87(23)	C
		(1)	3	-	4.18	0.221650(15)	0.325(10)	21.4(24)	
		(2)	1	1	4.30	0.2216463(63)	0.3319(56)	2.86(12)	
		(1)	4	-	6.00	0.221663(34)	0.314(29)	18.4(70)	
0.230	7	(2)	0	1	3.10	0.2216523(37)	0.3272(11)	2.767(19)	
		(1)	3	-	3.20	0.2216514(59)	0.3242(29)	21.08(63)	F
		(2)	1	1	3.37	0.221640(10)	0.3385(87)	3.01(20)	
		(1)	4	-	5.20	0.221651(24)	0.324(18)	21.1(45)	
		(2)	1	2	5.16	0.2216586(75)	0.3134(63)	2.46(12)	
		(1)	5	-	7.00	0.221665(36)	0.311(34)	17.5(85)	
0.235	8	(1)	4	-	4.27	0.2216457(81)	0.3286(48)	22.2(11)	K
		(2)	1	1	5.60	0.2216552(54)	0.3241(32)	2.701(63)	
		(2)	1	2	5.70	0.2216409(41)	0.3415(15)	3.100(29)	
		(2)	0	1	6.10	0.2216479(31)	0.32917(52)	2.8019(84)	
		(1)	5	-	6.16	0.221655(27)	0.321(22)	20.3(55)	
		(2)	2	2	6.34	0.2216452(39)	0.3241(24)	2.644(48)	
		(1)	3	-	6.60	0.2216604(40)	0.3187(12)	19.86(24)	
0.240	9	(1)	5	-	5.45	0.221640(10)	0.3330(67)	23.3(17)	Q
		(1)	3	-	5.70	0.2216599(32)	0.31895(59)	19.91(10)	
		(2)	0	1	6.50	0.2216494(27)	0.32876(36)	2.7948(57)	
		(1)	4	-	7.70	0.2216596(45)	0.3191(18)	19.94(38)	
		(2)	1	2	7.80	0.2216584(31)	0.31933(70)	2.599(11)	
0.250	10	(1)	4	-	7.00	0.2216580(35)	0.31996(87)	20.12(16)	M
		(2)	1	2	7.30	0.2216550(56)	0.3229(37)	2.672(72)	
		(1)	5	-	8.80	0.2216594(52)	0.3191(24)	19.94(51)	
		(2)	2	2	9.20	0.2216569(45)	0.3205(23)	2.621(42)	
0.260	11	(1)	4	-	6.30	0.2216590(30)	0.31954(47)	20.040(76)	N
		(1)	5	-	8.00	0.2216578(38)	0.3201(11)	20.15(23)	
		(2)	2	2	8.20	0.2216613(32)	0.31508(94)	2.510(16)	

FIG. 1. The results of fittings, A, B, C, F, K, Q, M and N are shown with the estimated values based on these results and the assumption of their statistical independence. (a), (b) and (c) show K_c, β and a_1, respectively. The values of a_1 of the fittings K and Q are not plotted in (c) because they are out of its horizontal axis but they are compatible with the estimate, $a_1 = 20.065(65)$. The error bars indicate the regions of one standard deviation. The solid lines and dotted lines show the estimated values and errors, respectively.

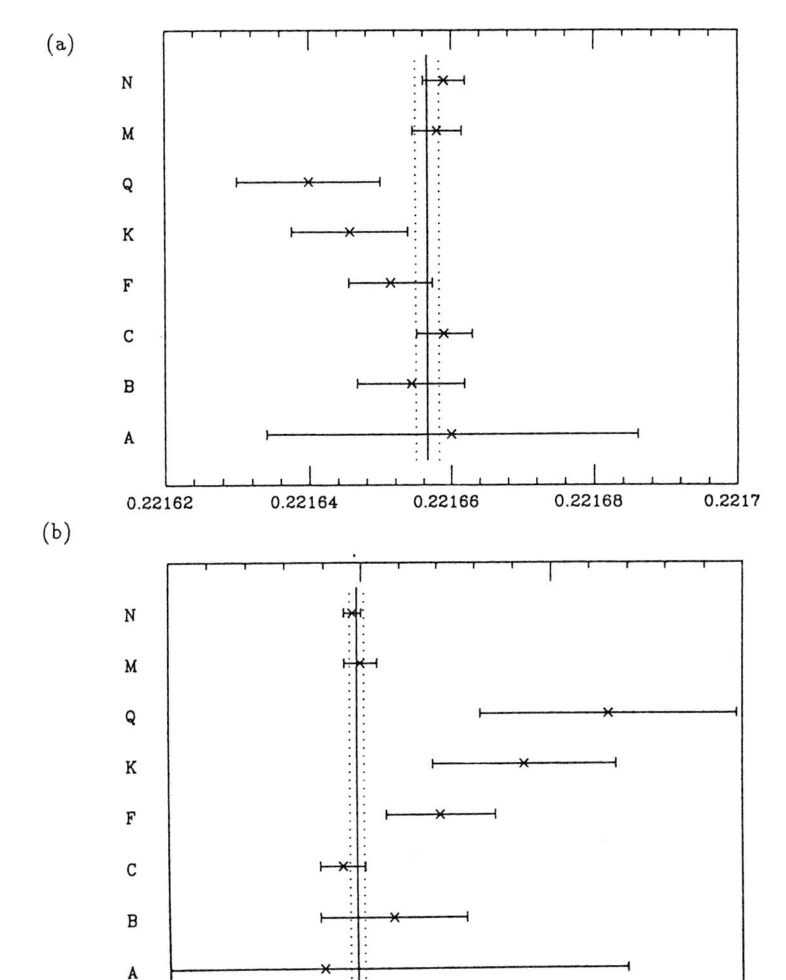

(c)

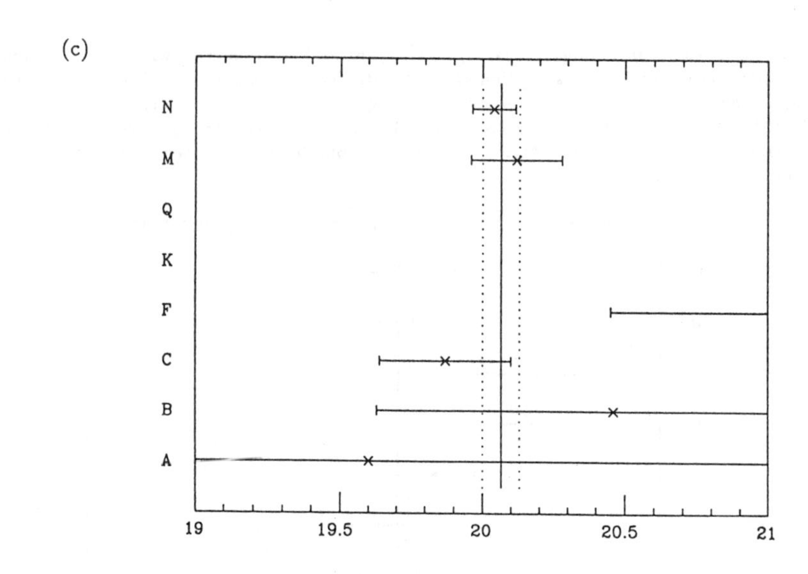

*Present Address: Computing and Information Systems Center, Japan Atomic Energy Research Institute, Tokai-mura, Naka-gun, Ibaraki 319-11, Japan.

[1]L. Fosdick, Phys. Rev. **116**(1959)565.
[2]P. Flinn and G. McManus, Phys. Rev. **124**(1961)54.
[3]N. Ito and Y. Kanada, Proceedings of Supercomputing '90 (IEEE Computer Society Press, 1990, Los Alamitos) p.753.
[4]N. Ito and Y. Kanada, Supercomputer **5** No. 3(1988)31
[5]N. Ito and Y. Kanada, Supercomputer **7** No. 1(1990)29
[6]H-O. Heuer, Comp. Phys. Comm. **59**(1990)387.
[7]R. H. Swendsen and J. S. Wang, Phys. Rev. Lett. **58**(1987)86.
[8]N. Ito and T. Chikyu, Physica **A166**(1990)193.
[9]N. Ito, Prog. Theor. Phys. **83**(1990)682.
[10]A. M. Ferrenberg and R. H. Swendsen, Phys. Rev. Lett. **61**(1988)2635.
[11]A. M. Ferrenberg, Springer Proceedings in Physics **53**, Computer Simulation Studies in Condensed Matter Physics III ed. D. P. Landau, K. K. Mon and H. B. Schitter (Springer-Verlag, 1991)p.30.
[12]H. Akaike, IEEE Trans. Autom. Contr. **AC19**(1974)716.
[13]N. Ito and M. Suzuki, J. Phys. Soc. Jpn. **60**(1991)No.6.
[14]Phase Transition and Critical Phenomena **3**, ed. C. Domb and M. S. Green (Academic Press, 1974).

CRITICAL DYNAMICS AND CONSERVATION LAWS

J. Kamphorst Leal da Silva and F. C. Sá Barreto
Departamento de Física - ICEx - UFMG - C. P. 702
30161 - Belo Horizonte - MG - Brazil

ABSTRACT

The kinetic Ising models with non-conserved and conserved order parameter are briefly reviewed. The effect on critical dynamics of global and local conservation laws is considered. Rigorous lower bounds for the dynamical exponent (z) are obtained from the initial response rate and the scaling hypothesis for the relaxation time. It is shown for several dynamical one-dimensional models that $z \geq 5$ even if the order parameter conservation occurs only globally. For one-dimensional models in which the maximal separation between two exchanged spins is proportional to the equilibrium correlation length, the inequality $z \geq 3$ is obtained.

1. INTRODUCTION

Universality is an important hypothesis in equilibrium second order phase transitions. It asserts that the nature of the critical point depends in large measure on general properties of a system, such as its dimensionality or the symmetries of its ordered state, but not very much on microscopic details of its hamiltonian. Therefore we can classify the physical systems in groups (the so called universality classes) such that all members of a group have the same critical exponents. Closely related to the concept of universality is the concept of scaling. According to the scaling hypothesis the divergence of the correlation length (ξ), which is considered as the only relevant length in the problem, is responsible for the singular behaviour of physical quantities. These two hypothesis are well established in equilibrium phase transition[1] .

In this work we are interested in critical dynamics. We study how a system initially out of equilibrium evolves in time to reach the equilibrium state. We are also interested how fluctuations of an equilibrium system decay. A characteristic time called the *relaxation time* (τ) for these phenomena can be defined. It is well known that this characteristic time diverges as the critical point is approached. According to the dynamical scaling hypothesis the static correlation length is still the relevant length and controls the relaxation time divergence. Thus we have the following scaling form for the relaxation time :

$$\tau_{\vec{q}} = \xi^z f(\vec{q}\xi) \quad , \tag{1.1}$$

where $\vec{q}$ is the appropriate critical wave vector and z is the critical dynamical exponent. The next important point is the classification of the physical systems in dynamical universality classes using the z exponent. It turns out that the z exponent depends beside the static properties on dynamical aspects, as the conservation laws entering the dynamics. These questions are nicely discussed by Hohenberg and Halperin[2]. Recently, the way as the conservation laws occur has been considered[3]. Models with global conservation laws would be in a universality class different of the models in which the conservation occurs in a local scale.

We consider the question of the universality classes in the simplest non trivial dynamical models, i.e. the kinetic Ising models. The first of these models has been proposed by Glauber[4] and consists of an Ising model with a stochastic dynamics in such a way that the magnetization is non-conserved. One important aspect of the Glauber model is that it can be solved exactly in one dimension. Thus the scaling hypothesis can be checked and the z exponent can be computed exactly ($z_G = 2$). Kawasaki[5] has proposed a dynamical model in which two nearest neighbor spins of opposite sign are exchanged. The magnetization is conserved for all times. The exact one dimensional exponent $z_K = 5$ has been evaluated for the isotropic ferromagnetic model[6]. This model is in a dynamic universality class different from the Glauber model because of the conservation of the order parameter. Note that the conservation law occurs *locally* since only nearest neighbour spins are exchanged. Models with *global*, rather than local, conservation laws have been considered recently. In these treatments the Kawasaki model has been generalized in order to include the exchange of spins separated by an *arbitrary* distance. It has been found[3] a lower z exponent which was related to the Kawasaki exponent z_K by $z = z_K - 2$. Therefore models which conserve the order parameter *only globally* were in a new universality class. This was supported by numerical simulations in two dimensions and dynamical renormalization group calculations in one dimension[7].

The main objective of this work is to consider the ferromagnetic kinetic Ising models with generalized Kawasaki dynamics in one dimension. Rigorous lower bounds for the z exponents are obtained from the initial response and the scaling hypothesis for the relaxation time. We obtain that several one dimensional models are in the same universality class of the nearest neighbour spins exchange model. We also obtain that for models in which the maximal distance between two exchanged spins is proportional to the equilibrium correlation length could be in a different universality class characterized by $z \neq 5$, in agreement with recent results[3]. One dimensional results could be especially important to clarify this question because critical dynamics in one dimension is very sensitive to any change. One example of this is the non universal behaviour found frequently in the literature[8]. Moreover, regarding problems and ambiguities with the dynamical renormalization group[9] a direct approach in one dimension seems to be worthwhile. The d-dimensional case ($d > 1$) is also briefly discussed. This work is organized as follow. In the next section we define the kinetic Ising models with non-conserved magnetization (Glauber models), we discuss the exact solution of $d = 1$ model and present in details the initial response approach. The dynamics of conserved magnetization models (Kawasaki models) is discussed in section 3. We consider firstly the standard Kawasaki model with local conservation of the magnetization. We also study the models with non-local conservation of the order parameter. In particular we consider models with only global conservation laws. The last section contains our concluding remarks.

2. DYNAMICS OF GLAUBER MODELS

A. The kinetic models

In the kinetic Ising model (KIM), introduced by Glauber[4], on each of the N sites of the lattice there is a spin represented as a stochastic function of time, $s_i(t)$, $(i = 1, 2, ...N)$. The reason for this stochastic representation is the following. Since at equilibrium the system is described by the Ising hamiltonian, it has no intrinsic dynamics. Therefore in order to obtain a dynamic model we must introduce some external perturbations. It is assumed that the Ising system is in contact with a thermal bath that induces flips of the spins. This bath is not treated explicitly. However we assume that there is a probability per unit time $w_i(s_1, s_2..s_i...s_N)$ that the i-th spin flips from the value s_i to $-s_i$ while all others are unaffected. The transition rates $w_i(s_1...s_N)$ are supposed independent of the previous history of the system (Markovian approximation). Although the transition rate $w_i(s_1...s_i...s_N)$ depends on the configuration $(s_1...s_N)$ it will be denoted as $w_i(s_i)$ from now on.

If $P(s_1, s_2, ...s_N; t)$ is defined as the probability to find the system in the configuration $(s_1, s_2, ...s_N)$ at time t then the master equation for the time evolution of $P(s_1, s_2, ...s_N; t)$ is

$$\frac{d}{dt}P(s_1...s_N;t) = -\sum_{i=1}^{N} w_i(s_i)P(s_1...s_N;t) + \sum_{i=1}^{N} w_i(-s_i)P(s_1... - s_i...s_N;t). \tag{2.1}$$

The correspondence between the KIM and the Ising model at equilibrium is made by determining the transition rates such that the stationary solution $P(s_1...s_N)$ of (2.1) is the equilibrium probability $P_{eq}(s_1...s_N)$. Thus we can write

$$\sum_{i=1}^{N} [-w_i(s_i)P_{eq}(s_1...s_i...s_N) + w_i(-s_i)P_{eq}(s_1...s_i...s_N)] = 0 \ . \tag{2.2}$$

A more explicit and restricted relation for $w_i(s_i)$ is obtained under the hypothesis of *detailed balance*, i.e., the sum (2.2) is zero term by term,

$$w_i(s_i)P_{eq}(s_1...s_i...s_N) = w_i(-s_i)P_{eq}(s_1... - s_i...s_N) \ . \tag{2.3}$$

When we consider the reduced Ising Hamiltonian

$$\begin{aligned} \mathcal{H} = -\frac{H}{k_B T} &= \frac{1}{2}\sum_{i,j} K_{i,j} s_i s_j \\ &= \sum_i E_i s_i \end{aligned} \tag{2.4}$$

and knowing that $P_{eq}(s_1...s_N) \sim \exp[\mathcal{H}(s_1...s_N)]$ we can write (2.3) in the form

$$\frac{w_i(s_i)}{w_i(-s_i)} = \exp(-2E_i s_i) \ . \tag{2.5}$$

This equation determines only partially the transition rate $w_i(s_i)$. Thus we can associate many kinetic models to the same equilibrium Ising model. However not all rates allowed by (2.5) correspond to physically meaningful models[10]. Obviously the transition rates $w_i(s_i)$ must be positive and symmetric under reversal of all spins as long as no magnetic field is present. Furthermore the rates should remain finite at all temperatures. When the master equation (2.1) is derived from a microscopic description of the interaction of the spins with the thermal bath, $w_i(s_i)$ is found to be proportional to time integrals of some correlation functions of the bath[11]. The divergence of any integral would imply the breakdown of the

Markovian approximation essential to (2.1). Similar arguments show that the rates should not vanish at any temperature. Even close to zero temperature reasonable heat baths admit nonvanishing rates because it is hard to see how arbitrary configurations of spins could freeze in.

Now, the time evolution of the order parameter will be considered. We define the functions $\langle s_k(t)\rangle$ as

$$\langle s_k(t)\rangle = \sum_{\{s\}} s_k P(s_1...s_k...s_N;t) \quad . \tag{2.6}$$

The equation for the evolution of this object is obtained by multiplying both sides of equation (2.1) by s_k and summing over all configurations $\{s\}$. After some algebra we obtain that :

$$\begin{aligned}\frac{d}{dt}\langle s_k(t)\rangle &= -2\sum_{\{s\}} s_k w_k(s_k) P(s_1...s_k...s_N;t) \\ &= -2\langle s_k(t) w_k[s_k(t)]\rangle \quad .\end{aligned} \tag{2.7}$$

Unfortunately the KIM has been solved exactly only in one dimension for some particular transition rates[4]. In order to illustrate the difficulties to find an exact solution let us consider the Glauber transition rates. From equation (2.5) we see that $w_i(s_i)$ can be written as

$$w_i(s_i) = \frac{\alpha}{2}[1 - s_i \tanh(E_i)] \quad , \tag{2.8}$$

where α fixes the scale of time. With this rate we obtain from (2.7) a equation of motion for $\langle s_i(t)\rangle$; namely

$$\frac{d}{dt}\langle s_i(t)\rangle = -\alpha\{\langle s_i(t)\rangle - \langle \tanh[E_i(t)]\rangle\} \quad . \tag{2.9}$$

Note that E_i depends on the neighbour spins of s_i. In dimension greater than one when $\langle \tanh(E_i)\rangle$ is developed higher-order correlation functions in the right hand side of (2.9) will appear. The equation of motion for these extra correlation functions depend on even higher-order ones. Thus an infinite hierarchy of coupled equations must be solved. No general solution of this problem has been found.

B. One-dimensional exact solution

In one dimension the problem is simpler because $\langle \tanh(E_i)\rangle$ can be expressed in terms of the one-point function $\langle s_i\rangle$. To illustrate this consider the homogeneous ferromagnetic Ising model. The hamiltonian is given by (2.4) with $K_{ij} = K > 0$. The static properties are easily obtained. For further use we note that the critical temperature is zero, the correlation length near $T = 0$ behaves as

$$\xi \sim \frac{1}{2}\exp(2K) \tag{2.10}$$

and the pair correlation function is given by

$$\langle s_k s_l\rangle_{eq} = \prod_{i=k}^{l-1} \tanh(K_i) \quad . \tag{2.11}$$

Since the local field $E_i = K(s_{i-1} + s_{i+1})$ takes only the values $2K, 0$ and $-2K$ we have that

$$\tanh(E_i) = \frac{\Gamma}{2}(s_{i-1} + s_{i+1})$$
$$w_i(s_i) = \frac{\alpha}{2}\{1 - \frac{\Gamma}{2}s_i(s_{i-1} + s_{i+1})\} \quad , \tag{2.12}$$

with $\Gamma = \tanh(2K)$. The N equations ($i = 1...N$)

$$\frac{d}{dt}\langle s_i(t)\rangle = -\alpha\langle s_i(t)\rangle + \frac{\alpha\Gamma}{2}\left[\langle s_{i-1}(t)\rangle + \langle s_{i+1}(t)\rangle\right] \tag{2.13}$$

have been solved by Glauber[4] . Here we calculate only the evolution of the order parameter $m(t) = (1/N)\sum_{i=1}^{N}\langle s_i(t)\rangle$ obtained by summing over i both sides of equation (2.13) and using the periodic boundary condition $s_{N+1} = s_1$. We have then

$$\frac{d}{dt}m(t) = -\frac{1}{\tau}m(t) \quad , \tag{2.14}$$

where the relaxation time τ is given by $\tau^{-1} = \alpha(1-\Gamma)$. Thus the system has an exponential relaxation from the initial finite magnetization to the zero equilibrium value. The z exponent can be easily evaluated because near the zero temperature $\Gamma \approx 1 - 2\exp(-4K)$. Using (2.10) for the correlation length we get

$$\tau \approx \frac{1}{2\alpha}\exp(4K) \approx \frac{2}{\alpha}\xi^2 \quad . \tag{2.15}$$

Hence $z_G = 2$. Glauber has also solved the equations for the pair correlation function. The complete solution of this model,i.e., the knowledge of $P(s_1...s_N;t)$, has been given by Felderhof[12].

C. The initial response

Let us now to establish rigorous lower bounds in arbitrary dimensions for the relaxation of the order parameter, as it has been made by Halperin[13].

Consider a function ϕ defined by

$$P(s_1...s_N;t) = P_{eq}(s_1...s_N)\phi(s_1...s_N;t) \quad . \tag{2.16}$$

Substituting (2.16) into the master equation (2.1) and using the detailed balance condition (2.3) we have that

$$\frac{d}{dt}\phi(s_1...s_N;t) = -D_s\phi(s_1...s_N;t) \quad , \tag{2.17}$$

where the operator D_s is defined by

$$D_s f(s_1...s_N;t) = \sum_{j=1}^{N} w_j(s_j)[f(s_1...s_j...s_N;t) - f(s_1... - s_j...s_N;t)] \quad . \tag{2.18}$$

Since the formal solution of equation (2.17) is given by

$$\phi(s_1...s_N;t) = \exp(-D_s t)\phi(s_1...s_N;0) \ , \tag{2.19}$$

the functions $\langle s_k(t)\rangle$ defined in (2.6) can be written as

$$\langle s_k(t)\rangle = \langle s_k \exp(-D_s t)\phi(s_1...s_N;0)\rangle_{eq} \ . \tag{2.20}$$

Here $\langle..\rangle_{eq}$ means the equilibrium thermal average.

In order to show that the evolution of the order parameter is related to the temporal spin correlation function let us consider a ferromagnetic Ising system in thermal equilibrium for $t < 0$ under a uniform and small magnetic field $B = b/(k_B T)$ with $T > T_c$. At $t = 0$ this field is switched off and the magnetization will decay to zero. The initial value of P, correct up to order of B, is easily shown to be

$$P(s_1...s_N;0) = P_{eq}(s_1...s_N)[1 + B\sum_j s_j + O(B^2)] \ . \tag{2.21}$$

Thus we find that $\phi(s_1...s_N;0) = 1 + B\sum_j s_j$ and equation (2.20) can be written as

$$\langle s_k(t)\rangle = B\sum_j \langle s_k s_j[t]\rangle_{eq} \ , \tag{2.22}$$

with $s_j[t] = \exp(-D_s t)s_j$. Therefore the evolution of the order parameter is given by the correlation functions $\langle s_k s_j[t]\rangle_{eq}$.

Let us return to the general problem. From (2.18) we can see that D_s is a real operator and using the detailed balance condition is easy to show that it has the following properties:

$$\langle f^* D_s g\rangle_{eq} = \langle g D_s f^*\rangle_{eq}, \quad \langle g^* D_s g\rangle_{eq} \geq 0 \ , \tag{2.23}$$

where g and f are arbitrary functions of $(s_1...s_N)$. Thus the eigenvalues ν_i of D_s are real and nonnegative. The correlation function

$$C_g(t) = \langle g^*[0]g[t]\rangle_{eq} \tag{2.24}$$

has a spectral representation of the form

$$C_g(t) = \int_0^\infty \varphi_g(\nu)\exp(-\nu t)d\nu \ , \tag{2.25}$$

with $\varphi_g(\nu) \geq 0$ for all ν. Assuming that the equilibrium average of g is zero we can define the relaxation time τ_g and the initial relaxation rate ν_g for the variable g by

$$\begin{aligned} \tau_g &= C_g(0)^{-1}\int_0^\infty C_g(t)dt \\ &= C_g(0)^{-1}\int_0^\infty \varphi_g(\nu)\nu^{-1}d\nu \\ \nu_g &= -C_g(0)^{-1}\frac{d}{dt}C_g(t)\Big|_{t=0} \\ &= C_g(0)^{-1}\int_0^\infty \varphi_g(\nu)\nu d\nu \ . \end{aligned} \tag{2.26}$$

Using the Schwartz inequality one obtains that

$$\tau_g \nu_g = \frac{\int_0^\infty \left(\sqrt{\frac{\varphi_g(\nu)}{\nu}}\right)^2 d\nu \int_0^\infty \left(\sqrt{\nu \varphi_g(\nu)}\right)^2 d\nu}{\left(\int_0^\infty \varphi_g(\nu) d\nu\right)^2} \geq 1 \ . \tag{2.27}$$

So we have the following inequality :

$$\tau_g \geq \nu_g^{-1}. \tag{2.28}$$

If $g = s_{\vec{q}} = (1/\sqrt{N}) \sum_j \exp(i\, \vec{q} \cdot \vec{r}_j) s_j$, τ_{s_q} will be the relaxation time of the order parameter for the appropriate $\vec{q}$. The equal-time correlation function is proportional to the static susceptibility, namely

$$C_{s_q}(0) = k_B T \chi_{\vec{q}} \ . \tag{2.29}$$

We have also that

$$\begin{aligned} \frac{d}{dt} C_{s_q}(t)\Big|_{t=0} &= \frac{d}{dt} \langle s_{-\vec{q}} \exp(-D_s t) s_{\vec{q}} \rangle_{eq} \Big|_{t=0} \\ &= -\langle s_{-\vec{q}} D_s s_{\vec{q}} \rangle_{eq} \\ \frac{d}{dt} C_{s_q}(t)\Big|_{t=0} &= -\frac{1}{N} \sum_{k,j} \exp[i\vec{q} \cdot (\vec{r}_j - \vec{r}_k)] \langle s_k D_s s_j \rangle_{eq} \ . \end{aligned} \tag{2.30}$$

It is easy to show that $D_s s_j = 2 s_j w_j(s_j)$ from the definition of D_s (2.18). So we must evaluate terms like $\langle s_k s_j w_j(s_j) \rangle_{eq}$. Using the detailed balance (2.3) we have that

$$\langle s_k s_j w_j(s_j) \rangle_{eq} = \delta_{k,j} \langle w_j(s_j) \rangle_{eq} \ . \tag{2.31}$$

Thus equation (2.30) can be written as

$$\frac{d}{dt} C_{s_q}(t)\Big|_{t=0} = -\frac{2}{N} \sum_j \langle w_j(s_j) \rangle_{eq} \ , \tag{2.32}$$

and the inequality (2.28) as

$$\tau_{s_q} \geq \frac{k_B T \chi_{\vec{q}}}{\frac{2}{N} \sum_{j=1}^N \langle w_j(s_j) \rangle_{eq}}. \tag{2.33}$$

In dimensions greater than one the denominator of (2.33) remains finite at the critical temperature[14]. Thus using the definition of the critical exponent γ

$$\chi \sim \xi^{\gamma/\nu} \tag{2.34}$$

and the scaling relation (1.1) for the relaxation time we obtain that $z_G \geq (\gamma/\nu)$.

Since in one dimension the critical temperature is zero, $\langle w_j(s_j) \rangle_{eq}$ vanishes at T_c. To illustrate this point consider the homogeneous ferromagnetic Ising model with the Glauber transition rates (2.12). In order to use inequality (2.33) we must evaluate $\langle w_j(s_j)\rangle_{eq}$. We can write that

$$\begin{aligned}\langle w_j(s_j)\rangle_{eq} &= \frac{\alpha}{2}[1 - \tanh(2K)\tanh(K)] \\ &\approx \frac{\alpha}{2}\xi^{-1} \quad \text{when } T \to 0 \ ,\end{aligned} \tag{2.35}$$

where (2.11) has been used to evaluate the correlation functions. In one dimension we have that $(\gamma/\nu) = 1$. Thus $z_G \geq 1 + (\gamma/\nu) = 2$ is obtained. Note that the lower limit coincides with the exact exponenent.

D. Nonuniversal behaviour

In recent years the question of the universality of the z exponent in one dimension has been discussed in the literature. It is well known that the dynamical exponent of the Ising model depends on the transition rates[10] which can be chosen in several forms once the detailed balance has been satisfied. In fact, each choice of the transition rates determines a different dynamical model. More surprisingly is the non-universality due to the spatial non-uniformity of the interactions. In this case the non-universality occurs not by comparing different dynamical models but within a fixed model. Droz et al.[8] have studied the critical dynamics of the one-dimensional ferromagnetic Ising model with two alternating different coupling constants $(J_1, J_2, J_1, J_2 \ldots)$. The Glauber transition rates generalized to this problem have been used. They have found that the z exponent depends on the constant couplings – microscopic parameters – in the following fashion $z_G = 1 + (J_1/J_2)$, $(J_1 \geq J_2)$. Thus z can vary continuously from 2 to ∞. This result has been obtained by solving exactly the equation for the magnetization and by using a very simple and nice argument about the movement of the domain walls[15]. This kind of behaviour has been also found in models of alternating interactions with arbitrary sign [16], Ising models with several different bonds[17], in the Fibonacci-chain quasicrystal[18] and by using the renormalization group techniques[9,18]. The nonuniversal z exponent has been found also in multi-spin-flips process with non-conserved magnetization[19] and in the disorder Ising ferromagnetic chain with two couplings constants[20,21]. The Q-state ferromagnetic Potts kinetic models in one dimension present the same non-universality. It has been discussed that the dynamical exponent depends on the transition rates, on the number of states (Q)[22–24] and on the spatial non-uniformity of the interactions[25]. Finally, this breakdown of the dynamic scaling hypothesis has also been found in ultrametric spaces[26] and in $d > 1$ Ising systems near the percolation threshold[27,28]. This nonuniversal behaviour occurs always near a zero temperature in structures where the dynamics involves thermal activation over barriers. For critical temperature different from zero we do not find the nonuniversal behaviour. For example in several Ising models with two different ferromagnetic bonds an universal z exponent has been found [29].

Now let us present in a few words two other approaches used in the study of critical dynamics of kinetic Ising models. The first one is the application of renormalization group methods[1] to dynamics. One example[2] is a generalization of the ϵ expansion, useful near the upper critical dimension. It describes the dynamics in terms of a Langevin equation and allows the evaluation of the time-dependent correlation functions. Other methods[30–32] are generalizations of the real-space renormalization group techniques. They are more efficient in low dimensions (frequently $d = 1$). It is worth mentioning that recently a correction, motivated by the nonuniversal behaviour of one-dimensional Ising models with several different coupling constants, has been introduced in the formalism[9]. The second one refers to simulations. The KIM are ideal for computer simulations because they have been formulated as Markov chains. In the Monte Carlo method[33], the evolution of the spins is described by a master equation. Since detailed balance is verified equilibrium configurations are generated in the long time limit. Thus we can calculate equilibrium properties easily. Here we are interested in the approach to equilibrium. Since the KIM cannot be solved exactly for $d > 1$ Monte Carlo is a powerful method to study it[34]. In the last years new algorithms with very small values of the z exponent have been presented[35–37]. They are characterized by the simultaneous update of several spins instead of the single spin-flip. This indicates that the dynamical exponent depends on the definition of the dynamical model.

3. DYNAMICS OF KAWASAKI MODELS

A. Models with local conservation laws

We consider now multi-spin -flip processes, in which the bath flips several spins simultaneously. This type of process is important because it allows the study of the dynamics where quantities of interest are conserved during the time evolution. For instance, one can describe transport phenomena caused by spatial inhomogeneity such as diffusion or heat conduction.

The standard Kawasaki model[5] is one in which neighboring Ising spins are exchanged $s_i, s_{i+\delta} \rightarrow s_{i+\delta}, s_i$ when $s_i \neq s_{i+\delta}$. As a result the total magnetization is conserved. In this case the master equation is

$$\frac{d}{dt}P(s_1 \ldots s_N;t) = -\left[\sum_{i,\delta} w_{i,i+\delta}(s_i, s_{i+\delta})\right] P(s_1 \ldots s_N;t) + \sum_{i,\delta} w_{i,i+\delta}(s_{i+\delta}, s_i) P(s_1 \ldots s_{i+\delta} \ldots s_i \ldots s_N;t) \quad , \tag{3.1}$$

where $w_{i,i+\delta}(s_i, s_{i+\delta})$ is the transition rate with which the spins i and $i+\delta$ are exchanged. If $s_i = s_{i+\delta}$ the transition rate is chosen to be zero.

Again the detailed balance condition must be imposed. Thus the following equation is obtained :

$$\frac{w_{i,i+\delta}(s_i, s_{i+\delta})}{w_{i,i+\delta}(s_{i+\delta}, s_i)} = \exp[-(s_i - s_{i+\delta})(E'_i - E'_{i+\delta})] \quad , \tag{3.2}$$

where $E'_i = E_i - K_{i,i+\delta} s_{i+\delta}$. The coupling between i-th spin and $(i+\delta)$-th is not present in (3.2) because the change of energy in the process is independent of $K_{i,i+\delta}$.

If we multiply both sides of (3.1) by s_k and sum over all configurations we find the evolution equation for $\langle s_k(t) \rangle$. The terms with $i+\delta \neq k \neq i$ drop out and it is easy to obtain,

$$\frac{d}{dt}\langle s_k(t)\rangle = -2\langle s_k(t) \sum_{\delta} \{w_{k-\delta,k}[s_{k-\delta}(t), s_k(t)] + w_{k,k+\delta}[s_k(t), s_{k+\delta}(t)]\}\rangle \quad . \tag{3.3}$$

Evidently if we sum (3.3) over all sites k the expression on the right is zero. The total magnetization is a constant of motion. This model has not been solved exactly even in one dimension due to the problem of higher-order correlation functions discussed in the preceding section. But one can obtain the exact z exponent using the linear response technique [6].

Let us consider now the initial response rate of the Kawasaki model. The function ϕ defined in (2.16) is introduced in the master equation (3.1). Then we can write an equation similar to (2.17). But we must define a new operator $D_s^{(K)}$ as

$$D_s^{(K)} f(s_1 \ldots s_N;t) = \sum_{j,\delta} w_{j,j+\delta}(s_j, s_{j+\delta})[\, f(s_1 \ldots s_j, s_{j+\delta} \ldots s_N;t) - f(s_1 \ldots s_{j+\delta}, s_j \ldots s_N;t)] \ . \tag{3.4}$$

Again properties (2.23) hold and the eigenvalues of $D_s^{(K)}$ are real and nonnegative[38]. Thus the general relations (2.24 - 2.28) apply in this case. Moreover, one finds

$$\begin{aligned} D_s^{(K)} s_j &= \sum_{\delta} \{ w_{j-\delta,j}(s_{j-\delta}, s_j)[s_j - s_{j-\delta}] + \ldots \\ &\quad + w_{j,j+\delta}(s_j, s_{j+\delta})[s_j - s_{j+\delta}]\} \\ \langle s_k s_j w_{j-\delta,j}(s_{j-\delta}, s_j)\rangle_{eq} &= [\delta_{k,j} - \delta_{k,j-\delta}]\langle w_{j-\delta,j}(s_{j-\delta}, s_j)\rangle_{eq} \\ \frac{d}{dt} C_{s_q}(t)\Big|_{t=0} &= -\frac{4}{N} \sum_{\delta,j} [1 - \cos(\vec{q}\cdot\vec{a}_\delta)]\langle w_{j,j+\delta}(s_j, s_{j+\delta})\rangle_{eq} \ . \end{aligned} \tag{3.5}$$

Here $\vec{a}_\delta$ are the vectors between neighboring spins. Thus the inequality for the relaxation time is given by

$$\tau_{s_q} \geq \frac{k_B T \chi_{\vec{q}}}{\frac{4}{N}\sum_{\delta,j}[1 - \cos(\vec{q}\cdot\vec{a}_\delta)]\langle w_{j,j+\delta}(s_j, s_{j+\delta})\rangle_{eq}} \ . \tag{3.6}$$

Let us consider again the one-dimensional homogeneous ferromagnetic model. The order parameter (the magnetization) is a conserved quantity. The inequality for the z exponent is obtained by expanding $\cos(\vec{q}\cdot\vec{a}_\delta)$ in the neighborhood of $\vec{q} = 0$ and writing the right side of (3.6) in terms of the scaling variable $q\xi$. We find that

$$\tau_{s_q} \geq \frac{k_B T \xi^{2+\gamma/\nu}}{2(qa\xi)^2 \langle w_j(s_j, s_{j+1})\rangle_{eq}} \ . \tag{3.7}$$

From (3.2) we obtain the Kawasaki rate, namely

$$w_{i,i+1}(s_i, s_{i+1}) = \frac{\alpha}{4}(1 - s_i s_{i+1})[1 - \frac{\Gamma}{2}(s_{i-1}s_i + s_{i+1}s_{i+2})] \ . \tag{3.8}$$

Here $\Gamma = \tanh(2K)$. Since $\langle w_{j,j+\delta}(s_j, s_{j+1})\rangle_{eq} \sim (\alpha/2)\xi^{-2}$ and assuming the scaling form (1.1) for the relaxation time the result $z_K \geq 5$ is obtained. Again the lower limit coincides with the exact exponent[6,10].

Let us make some comments about the initial response. When the inequalities (2.33) and (3.6) hold as equalities the relaxation times for $d \geq 1$ have the form prescribed by the conventional theory[2]. It assumes that the kinetic coefficients (simply the flip-rate per spin averaged over the system) remain finite at T_c. Thus the dynamics exponents are determined by the static ones,i.e., $z_G = \gamma/\nu$ for the Glauber dynamics and $z_K = 2 + (\gamma/\nu)$ for the Kawasaki dynamics. So, the conventional exponents are lower bounds to the exact ones. In one dimension we have nonconventional exponents because the kinetic coefficients vanish at zero temperature.

As in the Glauber models the standard Kawasaki model presents a nonuniversal behaviour near zero temperature. It is well known that the z exponent depends on the choice of the transition rates and on the spatial non-uniformity of the interactions[8,16,19,23,25].

B. Models with non-local conservation laws

The standard Kawasaki dynamics is based on a picture of correlated spin diffusion. However, this simple picture does not apply in all cases. When the spinodal decomposition in a binary fluid mixture including convection flow effects is considered a picture involving effective long-distance exchanges emerges[39]. Numerical simulations of spinodal decomposition in a one-component fluid have been made and a growth law for the ordering process different from a conserved order parameter system was found[40]. Recently, a long-range exchange model with a constant exchange probability in a region limited by a cutoff length has been introduced[41]. They have found an anomalous growth law if the exchange distance is comparable to the observed scale. Therefore it is worth the study of the generalized Kawasaki dynamics in which the exchange of opposite sign spins separated by an arbitrary distance occurs.

Let us consider first the one-dimensional case[42]. The dynamics is given by the following master equation :

$$\begin{aligned}\frac{\partial P(\{s\},t)}{\partial t} = \sum_{j=1}^{N}\sum_{r=1}^{n} M[&- w_{j,r}(s_j, s_{j+r})P(\{\ldots s_j \ldots s_{j+r} \ldots\},t) \\ &+ w_{j,r}(s_{j+r}, s_j)P(\{\ldots s_{j+r} \ldots s_j \ldots\},t)] \ .\end{aligned} \tag{3.9}$$

Here $w_{j,r}(s_j, s_{j+r})$ (with $s_j = -s_{j+r}$) stands for the non-normalized transition rate of the $\{s_1, \ldots, s_j, \ldots, s_{j+r}, \ldots, s_N\}$ configuration to the $\{s_1, \ldots, s_{j+r}, \ldots, s_j, \ldots, s_N\}$ one. If $s_j = s_{j+r}$, the transition rate is zero. Note that only the spins s_j and s_{j+r} have exchanged, maintaining constant the total magnetization. n is the maximal separation between two exchanged spins. M is an additional parameter useful to consider the several different dynamical models and responsible for the normalization of the transition rates. When $M = \delta_{r,R}$ where $\delta_{r,R}$ is the Kronecker delta, only spins separated by a fixed distance R are exchanged. The standard Kawasaki nearest neighbour spins exchange model is obtained by putting $M = \delta_{r,1}$. When $M = 1/n$ and $n = N$ we have the generalized Kawasaki model with exchange of spins separated by an arbitrary distance[3,7]. Now, a particular spin can be exchanged with several others. In this model the magnetization is not conserved in a local scale,and it is only conserved globally. Note that $\sum_r Mw_{j,r}$ is normalized. It means that the time scale of the heat bath (α) is the high temperature relaxation rate of a spin.

The transition rates are only partially determined by the detailed balance condition. We will consider the Kawasaki and the "exponential" transition rates. The Kawasaki rate for $r = 1$ is given by equation (3.8). For $r \geq 2$ the Kawasaki rates can be written as

$$\begin{aligned} w_{j,r}^{(1)}(s_j,s_{j+r}) = \frac{\alpha}{4}(1 - s_j s_{j+r})\{1 - c_1 s_j(s_{j-1} + s_{j+1} - s_{j+r-1} - s_{j+r+1}) \\ + c_3 s_j[s_{j-1}s_{j+1}(s_{j+r-1} + s_{j+r+1}) - s_{j+r-1}s_{j+r+1}(s_{j-1} + s_{j+1})]\} \end{aligned} , \tag{3.10}$$

where $c_1 = (1/8)[\tanh(4K) + 2\tanh(2K)]$, $c_3 = (1/8)[\tanh(4K) - 2\tanh(2K)]$ and α is the time scale defined by the heat bath. The "exponential" transition rates also often used[8] are defined by

$$w_{j,r}^{(2)}(s_j, s_{j+r}) = \frac{\alpha}{4}(1 - s_j s_{j+r})\sqrt{\frac{P_{eq}(\ldots s_{j+r} \ldots s_j \ldots)}{P_{eq}(\ldots s_j \ldots s_{j+r} \ldots)}} , \tag{3.11}$$

with $P_{eq}(\{s\}) \sim \exp(-\beta H)$.

The initial response formalism can be developed in the same lines as it has made above for the standard Kawasaki model. Now we must define an operator L instead of the D_K operator, namely

$$L\phi(\{s\},t) = \sum_{j,r} M w_{j,r}^{(b)}(s_j, s_{j+r})[\phi(\ldots s_j \ldots s_{j+r} \ldots, t) - \phi(\ldots s_{j+r} \ldots s_j \ldots, t)] , \tag{3.12}$$

where b can be 1 or 2 depending on the transition rate. All the other steps are easily done and we obtain that

$$\tau_{s_{\vec{q}}} \geq \frac{k_B T \chi_{\vec{q}}}{\frac{4}{N}\sum_{j=1}^{N}\sum_{r=1}^{n} M[1 - \cos(\vec{q}\cdot\vec{a}_r)]\langle w_{j,r}^{(b)}(s_j, s_{j+r})\rangle_{eq}} , \tag{3.13}$$

where $\chi_{\vec{q}}$ is the static susceptibility and $\vec{a}_r$ is the spacing vector between sites j and $j+r$. It is worth mentioning that this inequality becomes an equality in the high temperature limit (conventional theory)[2]. The physical reason is that in this limit ($T \to \infty$) a spin behaves independently of the others. Thus, the initial relaxation time is equal to the asymptotic one. In order to obtain a lower bound to the z exponent we must evaluate $\langle w_{j,r}^{(b)}(s_j, s_{j+r})\rangle_{eq}$ and assume the scaling hypothesis for the relaxation time (1.1). So, we must evaluate new static correlation functions for the one-dimensional ferromagnetic Ising model. It is very easy to obtain that

$$\langle s_i s_j s_k s_l\rangle = \tanh^{j-i}(K)\tanh^{l-k}(K) . \tag{3.14}$$

The sites considered in these relations satisfy the condition $l > k > j > i$. Note that this equilibrium result is valid for finite N in the open chain.

Let us consider the dynamical model defined by the "exponential rates" (3.11). Using (3.14), it is easily shown that $\langle w^{(2)}\rangle_{eq}$ is independent of r for $r > 1$, i.e.,

$$\begin{aligned} \langle w_{j,j+1}^{(2)}(s_j, s_{j+1})\rangle_{eq} &= \frac{\alpha}{4}\cosh^2(K)[1 - \Gamma(1)][1 - \Gamma(2)]^2 , \\ \langle w_{j,j+r}^{(2)}(s_j, s_{j+r})\rangle_{eq} &= \frac{\alpha}{4}\cosh^4(K)[1 - \Gamma(2)]^4 \quad \text{for } r > 1 , \end{aligned} \tag{3.15}$$

where $\Gamma(m) = \tanh^m(K)$. Near $T = 0$ we have that $\langle w^{(2)}_{j,j+1}\rangle_{eq} \sim (\alpha/2)\xi^{-2}$ and for $r > 1$ $\langle w^{(2)}_{j,j+r}\rangle_{eq} \sim \alpha\xi^{-2}$. Let us firstly consider the situation where $M = \delta_{r,R}$. In the limit $T \to 0$, we have that $\chi_{\vec{q}} \sim \xi$ near $\vec{q} \to 0$. Expanding $\cos(\vec{q}\cdot\vec{a}_r)$ for small $\vec{q}$ we obtain

$$\tau_{\vec{q}} \geq \frac{k_B T\xi^5}{2\alpha R^2(q\xi a)^2} \; . \tag{3.16}$$

Here a is the spacing lattice. Using the scaling form (1.1) for the relaxation time we have that $z \geq 5$.

Let us put $M = 1/n$ and consider n big but finite. For small $\vec{q}$ the expression (3.13) can be written as

$$\tau_{\vec{q}} \geq \frac{k_B T\chi_{\vec{q}}}{2q^2a^2\frac{1}{Nn}\sum_{j=1}^{N}\{\langle w^{(2)}_{j,j+1}\rangle_{eq} + \sum_{r=2}^{n} r^2\langle w^{(2)}_{j,j+r}\rangle_{eq}\}} \; . \tag{3.17}$$

Since $\langle w^{(2)}_{j,j+r}\rangle_{eq}$ is independent of r for $r > 1$, we can easily evaluate the two sums. Using the scaling variable $q\xi$, we can write that

$$\tau_{\vec{q}} \geq \frac{3k_B T\xi^5}{2\alpha n^2(qa\xi)^2} \; , \tag{3.18}$$

which also gives $z \geq 5$. We can evaluate (3.13) in the $T \to \infty$ limit to see how the high temperature relaxation time changes. In this limit we have that $k_B T\chi_{\vec{q}} = 1$ and $\langle w^{(2)}\rangle = \alpha/4$ for $r \geq 1$. Therefore we have that

$$\tau_{\vec{q}} = \frac{6\alpha^{-1}}{q^2a^2n^2} \; . \tag{3.19}$$

Comparing this result with the next neighbour exchange ($2/\{\alpha q^2a^2\}$), we see that α has been changed to $\alpha' = (\alpha n^2)/3$, meaning that the system now relaxes faster.

The exchange of spins separated by an arbitrary distance is described in our notation by $M = 1/n$ and $n = N$. Therefore, at the critical temperature ($T = 0$) and in the limit $n \to \infty$ we have from (3.18) that $\tau_{\vec{q}} \geq 0$. On the other hand, the high temperature limit, given by (3.19), for $n = N \to \infty$ we obtain $\tau_{\vec{q}} = 0$. In other words, the high temperature relaxation time is zero. This unphysical result can be avoided by introducing directly into the master equation (3.9) $M = n^{-2}M'$, with M' playing now the role of an additional constant. Alternatively we can change the time scale by $\alpha' = \alpha n^2$ and require it to be finite. In either case, the correct high temperature limit is obtained and we have that $z \geq 5$. Note that in order to obtain $z \geq 3$, in agreement with the results of the finite-size scaling[3] and renormalization group[7] calculations for the generalized Kawasaki model with exchange of spins separated by an arbitrary distance we must take $n \sim \xi$ in (3.18). This corresponds to a model in which the maximal flipping distance depends on the equilibrium correlation length. Thus for each temperature one has a different dynamical model. In particular, at very low or high temperatures the system could be frozen ($\tau_{\vec{q}} \to \infty$).

Now we consider the Kawasaki rates defined in equation (3.10). We can show that

$$\begin{aligned}\langle w^{(1)}_{j,j+1}(s_j, s_{j+1})\rangle_{eq} &= \frac{\alpha}{4}[1-\Gamma(1)][1-\tanh(2K)\Gamma(1)]\\ \langle w^{(1)}_{j,j+r}(s_j, s_{j+r})\rangle_{eq} &= \frac{\alpha}{4}[1-\Gamma(r)+c_1\{2\Gamma(r-1)+2\Gamma(r+1)-4\Gamma(1)\}\\ &\quad + c_3\{2\Gamma(r-1)+2\Gamma(r+1)-4\Gamma(3)\}] \quad \text{for } r\geq 2\ .\end{aligned} \tag{3.20}$$

In the limit $T \to 0$, it is easy to obtain from (3.19) that $\langle w^{(1)}\rangle_{eq} \sim \xi^{-2}$. Therefore using (3.13), we have that $z \geq 5$ if n is finite. When $n = N \to \infty$ or $n \sim \xi$ we have the same situations described in the last paragraph. The first situation is overcome by a similar procedure, giving $z \geq 5$. For $n \sim \xi$ we obtain $z \geq 3$.

In order to be more precise for exchange of spins separated by arbitrary distance we must consider that each possible exchange occurs only once. Therefore, in the master equation (3.9) we must change the upper limits of the sums. So, N must be changed to $N-1$ and n to $n-j$ in (3.9). These modifications are irrelevant and the above results remain valid.

In dimensions greater than one the inequality (3.13) is still valid, but now $\vec{a}_r$ is a d-dimensional vector and n is a d-folded index. We have not evaluate $\langle W(s_j, s_{j+r})\rangle_{eq}$, but we expect that it will be finite near T_c for n finite[13]. Note that $\langle W\rangle_{eq}$ is proportional to ξ^{-2} in one dimension only because the critical temperature is zero. In dimensions greater than one $T_c \neq 0$, so we have reasons to expect that $\langle W\rangle_{eq}$ will be nonzero[13] . For n finite, the exponent z must obey the inequality

$$z \geq 2 + \frac{\gamma}{\nu}\ . \tag{3.21}$$

When two spins separated by an arbitrary distance can be exchanged, we have again a zero relaxation time in high temperature. This problem can be overcome in the same line as for the one-dimensional case. Therefore inequality (3.21) is quite general. Except for $n \sim \xi$ when a smaller z exponent could be obtained[3].

4. CONCLUDING REMARKS

Critical dynamics of kinetic Ising models is a hard problem and known exact solutions are unfortunately restricted to particular one-dimensional cases. In two dimensions even with Monte Carlo simulation the problem is not simple. However it is a very interesting problem since it can illuminate the understanding of nonequilibrium systems.

We have described several one-dimensional nonhomogeneous kinetic Ising models with Glauber and Kawasaki dynamics. The main conclusion is that the critical dynamical exponent is nonuniversal and depends upon the nonhomogeneity. More precisely the z exponent is greater for nonhomogeneous systems than for isotropic ones. This nonuniversal behaviour is closely related with the zero critical temperature because when approaching the $T = 0$ limit, the transition rates can vanish faster than the inverse correlation length. This fact can provide an extra contribution to the critical slowing down. In addition, in Monte Carlo simulation the z exponent depends on the defined model.

We have also considered the homogeneous ferromagnetic Ising model with generalized Kawasaki dynamics in one dimension for two transition rates. We have obtained that $z \geq 5$ from the initial response and the scaling hypothesis for models in which the maximal separation of two exchanged spins is not proportional to the equilibrium correlation length. Based on physical grounds, we do not expect an exponent greater than 5. Therefore the exponent derived from the generalized Kawasaki model should be the same as the one obtained from the standard first neighbour spins exchange model. On the other hand, we have obtained that $z \geq 3$ for models in which the maximal distance of exchanged spins is proportional to the correlation length. Therefore these last models can be in a dynamical universality class different from the standard Kawasaki model one.

Moreover kinetic Ising models can be used to describe phase transitions in far-from-equilibrium steady states. One approach of constructing a model which has a non-equilibrium steady state is to consider the system connected to two different baths[43,44]. It would be interesting to look into such models where competing Kawasaki dynamics with local and non-local conservation laws are present.

REFERENCES

1. S. Ma, "Modern Theory of Critical Phenomena" (W. A. Benjamin, Inc., Massachusetts, 1976).

2. P. C. Hohenberg and B. Halperin, Rev. Mod. Phys. **49**, 435 (1976).

3. P. Tamayo and W. Klein, Phys. Rev. Lett. **63**, 2757 (1989).

4. R. J. Glauber, J. Math. Phys. **4**, 294 (1963).

5. K. Kawasaki in " Phase Transition and Critical Phenomena" vol. 2, eds. C. Domb and M. S. Green (Academic, N. Y., 1972).

6. W. Zwerger, Phys. Lett. A **84**, 269 (1981).

7. Y. Achiam, J. Phys. A **13**, 1825 (1980).

8. M. Droz, J. Kamphorst Leal da Silva and A. Malaspinas, Phys. Lett. A **115**, 448 (1986).

9. D. Kandel, Phys. Rev. B **38**, 486 (1988).

10. F. Haake and K. Thol, Z. Phys. B **40**, 219 (1980).

11. P. A. Martin, "Modèles en Mécanique Statistique des Processus Irréversibles" (Springer, Berlin, 1979).

12. B. U. Felderhof, Rep. Math. Phys. **1**, (1970).

13. B. I. Halperin, Phys. Rev. B **8**, 4437 (1973).

14. R. Abe, Prog. Theor. Phys. **39**, 947 (1968).

15. R. Cordery, S. Sarker and J. Toboshnik, Phys. Rev. B **24**, 5402 (1981).

16. J. H. Luscombe, Phys. Rev. B **36**, 501 (1987).

17. J. C. Angles d'Auriac and R. Rammal, J. Phys. A **21**, 763 (1988).

18. J. A. Ashraff and R. B. Stinchcombe, Phys. Rev. B **40**, 2278 (1989).

19. J. Kamphorst Leal da Silva, Phys. Lett. A **119**, 37 (1986).

20. D. Dhar and J. Barma, J. Stat. Phys. **22**, 259 (1980).

21. M. Droz, J. Kamphorst Leal da Silva, A. Malaspinas and A. L. Stella, J. Phys. A **20**, L387 (1987).

22. E. J. S. Lage, J. Phys. A **18**, 2289 (1985).

23. M. Droz, J. Kamphorst Leal da Silva, A. Malaspinas and J. Yeomans, J. Phys. A **19**, 2671 (1986).

24. P. O. Weir, J. M. Kosterlitz and S. H. Adachi, J. Phys. A **19** , L757 (1986).

25. M. Silvério Soares and J. Kamphorst Leal da Silva, J. Phys. A **22**, 4959 (1989).

26. D. Kutasov, A. Aharony, E. Domany and W. Kinzel, Phys. Rev. Lett. **56**, 2229 (1986).

27. R. Rammal, J. Physique **46**,1837 (1985).

28. C. K. Harris and R. B. Stinchcombe, Phys. Rev. Lett. **56**, 869 (1986).

29. M. Droz, J. Kamphorst Leal da Silva and O. F. Alcantara Bonfim, unpublished (1987).

30. G. F. Mazenko, M. J. Nolan and O. T. Valls, Phys. Rev. Lett. **41**, 128 (1978).

31. W. Kinzel, Z. Phys. B **29**, 361 (1978).

32. Y. Achiam and J. M. Kosterlitz, Phys. Rev. Lett. **41**, 128 (1978).

33. K. Binder in "Statistical Physics", ed. K. Binder (Topics in Current Physics, Springer, Berlin 1979).

34. E. Stoll, K. Binder and T. Schneider, Phys. Rev. B **8**, 3266 (1973).

35. R. H. Swendsen and J.-S. Wang, Phys. Rev. Lett. **58**, 86 (1987).

36. D. Kandel, E. Domany, D. Ron, A. Brandt and E. Loh Jr., Phys. Rev. Lett. **60**, 1591 (1988).

37. U. Wolff, Phys. Rev. Lett. **62**, 361 (1989).

38. L. P. Kadanoff and J. Swift, Phys. Rev. **165**, 310 (1968).

39. K. Kawasaki in "Synergetics", ed. H. Haken (Teubner, Stuttgart, 1973).

40. S. W. Koch, R. C. Desai and F.F Abraham, Phys. Rev. A **27** , 2152 (1983).

41. Y. Enomoto and K. Kawasaki, Mod. Phys. Lett. B **3**, 605 (1989).

42. J. Kamphorst Leal da Silva and F. C. Sá Barreto, to appear in Phys. Rev. A (1991).

43. A. DeMasi, P. A. Ferrari and J. L. Lebowitz, Phys. Rev. Lett. **55**, 1947 (1985); J. Stat. Phys. **44**, 589 (1986).

44. M. Droz, Z. Rácz and J. Schmidt, Phys. Rev. A **39**, 2141 (1989).

STATISTICAL MECHANICS OF DRIVEN LATTICE GAS MODELS

R. K. P. Zia

Center for Stochastic Processes in Science and Engineering
and
Physics Department
Virginia Polytechnic Institute and State University
Blacksburg, VA 24061 USA

ABSTRACT

The physics of fast ionic conductors motivates us to consider the collective behavior lattice gas (Ising) models when driven to non-equilibrium steady states by an external field. At generic temperatures, violation of the fluctuation-dissipation theorem leads to singular structure factors and long range correlations. The critical properties the familiar second order transitions are modified. Below T_c, interesting new transitions associated with interfaces are also found. Other related systems are discussed.

INTRODUCTION

Non-equilibrium phenomena, by comparison to equilibrium ones, are abundant in nature, appearing in a large variety of interesting contexts. Yet they are poorly understood. One way to start in on this vast field is to investigate the simplest possible models of many-particle systems, just as Ising introduced his model in order to understand ferromagnetism[1]. Translated into the lattice gas language[2], it has been applied to a large number of physical systems, e.g., liguid-gas[3], binary alloys[4], hydrogen in metals[5], etc. With the understanding of universality classes[6] based on renormalization group analysis[7], certain theoretical results from simple models may be compared with experimental data from complex physical systems *quantitatively*. In this spirit, we now devote our attention to one of the the simplest generalizations of Ising's model: a lattice gas driven to a non-equilibrium steady state by an external field[8].

THE KATZ-LEBOWITZ-SPOHN MODEL

Let us start with the familiar Ising model defined on a square lattice, in which each site can be either vacant or occupied by a single particle. A configuration is thus labelled by a set of occupation numbers $\{\rho_i\}$, with $\rho_i=0,1$, where i is the site index. The energy of a particular configuration is given by the usual

$$\{\rho\} \quad = \quad - J \, \Sigma \, \rho_i \, \rho_{i'} \quad , \qquad (1)$$

where the sum runs over nearest neighbor pairs. We concentrate on cases in which J is positive, corresponding to attraction between particles (i.e., a 'ferromagnetic' interaction, if the lattice gas is phrased in terms of Ising spins). Specifying the boundary conditions, e.g., doubly periodic (PBC), completes all that is necessary for studying the equilibrium properties of this system. For dynamical aspects, we need to provide transition rates for the system to evolve from one configuration into another. Here, we will use the simplest, particle hopping, from site to nearest- neighbor site, obeying the Metropolis rate[9]. Simulating coupling to a heat bath at inverse temperature β, these rates are simply $\min\{1,\exp(-\beta\Delta)\}$, where Δ is the change in energy due to the hop.

Finally, we drive the system into a non-equilibrium steady state by imagining all particles to have unit charge and imposing an external 'electric' field **E** parallel to one of the axes. The effect of this field is incorporated by a simple modification of the rates, namely, $\min\{1,\exp[-\beta(\Delta+\epsilon E)]\}$, where E is the magnitude of **E** and

$$\epsilon = \begin{cases} -1 & \\ \;\;0 & \text{for hops} \\ +1 & \end{cases} \quad \begin{matrix} \text{along} \\ \text{transverse to} \\ \text{against} \end{matrix} \quad \mathbf{E} \qquad (2)$$

Note that the 'charge' of the particles and a lattice constant have been absorbed into E. A physical system which *motivated* this model is the fast ionic conductor, in which ions diffuse in a crystalline lattice. Subjected to an external field, a steady state current can be established. Though these systems exhibit phase transitions[10], we caution the readers not to attempt quantitative comparisons, at this stage, between the properties of this model and experimental data of physical systems. Instead, it (which we call the KLS model) should be considered in the same spirit as Ising's in 1925.

Even for such simple systems, analytic solutions are next to impossible; the three-dimensional Ising model in equilibrium has yet to be solved exactly. To gain some insight, Monte Carlo studies were carried out[8,11-17]. From the typical configurations, it is tempting to conclude that non-equilibrium effects are minimal: the second order transition survives at all E, with $T_c(E)$ increasing monotonically, saturating at $1.4T_c(0)$. Above criticality, particles and holes are distributed randomly. Below T_c, condensation occurs, i.e., the system seperates into a particle-rich (dense) phase and a hole-rich (rarify) one. However, many 'surprises' appear on closer examination, at *all* temperatures. Here, we restrict ourselves to two topics: generic singularities above T_c and effects of boundary conditions below T_c. To discuss the critical properties, which are drastically different from the equilibrium ones, is outside our scope here. Interested readers may read a recent review[18].

GENERIC SINGULARITIES ABOVE T_c

Thermodynamic quantities in an equilibrium Ising model is known to display no singularities above the critical temperature. Thus, it is remarkable that a 'simple' generalization, to the KLS model, should create singularities in, say, correlation functions, at all $T > T_c$. In particular, the two-body correlation $G(r,t=0)$ displays power-law decays at large distances[13]. The crucial ingredients for the manisfestation of such behavior, in this case, are[19]

(i) a conserved order parameter,
(ii) non-equilibrium dynamics, and
(iii) spatial anisotropy associated with (ii).

Even for equilibrium systems, the first ingredient already produces power-law decays in $G(r=0,t)$ at large times, i.e., $t^{-d/2}$, where d is the spatial dimension of the system. Using the ideas of scaling and recalling that r scales like $\sqrt{t}$ in diffusive processes, we may naively expect $G(r,t=0) \rightarrow r^{-d}$ instead of the usual $\exp(-r/\xi)$. This expectation is indeed correct, if it weren't for the validity of detailed balance, which lead to the fluctuation-dissipation theorem, forcing the amplitudes of such terms to vanish[13,20]. In this sense, an equilibrium state is 'special', 'atypical', or 'constrained'. The role of (ii) is to lift this 'constraint', so that power-laws in r are again generic. The third ingredient is more subtle, necessary only for producing power-laws in the *two-body* correlation[19].

We should mention that much is known[21,22] about power-law decays in two-body correlations prior to driven diffusive systems. The KLS model distinguishes itself by having (a) only one relevant (macroscopic) degree of freedom, i.e., the local particle density, $\rho(\mathbf{x},t)$, and (b) a *translationally invariant* steady state. These two simplifications allow us to ask: what are the bare essentials needed for singular behavior in non-equilibrium systems.

In momentum space this singularity in the two-point function takes a starker form. We also find it more convenient to understand its origins in this language. Thus, we consider $S(\mathbf{k})$, the Fourier transform of $G(\mathbf{x},t{=}0)$. Known as the structure factor, this quantity is important for experimentalists since it is proportional to the intensity of scattered beams. For an equilibrium system, even if it had anisotropic inter-particle interactions or diffusion, $S(\mathbf{k})$ will be anisotropic, but it will approach a *single-valued* constant as $\mathbf{k} \to 0$ (i.e., at macroscopic scales). By contrast, the $S(\mathbf{k})$ in our non-equilibrium model, has a *discontinuity* singularity at $\mathbf{k} = 0$. Thus, $S(\mathbf{k} \to 0)$ is a non-trivial function of θ, the angle between $\mathbf{k}$ and $\mathbf{E}$.

This behavior in the KLS model, observed at a temperature 60% above criticality, can be found in in Table IIc of Ref.[8], where $S(\theta{=}0,k{\to}0)$ is about 4 times larger than $S(\theta{=}90°,k{\to}0)$. At 15% above T_c, this ratio grows to 9.3[23]. In the thermodyanmic limit, it is expected to diverge as T is lowered to T_c. We again emphasize that this singularity is present for *all* temperatures. A convenient way to describe this behavior is to use a mesoscopic, Langevin equation of motion for the order parameter. Because of (i), this approach begins with the continuity equation: $\partial\rho/\partial t + \nabla\cdot\mathbf{j} = 0$.

For *equilibrium* systems, a mesoscopic Hamitonian H can be used to specify the current $\mathbf{j}$:

$$j_\alpha = - \Lambda_{\alpha\gamma}\nabla_\gamma \frac{\delta H}{\delta\rho} + \eta_\alpha \quad , \tag{3}$$

where $\alpha,\gamma = 1,..,d$ refer to the Cartesian components of vectors Λ is an Onsager coefficient and η is a Gaussian distributed Lagevin noise. The first term models (deterministic) drift of ρ towards configurations with minimal energy, while the second models the random interactions of the system with a heat bath:

$$\langle\eta\rangle = 0 \quad \text{and} \quad \langle\eta_\alpha(\mathbf{x},t)\eta_\gamma(\mathbf{x}',t')\rangle = N_{\alpha\gamma}\delta(\mathbf{x}-\mathbf{x}')\delta(t-t') \ . \tag{4}$$

Due to detailed balance and the fluctuation-dissipation theorem[24] the noise correlation matrix $N_{\alpha\gamma}$ is equal to $2k_BT\Lambda_{\alpha\gamma}$, ensuring that the distribution evolves to $\exp(-\beta H)$. For an isotropic system, $\Lambda_{\alpha\gamma}$ is usually taken as $\lambda\delta_{\alpha\gamma}$. Linearizing (3), which suffices for T far from T_c, and following standard routes, we find $S(\mathbf{k})\equiv\langle\rho(\mathbf{k})\rho(-\mathbf{k})\rangle$:

$$S(\mathbf{k}) \quad \propto \quad \frac{k_\alpha N_{\alpha\gamma} k_\gamma}{k_\alpha \Lambda_{\alpha\gamma} k_\gamma + O(k^4)} \quad , \tag{5}$$

which reduces to the familiar Ornstein-Zernike form $1/(1+O(k^2))$ in the *equilibrium* case.

Imposing a driving field **E** has two effects on (3). A resistive term, $\mathbf{j}=\sigma\mathbf{E}$, must be added to (3), with the conductivity σ being some non-trivial function of the local density ρ. The simplest choice is $\sigma \propto \rho(1-\rho)$, modelling the excluded volume constraint of the lattice gas. As it turns out, this effect is important only studying critical behavior[18]. For our purposes here, the more significant effect of **E** is anisotropy. Not only does it introduces a preferred direction, its non-equilibrium character invalidates the fluctuation-dissipation theorem. Thus, in general, in addition to $\Lambda_{\alpha\gamma} \propto \delta_{\alpha\gamma}$, we have

$$N_{\alpha\gamma} \quad \propto \quad \Lambda_{\alpha\gamma} \quad . \tag{6}$$

The consequences of this equation on (5) is profound: $S(\mathbf{k} \to 0)$ is no longer single valued, though still *bounded*. To see the angular dependence simply, let us consider the d=2 example, assuming that both N and Λ are diagonal, with elements n_0, n_1, λ_0, and λ_1, where the subscript 0 goes with the θ=0 component, etc. In terms of these, (6) implies $n_0/n_1 \neq \lambda_0/\lambda_1$, so that

$$S(\mathbf{k} \to 0) \quad \propto \quad \left(\frac{n_0}{\lambda_0} - \frac{n_1}{\lambda_1} \right) \left(\frac{\lambda_0 \cos^2\theta}{\lambda_0 \cos^2\theta + \lambda_1 \sin^2\theta} \right) + \frac{n_1}{\lambda_1} \tag{7}$$

has a non-trivial θ dependent term here.

Note that the singularities discussed here are qualitatively very different from those due to critical fluctuations. In the latter case, $S(\mathbf{k} \to 0)$ is typically divergent. Another expression of this qualitative difference is to consider the angular average of our $S(\mathbf{k})$, which assumes the *analytic* Ornstein-Zernike form. In terms of correlations in **x** space, this singularity takes the form of an angular dependent amplitude for the r^{-d} term. Further, this amplitude, when integrated over the angles, vanishes, so that the 'angular averaged correlations' decays exponentially.

It may be instructive to see how this singularity translates into power-law decays in **x** space. Instead of laboring with the specific example, a more elegant and efficient way is to start with the general expression (5). Diagonalize Λ and rescale the momenta, so that $k\Lambda k$ becomes simply k^2. (Note that we have assumed Λ to be positive.) Now the numerator reads kMk, where $M = \sqrt{\Lambda}^{-1} N \sqrt{\Lambda}^{-1}$. The difference between an equilibrium and a non-equilibrium case lies in M being proportional to the unit matrix or not. To find the large **x** behavior of G(**x**), we need to keep only the lowest order k terms in S. Putting the kMk factor as $\nabla M \nabla$, we have

$$G(\mathbf{x} \to \infty) \quad \propto \quad \nabla M \nabla \int dk \; k^{-2} \exp(i\mathbf{k} \cdot \mathbf{x}) \quad \propto \quad \nabla M \nabla \; r^{-d+2} \quad . \qquad (8)$$

If $M \propto \delta$, as in equilibrium cases, then (8) gives us zero for $r>0$. (The reader should not be alarmed at this result; the higher order k terms we neglected are responsible for the exponential decay.) Therefore, we can safely consider only $\underline{M}$, the traceless part of M. Explicit calculation with this part provides us with $(\mathbf{x}\underline{M}\mathbf{x})r^{-d-2}$, giving us both the r^{-d} power and zero angular average. This behavior is confirmed in simulations[13].

Finally, note that these singularities in S typically give rise to others in multi-body correlation functions. A good example is the three-point function, which is divergent[17] in the limit of small, generic momenta. Yet, it vanishes when **k**=0 is approached in special directions[17,20]!

To end this section, we remark that special singularities exist even in the absence of interactions (or $T=\infty$ limit) for systems of low dimensions ($d \leq 2$), taking the form of anomalous diffusion[25].

EFFECTS OF BOUNDARY CONDITIONS AND INTERFACE PROPERTIES

Below criticality, condensation occurs, just as in equilibrium systems. For a half-filled lattice with PBC, typical configurations consist of strips of equal width, one that is particle-rich (dense) and the other, hole-rich (rarify). Unlike the equilibrium case, only strips oriented *parallel* to **E** have been observed. The two interfaces seperating these phases are equivalent, being mirror images of each othe r. In this section, we consider the effects on these interfaces, as well as the bulk, when various other boundary conditions are imposed.

1. Shifted Periodic Boundary Conditions

One motivation for considering other boundary conditions lies in a study of the roughness of these interfaces in non-equilibrium steady state. For an Ising model in equilibrium, the interfaces are always rough in d=2, smooth in d>3 and, for certain orientations in d=3, undergoes a transition -- the roughening transition[26]. It is natural to ask if driving affects any of these properties. So far, we have data only for d=2 systems[27], in which interfaces appear to be always *smooth*. Specifically, the statistical width of the interface, instead of diverging as $\sqrt{L}$ (on a $L \times L$ lattice), seems to increase at a rate slower than any power of L. Several studies[28] addressed the issue of capillary waves in driven systems. However, until recently[29], this behavior remained a puzzle.

In attempting to understand why interfaces in d=2 are smooth, we drew a connection with equilibrium crystal shapes and asked if energy densities associated with vicinal surfaces display a kink discontinuity in its orientation dependence[30]. A standard method to induce high index interfaces is[31] by imposing *shifted* periodic boundary conditions (SPBC). With SPBC, a particle leaving the first column at row R re-enters the lattice at column L at row $R+h$, where h is the 'shift'. We carried out Monte Carlo studies on the KLS system, with saturation E and $T \approx 0.6T_c$. Not only are the expected 'tilted' interface observed, several surprising new properties are discovered[32]:

* *bulk* densities and energies depends on the tilt angle, i.e., arctan(h/L), displaying kink singularities;
* the single strip (of say, the high density phase) splits into many strips, representing a multiple winding on a torus, when the tilt angle becomes large;
* on further increase of the tilt, there are other transitions where N strips merge into N-1 strips.

To emphasize the first point, we remind the reader that, in equilibrium systems, bulk properties like energy densities are hardly affected by boundary conditions. Typically, the effects of the surface survives in a boundary layer of thickness controlled by the correlation length. Far below criticality, this layer would be around a lattice spacing thick. In our case, the *entire bulk* is affected: particle and energy densities increase with h. This behavior can be understood qualitatively, since the field can draw particles out of the bulk when it has a component *normal* to the interface. Indeed, one edge of the particle-rich strip acts as a source for particles evaporating into the particle-poor region. At the opposite edge, the interface acts as a sink. In other words, a 'transverse' current is induced by the SPBC, leading to h dependent particle densities in the bulk. By symmetry, this dependence must be even under $h \to -h$. However, a convincing argument in favor of *linear* dependence on h is still lacking. When found, that would complete our understanding of a kink singularity: $|h|$.

Beyond bulk singularities, there are interesting features in the small h configurations. Since the two interfaces play opposite roles in the dynamics, they are no longer equivalent. On closer examination, there is a boundary layer near the absorbing edge in which the particle density is *higher* in the *dense* phase. Again, it is easy to have qualitative understanding, while quantitative answers are far from easy, in both theoretical and 'experimental' arenas[33]. Much work remains to be done.

When the tilt angle becomes too large, the transverse current increases. The absorbing edge behaves much like the interface in a growing crystal, leading eventually to the splitting transition. Leung[33] attempted to explain this transition in terms of a Mullins-Sekerka instability. While qualitatively successful, this line of investigation represents only the beginning of a fully predictive theory.

2. Open Boundary Conditions

Another motivation for considering other boundary conditions comes from experimental constraints. It is difficult to set up a steady state (*non-superconducting*) current in a loop, satisfying PBC. On the other hand, open systems, in which charges enter the sample from one end and leave at the other, are elementary. Thus, we considered a KLS system with *open* boundary conditions[34]. So far, we have data only for the simplest case: maximum 'feeding' of particles at one edge and 'bleeding' at the other. Above T_c, the system is more-or-less disordered, but with non-uniform profile due to the feed. Below T_c with saturation E, we observe yet another new type of ordering: dense and rarified phases alternate in 'fingers', creating a pattern reminiscent of a backgammon board. In the regime studied, the number of fingers depend on the aspect ratio of the system, in such a way that the tilt-angle of the interfaces remains constant. Clearly, there must be a relationship between these patterns and tilted interfaces in SPBC. Again, making qualitative arguments for such patterns is easy, while much more work is needed for a comprehensive theory.

In a recent study[35], J. Krug found other surpising effects in driven systems with open boundaries. He found "behavior reminiscent of second and first order phase transitions even in one-dimensional systems with short range interactions." A rich phase diagram, for particles with *repulsive* nearest neighbor interactions, emerges. Being non-equilibrium in nature, these transitions are found to be governed by a 'maximum current principle'. It is clear that, for driven systems, boundary conditions play a much more significant role than their counterparts in equilibrium systems.

Before ending this subsection, we mention that driving with density gradients in open boundary systems has a long history[36]. Interested readers are encouraged to read the articles cited, since covering these systems is beyond the scope of this paper. Here we only point out that the models described above differ in a major respect, i.e., the drive in KLS systems act *thoughout* the bulk. As a result, the current does not vanish in the thermodynamic limit. Above criticality, the drive due to the open boundaries will affect the bulk less and less in this limit[34], so that our system will presumably revert to a KLS model with PBC, as opposed to a system with equilibrium properties. Below criticality, the non-vanishing bulk current, acting in consort with long-range correlations, is expected to carry the information of the boundary drives throughout the entire bulk. Non-trivial ordering, such as backgammon patterns are expected to survive the thermodynamic limit in some form. In this sense, the KLS systems appear to offer abundant richness, waiting to be explored.

3. Varying Field Strengths at a Boundary

Closely related to boundary-driven KLS models is another[37], in which particles are simply allowed to enter and leave the system through the two boundaries *transverse* to **E**. To conserve particle number, a particle leaving an edge is brought back through the other. Thus, this model may be regarded as having a defective row, in which the drive is $E_0 \equiv 0$, in the ordinary KLS model with PBC. At low temperatures, phase seperation still occurs. But the interface is oriented *transverse* to **E**, in sharp contrast to the case in a translationally invariant KLS low temperature state.

A more important effect of this defect line is the destruction of the second order phase transition[37]. One way to understand this result is to consider the extreme case, in which $E_0 = -\infty$ at the defect row, modelling a 'brick wall' boundary. The system behaves just like a liquid-gas under the influence of a gravitational field so that it is effectively in equilibrium. Such a system, although displaying qualitatively different configurations at high and low temperatures, is known to have no phase transition[38].

In the context of a defective drive, it is clear that, as E_0 increases up to E, the KLS model is retrieved. Thus, we may study a phase diagram with one more paramter, E_0, and ask what are the effects on interfaces, phase transitions, etc.

CONCLUDING REMARKS

For convenience and due to limited scope, we have concentrated on the simplest model of driven diffusive lattice gases -- the KLS model with attractive interactions. Many variations and extensions exist. Beyond these, there are seemingly limitless other related non-equilibrium systems. As concluding paragraphs, we will direct the readers' attention to a very limited number of these.

One of the first variations of the proto-system studied is the fast rate limit[39], which led to, remarkably, solvable models. Another considered the effects of steric hinderance and used jumps to next neighbor sites[40]. Other variations are: systems with repulsive interparticle interactions[41]; models with annealed random drive[42]; quenched random noise, modelling impurities[43]; systems containing more than one species of particles[44], with extensions to the polarized lattice gas[45] and the 'brazil nut problem'[46]. Clearly, the KLS model is very far from being a realistic model for fast ionic conductors, which was one of the physical motivations. Yet, there are some successful applications, such as the fluctuations of an anchored Toom interface[47] and a model for electrophoresis[48]. Other natural candidates are, e.g., flux creep in high T_c superconductors[49], hot electron transport[50], and biased diffusion of atoms on surfaces[51].

Beyond variations and extensions of the KLS model, there is a host of related non-equilibrium steady state systems. Naturally, the more realistic ones tend to be more complex. We end by naming a few: liquids driven by shear[52] and temperature gradients[53]; models with competing Glauber and Kawasaki dynamics[54]; systems coupled to two temperature baths[55], some are possibly related to the KLS model with random drive[41] and others displaying truely bistable phases[56]; crystal growth[57], pattern formation[58] and kinetic roughening in interfaces[59]; statistics of sandpiles[60] and earthquakes[61].

In this short article, we presented a brief glimpse into the remarkable world of driven diffusive lattice gas models. They show a large variety of singular behavior, compared to their equilibrium cousins, without special tuning of control parameters. In this sense, they may be part of a wider class of phenomena known as self-organized criticality[60] and generic scale invariance[19]. By no means a comprehensive review[62], this article would have served its purpose, if it whetted the reader's appetite to look further into the physics of non-equilibrium systems.

ACKNOWLEDGEMENTS

This work is supported in part by a grant from the National Science Foundation through the Division of Materials Research.

REFERENCES

1. E. Ising, Z. Physik **31**, 253 (1925).

2. C.N. Yang and T.D. Lee, Phys. Rev. **87**, 404 (1952).

3. T.D. Lee and C.N. Yang, Phys. Rev. **87**, 410 (1952).

4. See, e.g., R.K. Pathria, **Statistical Mechanics** (Pergamon, Oxford, 1972) pp.399.

5. See, e.g., H. Wagner and H. Horner, Adv. Phys. **23**, 587 (1974).

6. L.P. Kadanoff, in **Proc. Intern. School of Physics "Enrico Fermi"**. ed. M. Green (Academic, NY 1971).

7. K.G. Wilson, Rev. Mod. Phys. **47C**, 773 (1975); D.J. Amit: **Field Theory, the Renormalization Group, and Critical Phenomena** (revised 2nd edition: World Scientific, Singapore 1984).

8. S. Katz, J.L. Lebowitz, and H. Spohn, Phys. Rev. **B28**, 1655 (1983) and J. Stat. Phys. **34**, 497 (1984).

9. N. Metropolis, A.W. Rosenbluth, M.M. Rosenbluth, A.H. Teller, and E. Teller, J. Chem. Phys. **21**, 1087 (1953).

10. See, e.g., W. Dieterich, P. Fulde, and I. Peschel, Adv. Phys. **29**, 527 (1980), and references therein.

11. J. Marro, J.L. Lebowitz, H. Spohn, and M.H. Kalos, J. Stat. Phys. **38**, 725 (1985),

12. J.L. Vallés and J. Marro, J. Stat. Phys. **43**, 441 (1986) and **49**, 89, (1987); J. Marro, J.L Vallés and J.M. González-Miranda, Phys. Rev. **B35**, 3372 (1987) and J. Marro and J.L. Vallés, J. Stat. Phys. **49**, 121, (1987).

13. M.Q. Zhang, J.-S. Wang, J.L. Lebowitz and J.L. Vallés, J. Stat. Phys. **52**, 1461 (1988).

14. K.-t. Leung, K.K. Mon, J.L. Vallés and R.K.P. Zia, Phys. Rev. Lett. **61**, 1744 (1988) and Phys. Rev. **B39**, 9312 (1989).

15. J.L. Vallés, K.-t. Leung and R.K.P. Zia, J. Stat. Phys. **56**, 43 (1989).

16. D.H. Boal, B. Schmittmann and R.K.P. Zia, Phys. Rev. **A43**, 5214 (1991).

17. K. Hwang, B. Schmittmann and R.K.P. Zia, Phys. Rev. Lett. **67**, 326 (1991).

18. B. Schmittmann, Int. J. Mod. Phys. **B4**, 2269 (1990).

19. G. Grinstein, J. Appl. Phys. **69**, 5441 (1991); G. Grinstein, D.H. Lee and S. Sachdev, Phys Re. Lett. **64**, 1927 (1990).

20. P.L. Garrido, J.L. Lebowitz, C. Maes and H. Spohn, Phys. Rev. **A42**, 1954 (1990).

21. A. Onuki, J. Stat. Phys. **18**, 475 (1978). See also [52,53].

22. H. Spohn, J. Phys. **A16**, 4275 (1983). See also [36].

23. K. Hwang, private communication.

24. G.E. Uhlenbeck and L.S. Ornstein, Phys. Rev. **36**, 823 (1930); R. Kubo, Rep. Progr. Phys. **29**, 255 (1966); R. Graham, in **Springer Tracts in Modern Physics** Vol. 66 (Springer, Berlin 1973); L.E. Reichl, **A Modern Course in Statistical Physics** (University of Texas Press, Austin 1980).

25. H.K. Janssen and B. Schmittmann, Z. Phys. **B63**, 517 (1986). See also D. Forster, D.R. Nelson and M.J. Stephen, Phys. Rev. **A16**, 732 (1977); H. van Beijeren, J. Stat. Phys. **63**, 47 (1991).

26. H. van Beijeren and I Nolden, "The Roughening Transition", in **Topics in Current Physics, 43**, ed. W. Schommers and P. von Blanckenhagen (Springer, Berlin, 1987) pp.259.

27. K.-t. Leung, K.K. Mon, J.L. Vallés and R.K.P. Zia, **Phys. Rev. Lett. 61**, 1744 (1988) and Phys. Rev. **B39**, 9312 (1989).

28. K.-t. Leung, J. Stat. Phys. **50**,405 (1988); A. Hernandez-Machado and D. Jasnow, Phys. Rev. **A37**,626 (1988); A. Hernandez-Machado, H. Guo, J.L. Mozos, and D. Jasnow, Phys. Rev. **A39**, 4783 (1989).

29. R.K.P. Zia and K.-t. Leung, submitted to J. Phys. A.

30. R.K.P. Zia, "Anisotropic Surface Tension and Equilibrium Crystal Shapes", in **Progress in Statistical Mechanics**, ed. C.K. Hu, (World Scientific, Singapore, 1988) pp. 303.

31. See, e.g., K.K. Mon, S. Wansleben, D.P. Landau and K. Binder, Phys. Rev. Lett. **60**, 708 (1988) and Phys. Rev. **B39**, 7089 (1989).

32. J.L. Vallés, K.-t. Leung and R.K.P. Zia, J. Stat. Phys. **56**, 43 (1989).

33. K.-t. Leung, J. Stat. Phys. **137**, 341 (1990).

34. D.H. Boal, B. Schmittmann and R.K.P. Zia, Phys. Rev. **A43**, 5214 (1991).

35. J. Krug, IBM preprint, to be published.

36. R.P. Smith, Acta Metal. **1**, 578 (1953); G.E. Murch, Phil. Mag. **A41**, 157 (1980). Some recent studies are: J.V. Andersen and O.G. Mouritsen, Phys. Rev. Lett. **65**, 440 (1990); H.C. Fogedby and A. Svane, Phys. Rev. **B42**, 1056 (1990).

37. J.V. Andersen and K.-t. Leung, Phys. Rev. **B43**, 8744 (1991).

38. J.V. Sengers and J.M.J. van Leeuwen, Int. J. Thermophys. **6**, 545 (1985); J.M.J. van Leeuwen and J.V. Sengers, Physica **138A**, 1 (1986); J.H. Sikkenk, J.M.J. van Leeuwen and J.V. Sengers, Physica **139A**, 1 (1986)

39. H. van Beijeren and L.S.Schulman, Phys. Rev. Lett. **53**, 806 (1984); J. Marro, J.L. Lebowitz, H. Spohn and M.H. Kalos, J. Stat. Phys. **38**, 725 (1985); H. van Beijeren, R. Kutner and H. Spohn, Phys. Rev. Lett. **54**, 2026 (1985); J. Krug, J.L. Lebowitz, H. Spohn and M.Q. Zhang, J. Stat. Phys. **44**, 535 (1986).

40. K.K. Mon and K. Hwang, priviate communications. See also [43].

41. K.-t. Leung, B. Schmittmann and R. K. P. Zia, Phys. Rev. Lett. **62**, 1772 (1989) and R. Dickman, Phys. Rev. **A41**, 2192 (1990). See also G. Szabo and A. Szolnoki, Phys. Rev. **A41**, 2235 (1990).

42. B. Schmittmann and R.K.P. Zia, Phys. Rev. Lett. **66**, 357 (1991).

43. V. Becker and H.K. Janssen, submitted to Europhys. Lett.

44. B. Schmittmann, K. Hwang and R.K.P. Zia, to be published.

45. M. Aertsens and J. Naudts, J. Stat. Phys. **62**, 609 (1991).

46. F.J. Alexander and J Lebowitz, J. Phys. **A23**, L375 (1990).

47. B. Derrida, J.L. Lebowitz, E.R. Speer and H. Spohn, Phys. Rev. Lett. **67**, 165 (1991).

48. Y. Shnidman, in **Mathematics in Industrial Problems, IV**, ed. A. Friedman (Springer, Berlin, 1991). See also T.A.J. Duke, Phys. Rev. Lett. **62**, 2877 (1989) and J. Chem. Phys. **93**, 9049 and 9055 (1990).

49. S. Doniach, in **High Temperature Superconductivity: Proceedings**, eds. K.S. Bendell, D. Coffey,D.E. Meltzer, D. Pines and J.R. Schrieffer (Addison Wesley, NY, 1990). See also B.I. Ivlev and N.B. Kopnin, J. Low Temp. Phys. **80**, 161 (1990).

50. See, for example, H.U. Baranger, J.-L. Pelouard, J.-F. Pône and R. Castagné, Appl. Phys. Lett. **51**, 1708 (1987) and references therein.

51. S.C. Wang and T.T. Tsong, Phys. Rev. **B26**, 6470 (1982). See also L.J. Whitman, J.A. Stroscio, R.A. Dragoset and R.J. Celotta, Science **251**, 1206 (1991).

52. D. Beysens and M. Gbadamassi, Phys. Rev. **A22**, 2250 (1980); D. Ronis and I. Procaccia, Phys. Rev. **A25**, 1812 (1982); A.-M.S. Trembley, E.D. Siggia and M.R. Arai, Phys. Rev. **A23**, 1451 (1981) and **A24**, 1655 (1981); D. Ronis and I. Procaccia, Phys. Rev. **A25**, 1812 (1982); J.W. Duffy and J. Lutsko, in **Recent Developments in Non-equilibrium Thermodynamics**, eds. J. Casas-Vazques, D. Jou and J.M. Rubi, (Spinger, Berlin, 1986); C.K. Chan and L. Lin, Europhys. Lett. **11**, 13 (1990); X.L. Wu, D.J. Pine and P.K. Dixon, Phys. Rev. Lett. **66**, 2408 (1991).

53. D. Ronis, I. Procaccia and I. Oppenheim, Phys. Rev. **A19**, 1324 (1979); R. Desai and M. Grant, Phys Rev **A27**, 2577 (1981); T.R. Kirkpatrick, E.G.D. Cohen and J.R. Dorfman, Phys. Rev. **A26**, 950 (1982). For other references and recent reviews, see, e.g., R. Schmitz, Phys. Rep. **171**, 1 (1988). B.M. Law and J.V. Sengers, J. Stat. Phys. **57**, 531 (1989).

54. A. DeMasi, P.A. Ferrari and J.L. Lebowitz, Phys. Rev. Lett. **55**, 1947 (1985) and J. Stat. Phys. **44**, 589 (1986); J.M. Gonzalez-Miranda, P.L. Garrido, J. Marro and J.L. Lebowitz, Phys. Rev. Lett. **59**, 1934 (1987).

55. Z. Cheng, P.L. Garrido, J.L. Lebowitz, and J.L. Vallés, Europhys. Lett. **14**, 507 (1991); H.W.J. Blöte, J.R. Heringa, A. Hoogland and R.K.P. Zia, J. Phys. **A23**, 3799 (1990)..H.W.J. Blöte, J.R. Heringa, A. Hoogland and R.K.P. Zia, Int. J. Mod. Phys. **B5**, 685 (1991); H. Shinkai, H.W.J. Blöte, J.R. Heringa, A. Hoogland and R.K.P. Zia, submitted to Phys. Rev. A (1991).

56. C.H. Bennett and G. Grinstein, Phys. Rev. Lett. **55**, 657 (1985); Y. He, C. Jayaprakash and G. Grinstein, Phys. Rev. **A42**, 3348 (1990).

57. See, e.g., **Crystal Growth**, ed. B.R. Pamplin, (Pergamon, Oxford, 1980).

58. See, e.g., J.S. Langer, in **Chance and Matter**, ed. J. Souletie (North-Holland, Amsterdam, 1987); O.G. Mouritsen, Int. J. Mod. Phys. **B4**, 1925 (1990).

59. M. Kardar, G. Parisi and Y.C. Zhang, Phys. Rev. Lett. **58**, 2087 (1987). For further references, see J. Krug and H. Spohn, in **Solids Far.From Equilibrium: Growth, Morphology and Defects**, ed. C. Godreche, (Cambridge, 1991)

60. P. Bak, C. Tang and K. Wiesenfeld, Phys. Rev. **A38**, 364 (1988).

61. J.M. Carlson and J.S. Langer, Phys. Rev. Lett. **62**, 2632 (1989).

62. A comprehensive review on driven systems will be published, in **Phase Transitions and Critical Phenomena**, eds. C. Domb and J.L. Lebowitz (Academic, NY), by B. Schmittmann and R.K.P. Zia.

CHAOS, NEURAL NETWORK, AND SELF-ORGANIZED CRITICALITY

RECURRENCE OF KAM TORI AND THEIR NOVEL CRITICAL BEHAVIOR IN NONANALYTIC TWIST MAPS

Bambi Hu and Jicong Shi
Department of Physics, University of Houston, Houston, Texas 77204-5504

ABSTRACT

A novel type of critical phenomena is discovered in a class of nonanalytic twist maps with a variable degree of inflection z. It is found that when $z > 3$ a KAM torus can reappear after it has disappeared. An "inverse residue criterion" is introduced to locate the precise reappearance point. We have also studied the local and global scaling behaviors of these tori. The critical exponents, the singularity spectrum and the generalized dimension all vary with z when $2 \leq z < 3$ but are independent of z when $z \geq 3$. In this sense the degree of inflection plays a role quite similar to that of dimensionality in phase transitions with $z = 2$ and 3 corresponding respectively to the lower and upper critical dimensions. The resemblance to phase transitions is remarkable.

I. INTRODUCTION

In the context of the standard map the picture for the breakup of KAM tori is relatively simple. Each rotational torus is characterized by an irrational winding number. These tori serve as barriers to locally stochastic motion. However, as the perturbation strength is increased, more and more of them disappear. At a critical perturbation value, the last KAM torus characterized by the "golden-mean" winding number breaks up and global stochasticity sets in. A "residue criterion" was introduced by Greene[1] to determine the breakup point.

In many respects the breakup of KAM tori is analogous to a phase transition. By utilizing concepts and techniques used in the study of phase transitions, remarkable progress has been made in the understanding of the breakup of KAM tori.[2,3]

In phase transitions diverse physical system can be divided into equivalence classes according to certain criteria. These universality criteria are well known in phase transitions. However, much less is known about them in chaotic transitions.[4]

To gain a better understanding of universality in the breakup of KAM tori, we propose to study a class of nonanalytic twist maps:

$$T : \quad \begin{cases} r_{i+1} = r_i - kg(\theta_i), \\ \theta_{i+1} = \theta_i + r_{i+1} \quad (mod\ 1), \end{cases} \tag{1}$$

where $\theta_i \in [-\frac{1}{2}, \frac{1}{2})$ and $r_i \in [0,1]$. The function $g(\theta) = \theta(1 - |2\theta|^{z-1})$ is nonanalytic (C^1) and is so designed that it has a variable degree of inflection z at the inflection point $\theta = 0$. The motivation for studying such a function is that in the circle map, which can be regarded as the dissipative limit of the standard map, the degree of inflection serves as a universality criterion. The sine function in the circle map possesses a cubic degree of inflection ($z = 3$). To generalize it to an arbitrary degree of inflection, $g(\theta)$ was invented.[5,6] Although the substitution of this function in the circle map produced no particularly surprising results, many novel features have been discovered in the conservative case.

For area-preserving twist maps, each orbit is characterized by a winding number ω. ω is a rational number for a periodic orbit and an irrational number for a KAM torus. A KAM torus with an irrational winding number ω can be approximated by a sequence of periodic orbits whose winding numbers are the successive convergent in the continued fraction expansion of ω. The particular KAM torus we will focus our attention on is the one whose winding number is the inverse of the "golden-mean," $\omega = (\sqrt{5} - 1)/2$. Its convergents are $\omega_n = F_n/F_{n+1}$, where F_n is a Fibonacci number satisfying $F_{n+1} = F_n + F_{n-1}$ with $F_0 = 0$, $F_1 = 1$. This KAM torus is the most "robust" and is very often the last one to break up. In the following, we will study the behavior of this critical torus as we vary the degree of inflection.

Since $g(\theta)$ is an odd function, the map (1) is reversible and possesses four symmetry lines:

$$a. \quad \theta = -1/2 \tag{2}$$

$$b. \quad \theta = 0 \tag{3}$$

$$c. \quad \theta = (r-1)/2 \tag{4}$$

$$d. \quad \theta = r/2. \tag{5}$$

For a rational winding number, each orbit (elliptic or hyperbolic) has two points on two of these four lines. One line will be mapped into another at the half-way point of the orbit. Thus finding a periodic orbit is greatly simplified.

II. RECURRENCE OF KAM TORI

The following is a summary of our main findings.

(1) $k < 0$ or $z \leq 2$

For $k \neq 0$ the residue tends to infinity and the action difference[7] tends to a nonzero constant as the period tends to infinity. It suggests that there are no KAM tori when $k \neq 0$. As in the sawtooth map, $k_D = 0$ is the trivial critical point of the breakup of the KAM torus. This is similar to a zero-temperature phase transition.

(2) $2 < z \leq 3$

In this case $k_D \neq 0$; however, no reappearance of KAM tori has been observed. k_D tends to zero monotonically as z decreases from 3 to 2. The phase

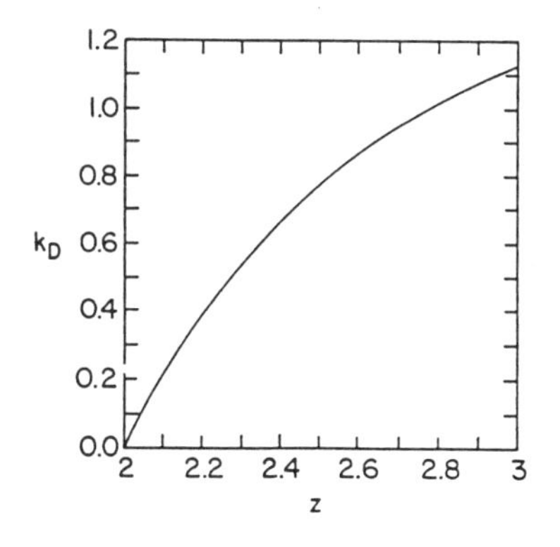

Figure 1: The critical value k_D as a function of z for $2 \leq z \leq 3$. The area below (above) the critical line indicates the existence (non-existence) of the "golden-mean" KAM torus.

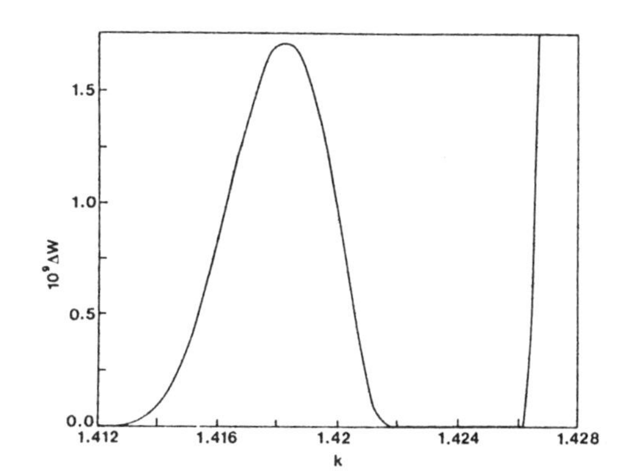

Figure 2: The action difference ΔW as a function of k for $z = 4$ and $(Q_i, P_i) = (1597, 987)$.

diagram is shown in Fig. 1. The area below (above) the line indicates the existence (non-existence) of the KAM torus.

(3) $z > 3$

Reappearance of KAM tori[8–10] has been observed. We found[11] that there is more than one value of k which satisfies Greene's criterion. For example, for $z = 4$, two disappearance points $k_D^{(1)} = 1.4129353$ and $k_D^{(2)} = 1.4261557$ have been found. If a KAM torus disappears at two points, there must be a point k_R, $k_D^{(1)} < k_R < k_D^{(2)}$, at which it reappears. To determine k_R, we introduce an "inverse residue criterion" for the reappearance of a KAM torus:

A KAM torus with a winding number ω that has disappeared at k_D will reappear at k_R if

$$\lim_{i \to \infty} R_i^{\pm}(k) = \begin{cases} \pm\infty, & k < k_R; \\ R^{\pm}, & k = k_R; \\ 0^{\pm}, & k > k_R. \end{cases} \tag{6}$$

$R_i^{\pm}(k)$ are the residues of the minimax (+) and minimizing (−) orbits with winding numbers $\omega_n = P_n/Q_n$ at a given value of k. $|R^{\pm}| < 1$ are two constants. This "inverse residue criterion" enables us to make a precise determination of the reappearance point. It is complementary to the "residue criterion" for the determination of the disappearance point. For example, for $z = 4$, $k_R^{(1)} = 1.42173415$; and for $z = 6$, $k_R^{(1)} = 1.29946540$, $k_R^{(2)} = 1.45296340$. The superscript "i" in $k_{D,R}^{(i)}$ refers to the ith time the KAM tours disappears (D) or reappears (R). We have also computed the action difference and found $\Delta W \longrightarrow 0$ if $k < k_D^{(1)}$ or $k_R^{(1)} < k < k_D^{(2)}$; and $\Delta W \longrightarrow$ *constant* > 0 if $k_D^{(1)} < k < k_R^{(1)}$ or $k > k_D^{(2)}$.

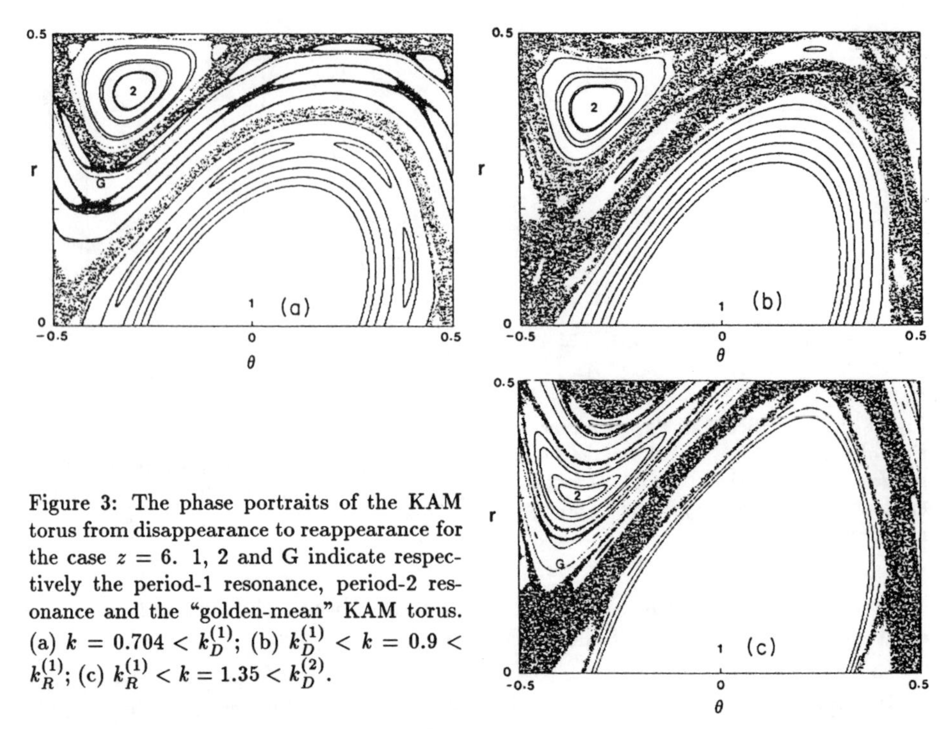

Figure 3: The phase portraits of the KAM torus from disappearance to reappearance for the case $z = 6$. 1, 2 and G indicate respectively the period-1 resonance, period-2 resonance and the "golden-mean" KAM torus. (a) $k = 0.704 < k_D^{(1)}$; (b) $k_D^{(1)} < k = 0.9 < k_R^{(1)}$; (c) $k_R^{(1)} < k = 1.35 < k_D^{(2)}$.

Fig. 2 shows a typical case. Fig. 3 shows the evolution of the phase portrait from disappearance to reappearance of the KAM torus for the case $z = 6$. In Fig. 3(a) $k < k_D^{(1)}$, the chaotic regions near the period-1 and the period-2 resonances are separated by the KAM torus. In Fig. 3(b) $k_D^{(1)} < k < k_R^{(1)}$, the KAM torus has disappeared and the chaotic regions become connected. In Fig. 3(c) $k_R^{(1)} < k < k_D^{(2)}$, the KAM torus has reappeared and the chaotic regions become separated again. These results together with the critical behavior to be discussed later suggest that $k_R^{(i)}$ are indeed the points at which the KAM torus reappears. Moreover, a KAM torus can recur more than once. For example, for $z = 6$, we have observed that the KAM torus has recurred at least twice. The residues are no longer monotonic functions of k. They tend to infinity right after the KAM torus has disappeared, and become finite again as it reappears. Since we cannot ascertain the existence of a "final" disappearance, the KAM torus can conceivably recur infinitely many times. We have also observed an exchange of stability before the first breakup of the KAM torus. Fig. 4 shows the variation of the residues with k. It is evident here that stability exchange does not necessarily entail reappearance. When z is a fraction, the dependence of the residues on k is quite complicated. A "bifurcation" of the regions in which the KAM torus exists has also been observed as z varies.

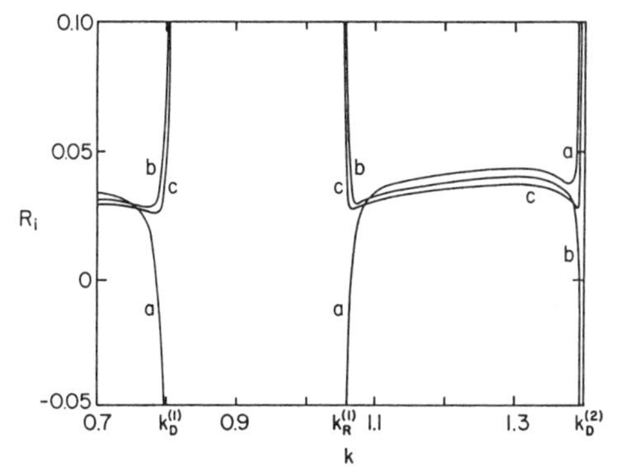

Figure 4: The residues R_i of the initially elliptic orbits for $z = 5$. (a) $(Q_i, P_i) = (233, 144)$; (b) $(377, 233)$; (c) $(610, 377)$. The residues of the initially hyperbolic orbits are symmetrically located and are not shown. Stability exchange occurs before the first breakup of the KAM torus for (a), but not for (b) and (c).

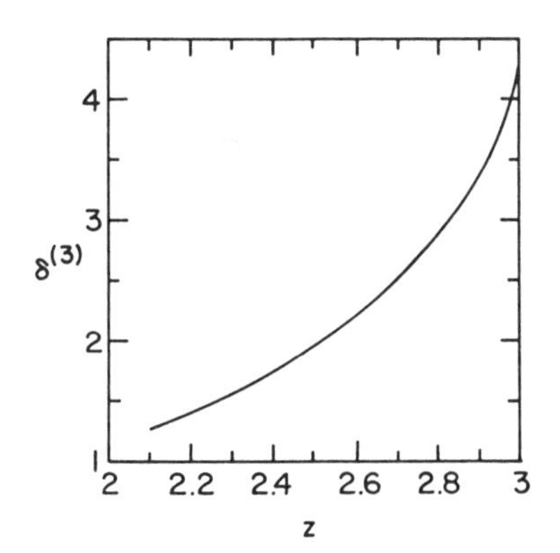

Figure 5: The parameter scaling exponent $\delta^{(3)}$ as a function of z for $2 \leq z \leq 3$.

III. LOCAL SCALING BEHAVIOR

The local critical behavior of a KAM torus in the standard map has been well understood by the method of renormalization.[2,3] One fixed point of the renormalization operator, called the simple fixed point, corresponds to the linear map. The other one, called the standard fixed point, corresponds to the analytic critical area-preserving twist map.[3] It is known from numerical studies that a broad class of area-preserving twist maps belongs to the same universality class for all quadratic irrationals with the same periodic pattern in the continued fraction expansion. Our results, however, indicate that the critical behavior is much more complicated than expected. We found that the critical exponents of the KAM torus vary with z when $2 \leq z < 3$ but are independent of z when $z \geq 3$. There is a crossover from the simple fixed point to the standard fixed point.

To discuss the critical behavior of the map (1), we first summarize the definitions of various critical exponents. For a pair of minimizing and minimax orbits with a winding number P_n/Q_n, the distance between two neighboring orbit points is $d_n = |\theta_n^{max} - \theta_n^{min}|$, where the superscripts "max" and "min" denote respectively the minimax and minimizing orbits. Period-1 scaling is defined by

$$\frac{d_n}{d_{n+1}} \sim (\frac{Q_n}{Q_{n+1}})^{-x^{(1)}},$$

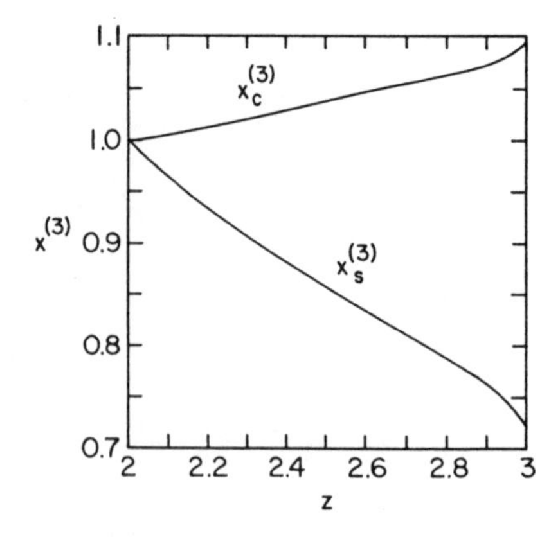

Figure 6: The critical exponents on the dominant symmetry line $x_s^{(3)}$ and the nondominant symmetry line $x_c^{(3)}$ as a function of z for $2 \leq z \leq 3$.

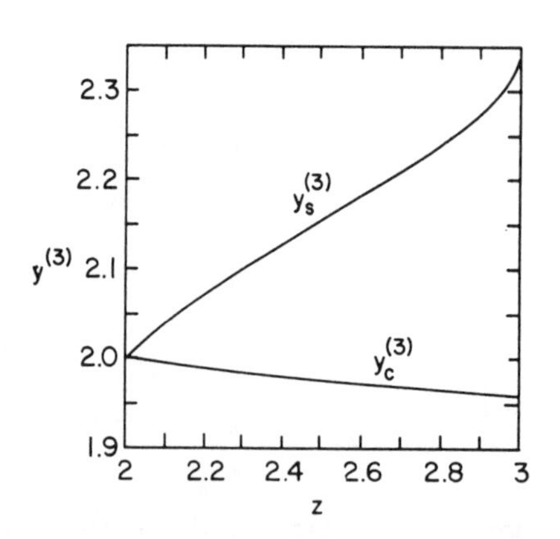

Figure 7: The critical exponents on the dominant symmetry line $y_s^{(3)}$ and the nondominant symmetry line $y_c^{(3)}$ as a function of z for $2 \leq z \leq 3$.

$$\frac{r_n - r_{n-1}}{r_{n+1} - r_n} \sim (\frac{Q_n}{Q_{n+1}})^{-y^{(1)}}; \tag{7}$$

and period-3 scaling

$$\begin{aligned} \frac{d_n}{d_{n+3}} &\sim (\frac{Q_n}{Q_{n+3}})^{-x^{(3)}}, \\ \frac{r_n - r_{n-3}}{r_{n+3} - r_n} &\sim (\frac{Q_n}{Q_{n+3}})^{-y^{(3)}}. \end{aligned} \tag{8}$$

Due to its better convergence, period-3 scaling is used. For convenience, the exponents on the dominant symmetry line are denoted by $x_s^{(i)}$ and $y_s^{(i)}$, and the exponents on the symmetry line l (l=a,b,c,d) are denoted by $x_l^{(i)}$ and $y_l^{(i)}$. In the standard map[2] it was found that $x_s^{(1)} = 0.721$, $y_s^{(1)} = 2.329$, $x_a^{(3)} = 1.093$, $y_a^{(3)} = 2.329$, and $x_s^{(1)} + y_s^{(1)} = x_a^{(3)} + y_a^{(3)} = 3.05$. The convergence rate of the parameter defines another critical exponent. Again period-3 scaling is used

$$\delta^{(3)} = \frac{k_{n+1} - k_n}{k_{(n+3)+1} - k_{n+3}}. \tag{9}$$

$\delta^{(3)}$ is related to the usual δ by $\delta^{(3)} = \delta^3$. In the standard map $\delta^{(3)} = (1.628)^3$.

For $k < 0$ or $z \leq 2$ the critical point is $k_D = 0$. The system is integrable, and the critical exponents can be calculated analytically: $x = 1$ and $y = 2$. This is just the critical behavior of a linear system. For $2 < z < 3$, we found that the critical exponents at the critical point k_D vary with z (see Table I and Figs. 5-7). For $z = 3$ the critical behavior is the same as that of the standard map. Therefore, the critical behavior changes from that of the linear map to

Table I. The critical points k_D, the critical exponents and the endpoints of the $f(\alpha)$ curves for $2 \leq z \leq 3$. $x_i^{(3)}$ and $y_i^{(3)}$ are the exponents on the dominant symmetry line for $i = b$ and the nondominant symmetry line for $i = a$ respectively.

z	k_D	$\delta^{(3)}$	$x_a^{(3)}$	$y_a^{(3)}$	$x_a^{(3)} + y_a^{(3)}$	$x_b^{(3)}$	$y_b^{(3)}$	$x_b^{(3)} + y_b^{(3)}$	α_{min}	α_{max}
2.0	0	—	1	2	3	1	2	3	1	1
2.1	0.219	1.27	1.021	1.978	2.999	0.965	2.036	3.001	0.979	1.036
2.2	0.391	1.42	1.032	1.969	3.001	0.935	2.067	3.002	0.969	1.070
2.3	0.5375	1.55	1.040	1.963	3.003	0.907	2.098	3.005	0.962	1.103
2.4	0.6617	1.75	1.047	1.958	3.005	0.881	2.127	3.008	0.955	1.135
2.5	0.76828	1.97	1.053	1.956	3.009	0.858	2.155	3.013	0.950	1.166
2.6	0.86037	2.23	1.060	1.954	3.014	0.835	2.181	3.016	0.943	1.198
2.7	0.94034	2.53	1.068	1.952	3.020	0.810	2.211	3.021	0.936	1.235
2.8	1.010114	2.88	1.075	1.950	3.025	0.788	2.239	3.027	0.930	1.269
2.9	1.071375	3.36	1.084	1.950	3.034	0.763	2.271	3.034	0.923	1.311
3.0	1.125454	4.25	1.100	1.947	3.047	0.721	2.330	3.051	0.909	1.387

Table II. The disappearance (k_D) and reappearance (k_R) points and the critical exponents for $z > 3$.

z	k	$x_s^{(3)}$	$y_s^{(3)}$	$x_s^{(3)} + y_s^{(3)}$	$x_n^{(3)}$	$y_n^{(3)}$	$x_n^{(3)} + y_n^{(3)}$
3.8	$k_D^{(1)} = 1.38253450$	0.7284	2.3261	3.0545	1.0990	1.9448	3.0438
	$k_R^{(1)} = 1.38760367$	0.7226	2.3295	3.0521	1.1018	1.9512	3.0530
	$k_D^{(1)} = 1.41293530$	0.7219	2.3287	3.0506	1.1045	1.9419	3.0464
4	$k_R^{(1)} = 1.42173415$	0.7223	2.3281	3.0504	1.1015	1.9466	3.0481
	$k_D^{(2)} = 1.42615570$	0.7203	2.3329	3.0533	1.0986	1.9425	3.0412
	$k_D^{(1)} = 0.80993000$	0.7234	2.3281	3.0515	1.1030	1.9432	3.0462
5	$k_R^{(1)} = 1.05287350$	0.7216	2.3288	3.0504	1.1040	1.9432	3.0472
	$k_D^{(2)} = 1.39647420$	0.7221	2.3285	3.0505	1.1044	1.9422	3.0466
	$k_D^{(1)} = 0.70400046$	0.7218	2.3289	3.0507	1.1046	1.9421	3.0467
	$k_R^{(1)} = 1.29946540$	0.7227	2.3284	3.0511	1.1037	1.9432	3.0470
6	$k_D^{(2)} = 1.43731867$	0.7228	2.3304	3.0532	1.1121	1.9327	3.0448
	$k_R^{(2)} = 1.45296340$	0.7229	2.3303	3.0533	1.1121	1.9326	3.0447
	$k_D^{(3)} = 1.51257548$	0.7221	2.3293	3.0513	1.1024	1.9454	3.0478
11	$k_D^{(1)} = 0.27830553$	0.7211	2.3303	3.0515	1.1021	1.9462	3.0483

that of the standard map as z varies from 2 to 3. The sum of the exponents, $x+y$, which is a more useful quantity in the study of transport, shows a slightly increasing trend. However, the increase is too small to exclude the possibility that it is in fact a constant. Since δ decreases with z, higher-period orbits are needed to compute the exponents. As z is close to 2, the convergence of δ_n towards δ is slowed down, and the exponents are very hard to compute. When $z \geq 3$, the exponents at the disappearance and reappearance points are equal and the same as those of the standard map. They are independent of z (see Table II). In this sense the degree of inflection z plays a role quite similar to that of dimensionality in phase transitions with $z = 2$ and 3 corresponding respectively to the lower and upper critical dimensions.

IV. GLOBAL SCALING BEHAVIOR

At the critical point, the KAM torus loses its analyticity and breaks up into a Cantor set called a "cantorus". The full complexity of its scaling structure can be described only by a spectrum of critical exponents α and their densities $f(\alpha)$, of which the exponents x and y comprise only a part.[11,12] To compute $f(\alpha)$ we divide the torus into pieces labeled by an index i which runs from 1 to N. The size of the ith piece is l_i and the density of points in l_i is p_i. α is defined by $p_i = l_i^{\alpha}$. It lies in an interval $[\alpha_{min}, \alpha_{max}]$, the endpoints of which are determined by the local scaling behavior. $f(\alpha)$ is the Hausdorff dimension of the set of points having exponent α.

To calculate $f(\alpha)$ of the cantorus, we follow the formalism described in Ref. 12. A periodic orbit with a winding number $\omega_n = F_{n-1}/F_n$, is divided into $N = F_n$ pieces consisting of two nearby orbit points. The Euclidean distance between these two points $l_i(n) = \sqrt{(r_i - r_{i+F_{n-1}})^2 + (\theta_i - \theta_{i+F_{n-1}})^2}$ defines a natural scale for the partition with a measure $p_i = 1/F_n$. Then a partition function can be defined

$$\Gamma_n(q,\tau) = \sum_{i=1}^{F_n} \frac{p_i^q}{l_i^\tau} = F_n^{-q} \sum_{i=1}^{F_n} l_i^{-\tau}. \tag{10}$$

For an asymptotically recursive structure, the partition function is of order unity when

$$\tau = (q-1)D_q. \tag{11}$$

D_q is a set of generalized dimensions, of which D_0 is the Hausdorff dimension, D_1 the information dimension and D_2 the correlation dimension. The spectrum $f(\alpha)$ is defined by the Legendre transformation of $\tau(q)$,

$$\alpha(q) = \frac{d\tau}{dq}, \tag{12}$$

$$f(q) = q\alpha(q) - \tau. \tag{13}$$

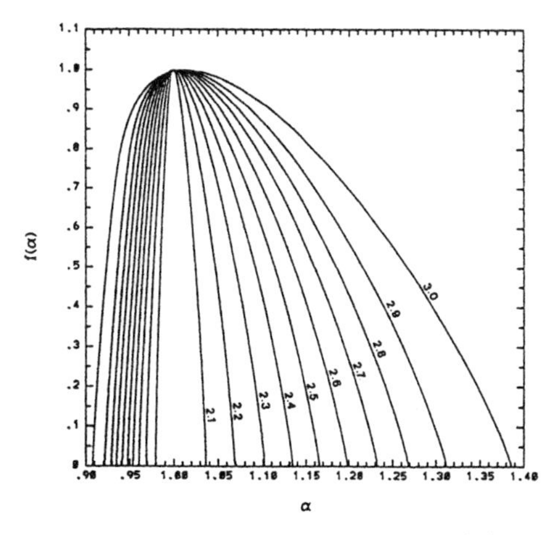

Figure 8: The singularity spectrum $f(\alpha)$ of the critical KAM torus for $2 \leq z \leq 3$. The number on the curve indicates the value of z.

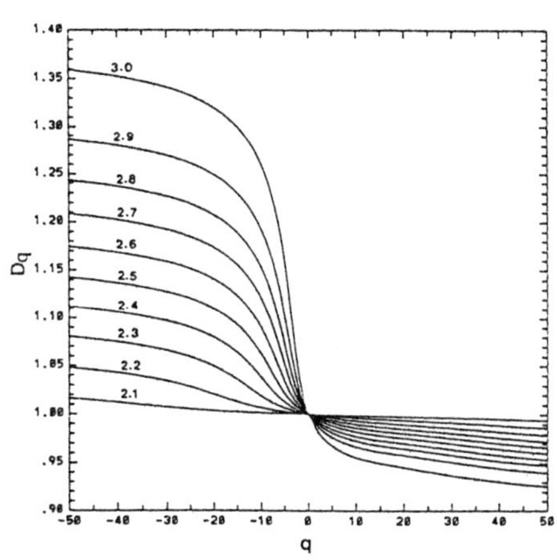

Figure 9: The generalized dimensions D_q of the critical KAM torus for $2 \leq z \leq 3$. The number on the curve indicates the value of z.

Eliminating q gives the function $f(\alpha)$. To improve the convergence of $f(\alpha)$ as $\omega_n \to \omega$, one can employ the usual ratio trick

$$\frac{\Gamma_n(q,\tau)}{\Gamma_{n-1}(q,\tau)} = 1 \tag{14}$$

For the map (1), we will use period-3 ratio due to its better convergence

$$\frac{\Gamma_n(q,\tau)}{\Gamma_{n-3}(q,\tau)} = 1. \tag{15}$$

From Eq. (12), q and α can be calculated as functions of τ

$$q = \frac{1}{ln(\frac{F_{n-1}}{F_n})}\left[ln(\sum_{i=1}^{F_{n-1}} l_i^{-\tau}(n-1)) - ln(\sum_{i=1}^{F_n} l_i^{-\tau}(n))\right] \tag{16}$$

$$\alpha = ln(\frac{F_{n-1}}{F_n})\left[\frac{\sum_{i=1}^{F_n} l_i^{-\tau}(n) ln(l_i(n))}{\sum_{i=1}^{F_n} l_i^{-\tau}(n)} - \frac{\sum_{i=1}^{F_{n-1}} l_i^{-\tau}(n-1) ln(l_i(n-1))}{\sum_{i=1}^{F_{n-1}} l_i^{-\tau}(n-1)}\right]^{-1} \tag{17}$$

We have calculated $f(\alpha)$ at the disappearance and reappearance points for a variety of values of z. The $f(\alpha)$ curves are the same as that of the standard map[13] and are independent of z for $z \geq 3$, but they vary with z for $2 \leq z < 3$. Fig. 8 plots the $f(\alpha)$ curves for $2 \leq z \leq 3$. It shows that as z varies from 3 to 2 the $f(\alpha)$ curves shrink from that of the standard map to that of the linear case. Fig. 9 plots D_q versus q for $2 \leq z \leq 3$. For $z = 2$, the critical case has a trivial scaling with a single dimension $D = 1$. For $z = 3$, the critical case is the same as that of the standard map which has a set of fractal dimensions lying in the interval $[D_{-\infty}, D_{\infty}]$. As z varies from 3 to 2, the critical KAM torus changes smoothly from a cantorus characterized by the standard fixed point to a smooth torus characterized by the simple fixed point.

The endpoints of the $f(\alpha)$ curves can be determined exactly in terms of the local critical exponents. In the partition function (10), as $q \to -\infty$, the dominant contribution comes from the largest l_{max}, corresponding to the most rarefied region of the cantorus; as $q \to \infty$, the dominant contribution comes from the smallest l_{min}, corresponding to the most concentrated region of the cantorus. The most rarefied region is on the dominant symmetry line, and the most concentrated region is on one of the nondominant symmetry lines. For $2 < z < 3$, they are $\theta = 0$ and $\theta = -1/2$. Therefore, from Eq. (15),

$$\lim_{q \to -\infty} \frac{\Gamma_n}{\Gamma_{n-3}} = \lim_{q \to -\infty} \left(\frac{F_n}{F_{n-3}}\right)^{-q} \left(\frac{d_n(0)}{d_{n-3}(0)}\right)^{-\tau}, \tag{18}$$

$$\lim_{q \to \infty} \frac{\Gamma_n}{\Gamma_{n-3}} = \lim_{q \to \infty} \left(\frac{F_n}{F_{n-3}}\right)^{-q} \left(\frac{d_n(-\frac{1}{2})}{d_{n-3}(-\frac{1}{2})}\right)^{-\tau}. \tag{19}$$

Together with Eq. (8), we obtain,

$$\alpha_{max} = D_{-\infty} = \frac{1}{x_b^{(3)}}, \tag{20}$$

$$\alpha_{min} = D_{\infty} = \frac{1}{x_a^{(3)}}. \tag{21}$$

In Table I, the variation of α_{max} and α_{min} with z is listed for $2 \leq z \leq 3$. Excellent agreement with those obtained from the $f(\alpha)$ curves can be seen.

V. CONCLUDING REMARKS

The study undertaken here has revealed surprisingly novel behaviors of KAM tori. The resemblance to phase transitions is really amazing. The degree of inflection plays a role very similar to that of dimensionality in phase transitions. Recurrence of KAM tori is another new phenomenon that has analogies in, for example, the ANNNI model in commensurate-incommensurate phase transitions. This phenomenon may be relevant to physical systems exhibiting a sequence of metal–insulator transitions.[8] All these unexpected new features suggest that the behavior of KAM tori may indeed be far more complicated than we have been accustomed to believe. It is thus worthwhile to conduct a more thorough study. Such a study will lead to not only a more complete understanding of KAM tori but also their application to real physical systems.

REFERENCES

1. J. M. Greene , J. Math. Phys. **20**, 1183 (1979).
2. L. P. Kadanoff, Phys. Rev. Lett. **47**, 1641(1981); S. J. Shenker and L. P. Kadanoff, J. Stat. Phys. **27**, 631 (1982).

3. R. S. MacKay, Physica D **7**, 283(1983).

4. B. Hu and J. M. Mao, Phys. Rev. A **25**, 3259(1982); B. Hu and I. Satija, Phys. Lett. A **98**, 143(1983).

5. S. Ostlund, D. Rand, J. Sethna and E. Siggia, Physica D **8**, 303(1983).

6. B. Hu, A. Valinia and O. Piro, Phys. Lett. A (1990).

7. J. N. Mather, Topology **21**, 457 (1982); Erg. Theory Dyn. **4**, 301 (1984).

8. H. J. Schellnhuber and H. Urbschat, Phys. Rev. Lett. **54**, 588(1985).

9. J. Wilbrink, Physica D **26**, 385 (1987).

10. J. A. Ketoja and R. S. MacKay, Physica D **35**, 318(1989).

11. B. Hu, J. Shi and S. Y. Kim, J. Stat. Phys. **62**, 631(1991); Phys. Rev. **43**, 4249(1991).

12. T. C. Halsey, M. H. Jensen, L. P. Kadanoff, I. Procaccia and B. I. Shraiman, Phys. Rev. A **33**, 1141(1986).

13. A. H. Osbaldestin and M. Y. Sarkis, J. Phys. A **20**, L953(1987).

14. J. Shi and B. Hu, Phys. Lett. A (to appear).

MAPPINGS IN HIGHER DIMENSIONS

M.P. Bellon, **J-M. Maillard**, C-M. Viallet

ABSTRACT

We introduce non trivial two-dimensional and three-dimensional mappings. These mappings are birational transformations, the iterates of which give a non linear representation of infinite Coxeter groups. These mappings originate from integrable models in statistical mechanics. They exhibit a number of remarkable properties. Some have algebraic invariants allowing for orbits lying on some smooth manifolds. Other give examples of mappings the iterates of which are Hofstadter- like patterns and exemplify onset of order from disorder (Saturn's ring).

These examples are not avatars of the well-known trace mappings.

PACS: 05.50, 05.20, 02.10, 02.20
AMS classification scheme numbers: 82A68, 82A69, 14E05, 14J50, 16A46, 16A24, 11D41
Key-words: Iterated mappings, Dynamical systems, Integrable models, Coxeter groups, Birational transformations, Cremona transformations, Inversion relations, Automorphisms of algebraic varieties, Deformations of algebraic varieties, Elliptic curves.

work supported by CNRS

Postal address: Laboratoire de Physique Théorique et des Hautes Energies
Université de Paris 6, Tour 16, 1er étage, boîte 126.
4, Place Jussieu/ F-75252 PARIS Cedex 05

It is not necessary to recall the attention paid to investigations of different ways in which chaos can arise in strongly deterministic dynamical systems (see for instance James Gleick "Chaos" 1987 The Viking Press, New York) As models for numerical analysis one or two dimensional maps are often considered : it is not either necessary to recall the success of all kinds of Feigenbaumology [1] or the celebrated Hénon map [2]

$$x_{n+1} = y_n + 1 - Ax_n^2$$
$$y_{n+1} = Bx_n$$

However very few examples of higher dimensional map are available in the literature for which "something interesting" can be said. One of these very few example is the trace-map which is linked to systems which are neither periodic nor random (i.e. neither crystalline nor amorphous) and for which a recursion formula expresses the existence of an exact renormalization group structure in the model [3].

Recalling L. Kadanoff [4] "It is easy, too easy, to perform computer simulations upon all kinds of models and to compare the results with each other and with real-world outcomes. But without organizing principles, the field tends to decay into a zoology of interesting specimens and facile classifications".

We propose here a completely new set of mappings : they are n-dimensional maps (n arbitrary) depending on m-parameters (m arbitrary) and these mappings remarkably **preverse algebraic varieties of various codimensions**. The construction of these mappings could not have easily been found ex nihilo. Considerations about exactly solvable models in lattice statistical mechanics are essential in their building up. We will not explain here the very origin of these mappings and let the reader to refer to parallel publications [5], [6], [7], [8]. We will only give some examples and comment the associated structures and results.

Let us just mention that these mappings are **birational** transformations **generated by a finite set of involutions** $I_1, I_2, ...I_r$. For $r = 2$ the two involutions generate a (generically) infinite order birational transformation $I_1 I_2$ and we recover the well-known iteration of a mapping. We do have examples with more than two involutions were the number of infinite order birational transformations no longer reduces to one, which have no relation between them and which remarkably preserve non trivial algebraic varieties [8].

In the simplest cases this infinite set of birational transformations can be seen as a non linear representation of **affine Weyl group** (Coxeter groups [7,8]). They are of paramount relevance in the analyzis of integrable models in Statistical Mechanics, or Field Theory (quantum groups...) and we do believe they form of class of transformations large enough to be a powerfull tool in the analyzis of all kinds of physical problems (dynamical problems, hydrodynamics...) because these higher dimensional mappings do correspond to non trivial group structure, because they can have many parameters, because they enable to recover the well-known "zoology" of behaviour from complete integrability to chaotical strange-attractor structures (Hofstadter butterfly...) and finally because they can present order from disorder phenomena and lead to an interesting and natural deformation theory of these structures.

Let us now give some examples

A two-dimensional map

Let us introduce the following (bi)rational mappings:

$$J : u \rightarrow 1/u \quad , v \rightarrow 1/v$$

$$I : u \rightarrow \frac{-u - u^2 + 2v^2}{1 + u + 2v - u^2 - 2uv - v^2}, \quad v \rightarrow \frac{u^2 + uv - v^2 - v}{1 + u + 2v - u^2 - 2uv - v^2}$$

I and J are both involutions and one can even see that they are conjugated by a collineation :

$$u \longrightarrow \frac{1 - v}{1 + 2u + 3v}, \quad v \longrightarrow \frac{1 - u}{1 + 2u + 3v}$$

The iteration of the birational transformation IJ remarkably reads to orbits lying on **algebraic curves** (see figure 1). These curves actually form a (linear) pencil of curves given by the equation

$$\Delta = \frac{(2v^2 + 2uv - u^2 - 2u^3 - 2vu^2 + v^2u)(u - v^2)^2}{(u + v)^4(1 - u)(1 - v)^2} \tag{1}$$

where Δ is some constante.

Some arguments of algebraic geometry (existence of an infinite set of automorphisms of a curve) enable to argue that these curves are necessarily of genus 0 or 1 [9]. These curves are actually genus one curves except for a **finite number of values of** $\Delta(\Delta = 3/16...)$ where one has genus 0 curves (one can actually proove that this is the only possible situation).

Moreover subvarieties made out of points having **finite orbits** are **automatically stable** by I and J. Such subvarieties, $(IJ)^r = e$, are actually curves of the pencil. For instance $r = 3$ corresponds to the component $u - v^2 = 0$ of the curve $\Delta = 0$. $r = 6$ corresponds to the whole curve $\Delta = 0$. $r = 5$ corresponds to the curve $\Delta = -1$. For higher values of r, there will be different curves of points of order r corresponding to fixed values of Δ. For $r = 7$, there are two curves with $\Delta = \frac{1}{2}(7 \pm 3\sqrt{5})$. For $r = 8$, we have two values $\Delta = \frac{1}{2}(3 \pm \sqrt{7})$. For $r = 9$, the values of Δ are the three roots of the polynomial $x^3 + 3x^2 - 6x + 1$. For $r = 10$, we have naturally the solution of $r = 5$ and

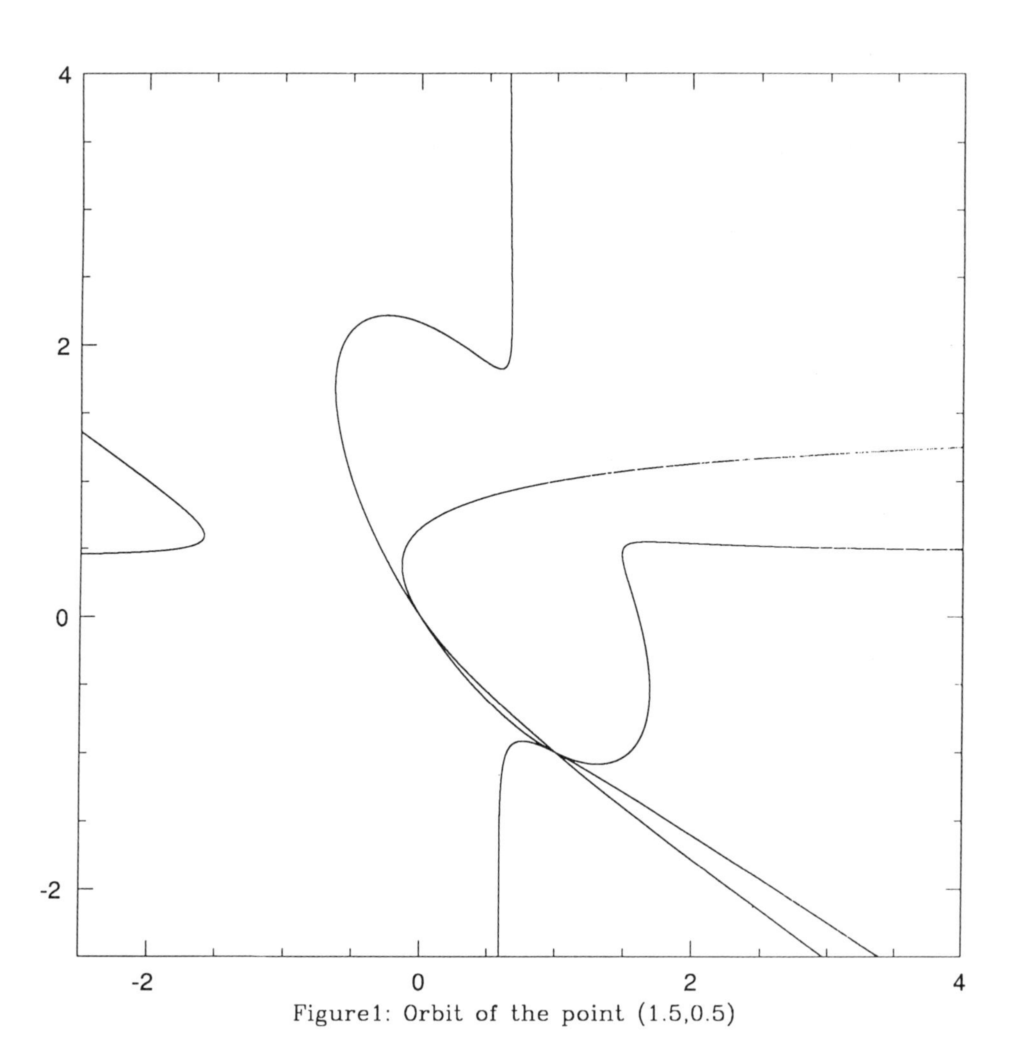

Figure1: Orbit of the point (1.5,0.5)

in addition the values of Δ corresponding to the three roots of $x^3 - 9x^2 + 7x - 1$. Finally the point of order $r = 11$ lie on the curves with Δ solution of the fifth degree equation $x^5 - 13x^4 + 55x^3 - 17x^2 - 4x + 1$.

Introducing an elliptic parametrization [10] IJ is translation $\theta \to \theta + \lambda$ of the uniformizing (spectral) parameter. We have here a situation equivalent to translation on the circle S^1 with a shift (generically) not commensurate with 2π. It is thus natural to densify (generically) the elliptic curve (1) by iteration of IJ and it is then also natural to be able to extend this infinite **discrete** group action into a **continuous** group $\theta \to \theta + d\theta$ which enables to write down differential equations for a running point of curve (1):

$$\frac{du}{dt} = \phi(u,v)(9u^2v^2 - 2v^2u - 9uv^3 + 3u^3v + u^2v^3 - 9u^3v^3 + 3uv^4 - 6u^4v - 2u^3v^3 + 6vu^2 - 4v^3 - 6v^4 + 3u^4 + 3v^5 - 4u^4v^2 + 6u^2v^4)(v-1)$$

$$\frac{dv}{dt} = \phi(u,v)(-2v^2u - u^2v^2 - 3uv^3 + 4u^3v + 4u^2v^3 - u^3v^2 + 2uv^4 - 3u^4v - u^3v^3 - vu^2 + 3u^3 - 2v^4 + 2u^4 - 2u^5 - 2u^4v^2 + 3u^2v^4)(u-1).$$

where $\phi(u,v)$ is an arbitrary function.

The limiting density of points on (1) is of course given by $d\theta = cste$ which can be rewritten in terms of u and v.

Deformation of a two dimensional mapping

Let us consider another involution I which, combined with J, also lead to a linear pencil of elliptic curves :

$$I : u \to \frac{-u + uv - u^2 + v^2}{1 + u + v - uv - v^2 - u^2}, \quad v \to \frac{-u + uv - v^2 + u^2}{1 + u + v - uv - u^2 - v^2}$$

I and J can be seen to be conjugated by a collineation (which an involution):

$$u \to \frac{1 + (\omega + \omega^4)u + (\omega^2 + \omega^3)v}{1 + 2u + 2v}, \quad v \to \frac{1 + (\omega^2 + \omega^3)u + (\omega + \omega^4)v}{1 + 2u + 2v}$$

where ω is a fifth root of unity.

One can envisage a great number of deformations preserving the involutive character of I and J. For simplicity we will keep I fixed and deform J into $\widetilde{J}_a$:

$$\widetilde{J}_a = u \to a/u, \quad v \to a/v$$

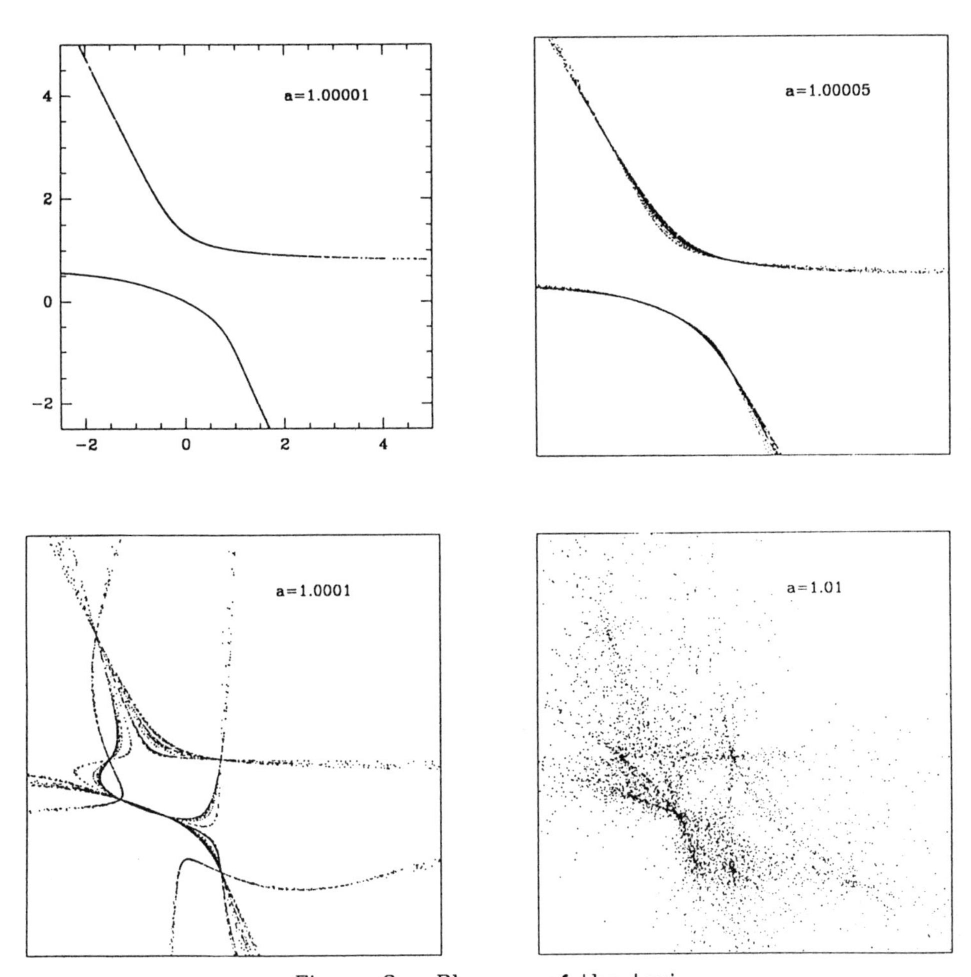

Figure 2 : Blow up of the tori

Figures 2 show how the initial invariant curve deforms with increasing values of a. For very small deformations ($a = 1 + 5 \cdot 10^{-5}$), the curve is slightly deformed. For $a = 1 + 10^{-4}$, the situation is drastically modified. The trajectories seem to wander between curves of the linear pencil previously described, with different values of the invariant Δ.

A chaotic two-dimensional mapping

I is now replaced by

$$I : u \longrightarrow \frac{-u + 2v^2 - u^2}{1 + 2u + 2v - u^2 - v^2 - 3uv}, v \longrightarrow \frac{-v - v^2 + 2u^2}{1 + 2u + 2v - u^2 - v^2 - 3uv}$$

I is also conjugate of J by a collineation.

The orbits of IJ are given on figure 3. There is clearly no curve. There is an extremely fast accumulation of numerical errors in the iteration. This is the reason why an iteration with 200 digits and another one with 300 digits has been performed. There is no discrepancy up to 20000 iterations for the orbit of figure 3. Figure (3) gives 20000 points and Figure (4) gives a magnification near the $u = 1$ axis. This last figure shows that this "Hoffstadter-butterfly like" orbit has a remarkable structure at an infinite set of special points $u = 1$, $v = \sin((n+1)\alpha)/\sin(n\alpha)$ with $\tan\alpha = \sqrt{7}$. This will be detailed elsewhere [10].

Mappings for the product of two two-dimensional space

In the previous examples we have introduced various birational transformations $R = IJ$:

$$u \rightarrow R_u(u,v), \quad v \rightarrow R_v(u,v)$$

Let us consider the action of R on a double copy of the parameter space $((u,v),(\bar{u},\bar{v}))$ defined as follows : one acts with R on (u,v) and with its inverse R^{-1} on $(\bar{u},\bar{v})$

$$\bar{u} \rightarrow (R^{-1})_u(\bar{u},\bar{v}), \quad \bar{v} \rightarrow (R^{-1})_v(\bar{u},\bar{v})$$

We look at the orbits of this particular action of R in the $(u,\bar{u})$ plane.

Remarkably these orbits are (algebraic and even elliptic) curves for the first two examples of two-dimensional mappings if the two initial point (u_o, v_o) and $(\bar{u}_o, \bar{v}_o)$ belong to the **same curve** of the linear pencil.

This can easily be understood recalling the elliptic parameters θ and $\bar{\theta}$ on the two copies : the curve $\theta + \bar{\theta} = cste$ is clearly invariant under the previous $R \times R^{-1}$ action.

In contrast the same $R \times R^{-1}$ action lead to drastically different orbits for the third "chaotic" example (see figure 5).

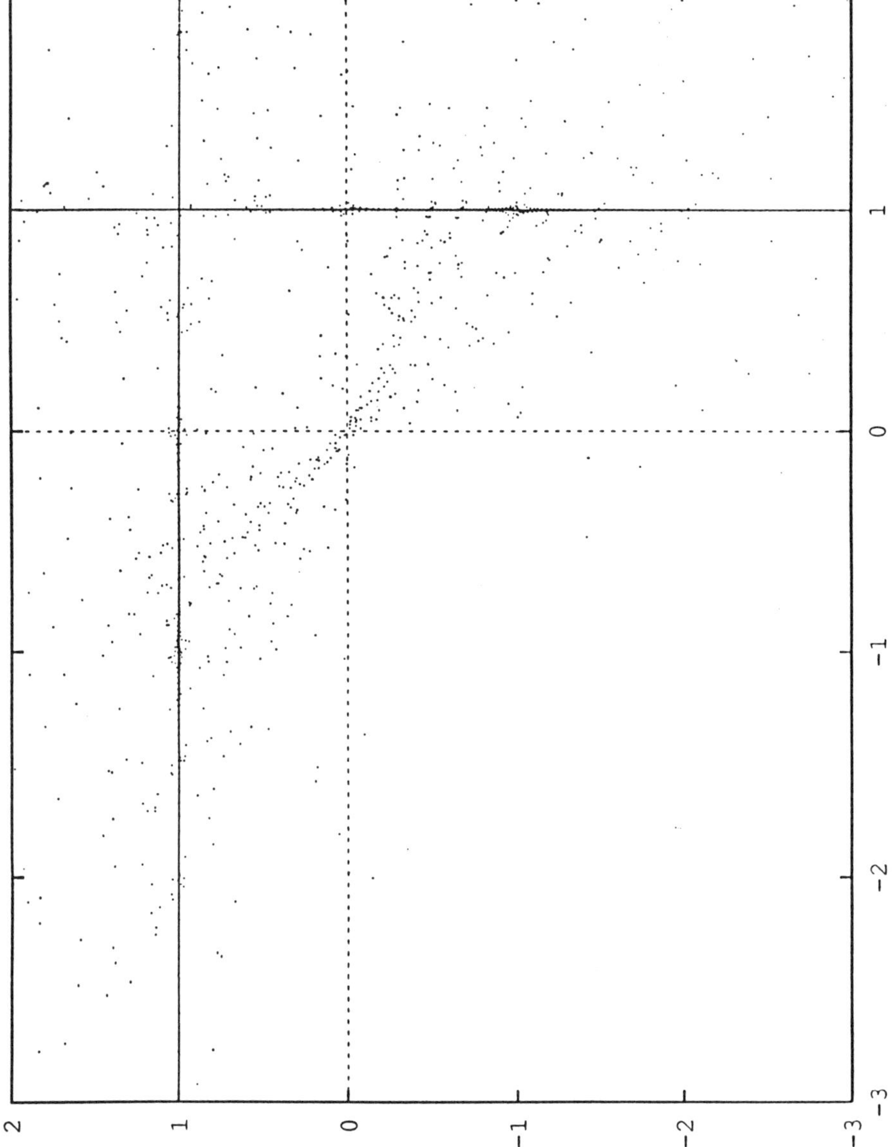

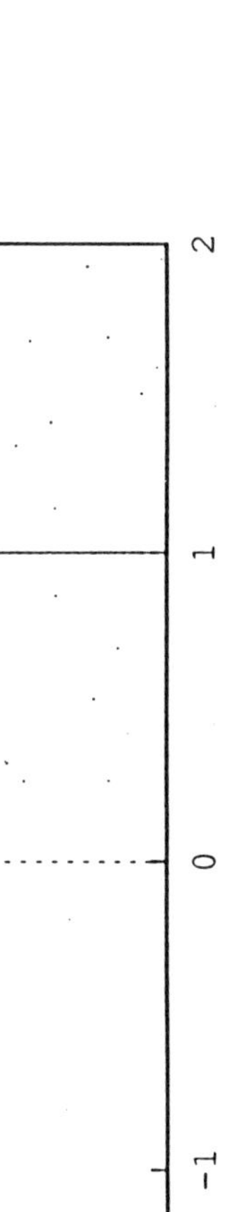

Figure 3: A chaotic orbit

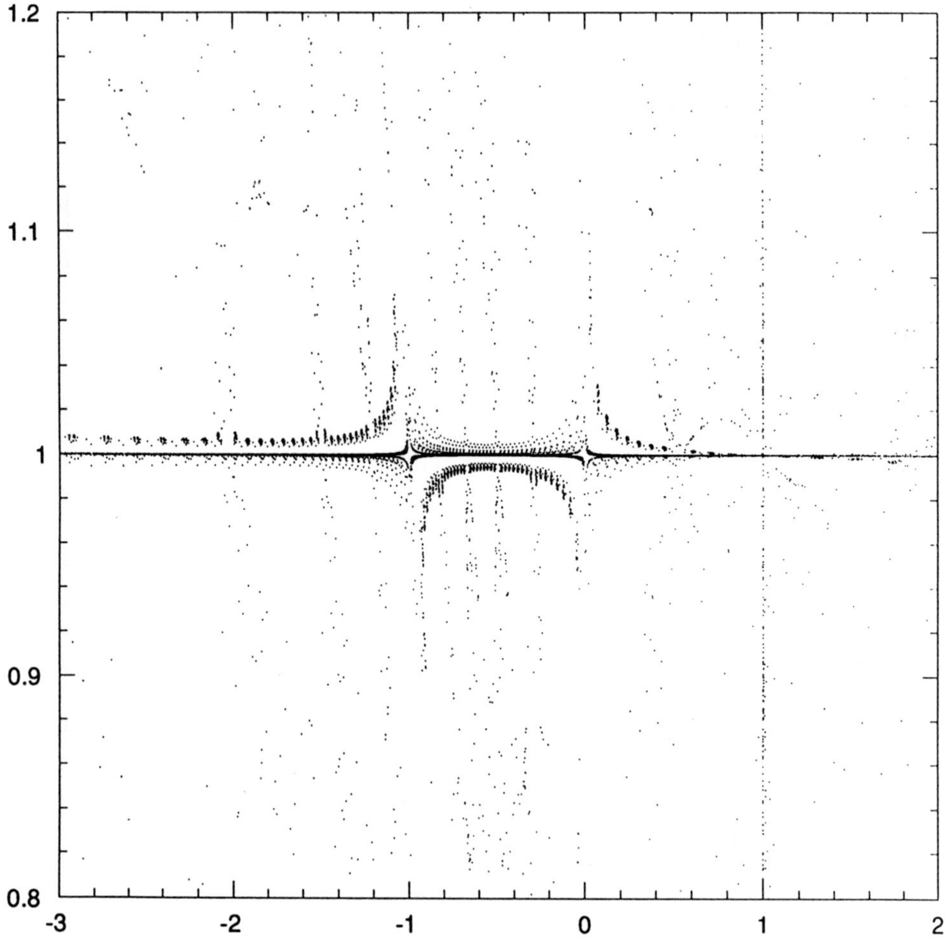

Figure 4: Magnification near u = 1

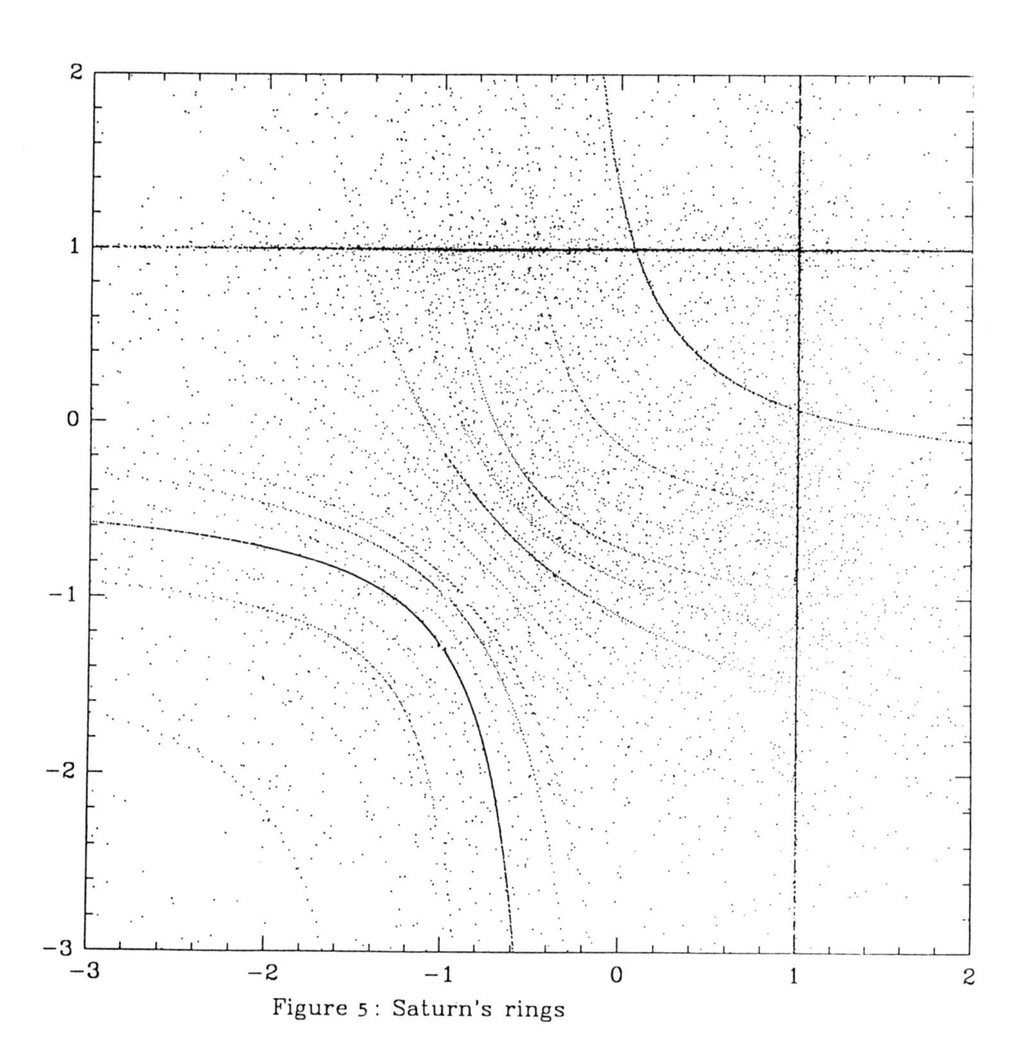

Figure 5: Saturn's rings

Figure (5) shows a remarkable structure : when one observes the accumulation of the points, one sees that the hyperbolae appear one after the other over an already constituted background. A very heuristic explanation of this emergence of order from disorder is the following : with the same notations as in [1], the two lines $u = 1$ and $v = 1$ are exchanged under the action of IJ, and contain an infinite set of remarkable attractive-repulsive points $(\sin((n+1)\alpha)/\sin(n\alpha)$, with $\tan(\alpha) = \sqrt{7})$. When the point (u, v) roves about the plane, it eventually passes nearby one of these highly unstable points. This stabilizes for a while the trajectories near their exact values. When restricted to the two lines $u = 1$ and $v = 1$, one can easily see that transformation IJ is an homography of infinite order. Orbits in the $u, \bar{u}$ plane are then hyperbolae as shows a calculation similar to the one of [11].

More non trivial mappings

Recalling the logistic map

$$x_{n+1} = rx_n(1 - x_n)$$

One has a particular value or r, that is $r = 4$, for which the mapping becomes completely ergodic. This can be understood easily introducing the rational parametrization $x_n = \sin^2\theta_n$ which leads $\theta_{n+1} = 2\theta_n$. Such a mapping clearly densify the circle S^1 for almost all initial value of θ_o. This can be easily generalized to mappings with elliptic parametrization for instance

$$x_{n+1} = r\ x_n(1 - x_n)\frac{(1 - k^2 x_n)}{1 - k^2 x_n^2}$$

which also reads $\theta_{n+1} = 2\theta_n$ with the parametrization $x_n = sn^2\theta_n$ where sn is the elliptic sinus of modulus k.

Such a phenomenon can obviously be transposed on the previously detailed examples of linear pencil of elliptic curves.

For simplicity let us consider a very simple example (associated to the six vertex model) for which a rational parametrization occur [12]: with the usual canonical notations

$$a = \sin\theta$$

$$b = \sin(\lambda + \theta)$$

$$c = \sin\lambda$$

the previous group generated from two involutions I and J reads $\theta \to \pm(\theta \pm n\lambda)$ and "ergodify" the circle S^1 (see [6]).

Let us consider this second way ($\theta \to 2\theta$) to densify the circle S^1. Introducing the notations $x = \frac{a}{c}$ and $y = \frac{b}{c}$ lead to consider the following two-dimensional map:

$$x \to -2xy - \frac{x}{y}(y^2 + x^2 - 1)$$
$$y \to y^2 - x^2$$

The orbits of this two-dimensional map similarly to the situation of figure (1) will lie on the linear pencil:

$$\frac{a^2 + b^2 - c^2}{ab} = cste = \frac{x^2 + y^2 - 1}{xy}$$

Similar calculations leading to new non trivial mappings can be done as soon as we get linear pencil for the iterates of IJ.

This can even be generalized combining, in a random way, the shift $\theta \to \theta + \lambda$ and transformation $\theta \to 2\theta$. This will lead to an affine (discrete) group which is the semi-direct product of two Z groups ($\theta \to \theta + n\lambda$ and $\theta \to 2^m\theta$). The corresponding transformations are rational transformations for $m \geq 0$.

A nine-dimensional mapping

Many other examples can be given such as three-dimensional mappings the iterates of which are (algebraic, elliptic) curves [5] or four-dimensional mappings the orbits of which lie on (algebraic) surfaces [7]. All kinds of behaviour can be observed (more or less chaotic, contracting towards algebraic subvarieties,...). The case where **finite order** orbits occur is highly interesting and deserves a particular attention [5,7,10] but will not be detailed here.

Let us finally give a more amazing example of birational transformations generically without relations between them, in nine dimensions which preserve algebraic varieties. This example comes from the quest for solutions of the so-called tetrahedron equations [11], [12], [13].

Let us first introduce the following notations for the entries of the 4×4 matrix

$$m = \begin{pmatrix} a & d_1 & d_2 & d_3 \\ d_1 & b_1 & c_3 & c_2 \\ d_2 & c_3 & b_2 & c_1 \\ d_3 & c_2 & c_1 & b_3 \end{pmatrix}$$

Let us denote $a', d_1', d_1' \cdots b_3'$ the entries of the inverse of the previous 4×4 matrix.

We introduce a first involution I_1 which corresponds to associate to each entry of the initial 4×4 matrix m the entry of the inverse matrix m^{-1}

$$I_1 : a \to a', \; d_1 \to d_1', \; d_2 \to d_2', \cdots b_3 \to b_3'$$

This is of course a birational transformation.

We also introduce three other involutions corresponding to conjugate I_1 by some permutations P of the ten (homogeneous) variables $a, \cdots b_3$ namely

$$P : d_1 \to d_2 \to c_1 \to c_2 \to d_1, \; d_3 \longleftrightarrow c_3, \; b_1 \longleftrightarrow b_2$$

a and b_3 being fixed. P is a transformation of order four.

The three involutions I_2, I_3, I_4 are $I_2 = P I_1 P^{-1}$, $I_3 = P^2 I_1 P^{-2}$, $I_4 = P^3 I_1 P^{-3}$. Almost equivalently we could have also defined P_i simply by the exchange of c_i and d_i and we could have introduced three involutions I_2, I_3, I_4 by $I_2 = P_1 I_1 P_1^{-1}, I_3 = P_2 I_1 P_2^{-1}, I_4 = P_3 I_1 P_3^{-1}$.

The algebraic variety given by the equation

$$a(b_1 + b_2 + b_3) + b_1 b_2 + b_2 b_3 + b_1 b_3 - (c_1^2 + c_2^2 + c_3^2 + d_1^2 + d_2^2 + d_3^2) = 0$$

is actually invariant under the four previous involutions which can be seen as transformations on the ten (homogeneous) variables $a, \cdots b_3$ or on the nine (non-homogeneous) variables $\frac{d_1}{a}, \frac{d_2}{a}, \cdots, \frac{b_3}{a}$.

Let us first study the subgroup generated by the iteration of (the generically infinite order generator) $I_1 I_2$.

For an initial point symmetric under the exchange of 2 and 3, we get remarkably **a curve**!!. Other initial points lead to orbits remarkably lying on higher dimensional non trivial varieties (see figure 6 and figure 7).

To sum up we are able to exhibit in a systematic way non trivial examples of mappings which have non trivial structures and do not reduce to the few well-known examples of the literature (mappings of the interval [1], Hénon mapping [2], trace-map [3]...). We do think these transformations should be in the next future a useful tool to analyze various type of physical problems. They have already been seen to be a powerful tool in statistical mechanics to analyze the exactly solvable models [9], [16].

Acknowledgments

We would like to thank Dr. J.C. Anglès d'Auriac for performing for us the multi-precision calculations of figure (3) and (4).

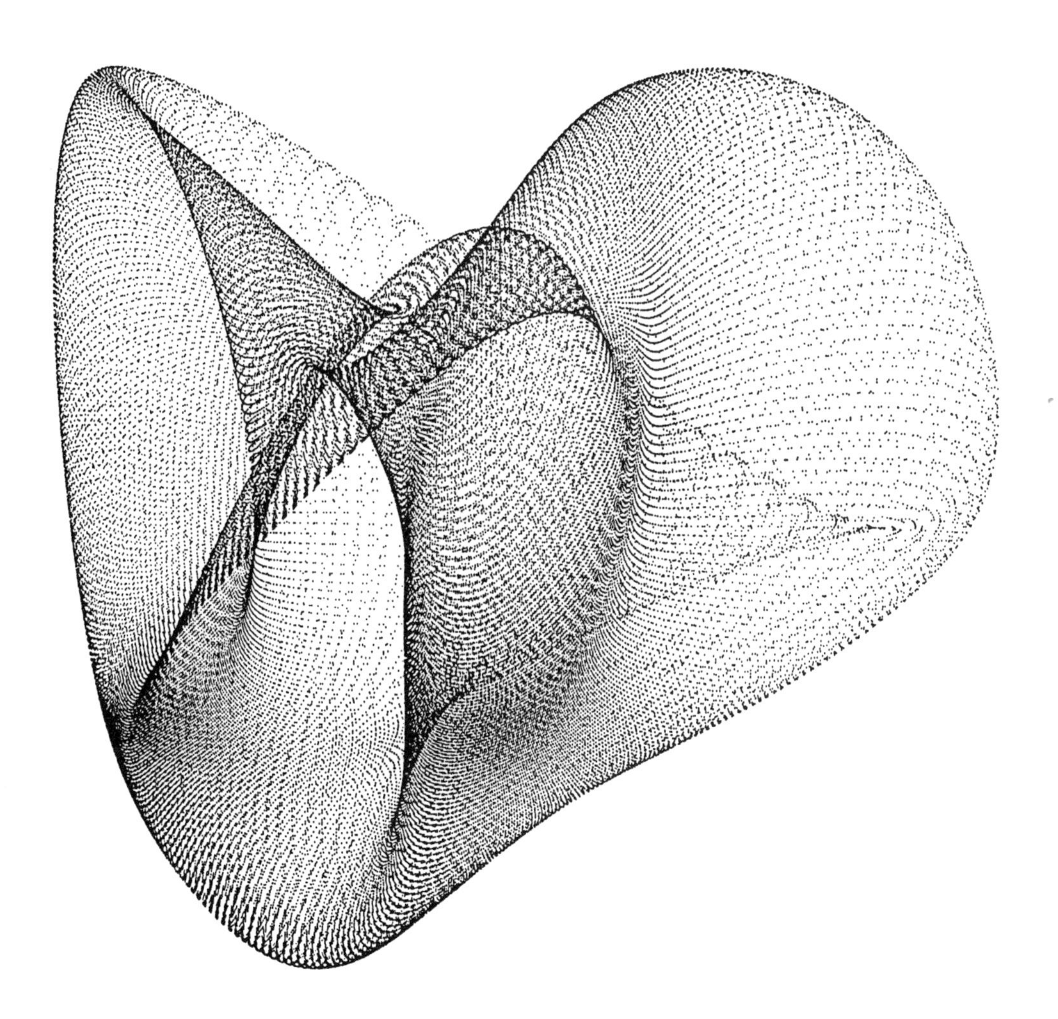

Figure 6

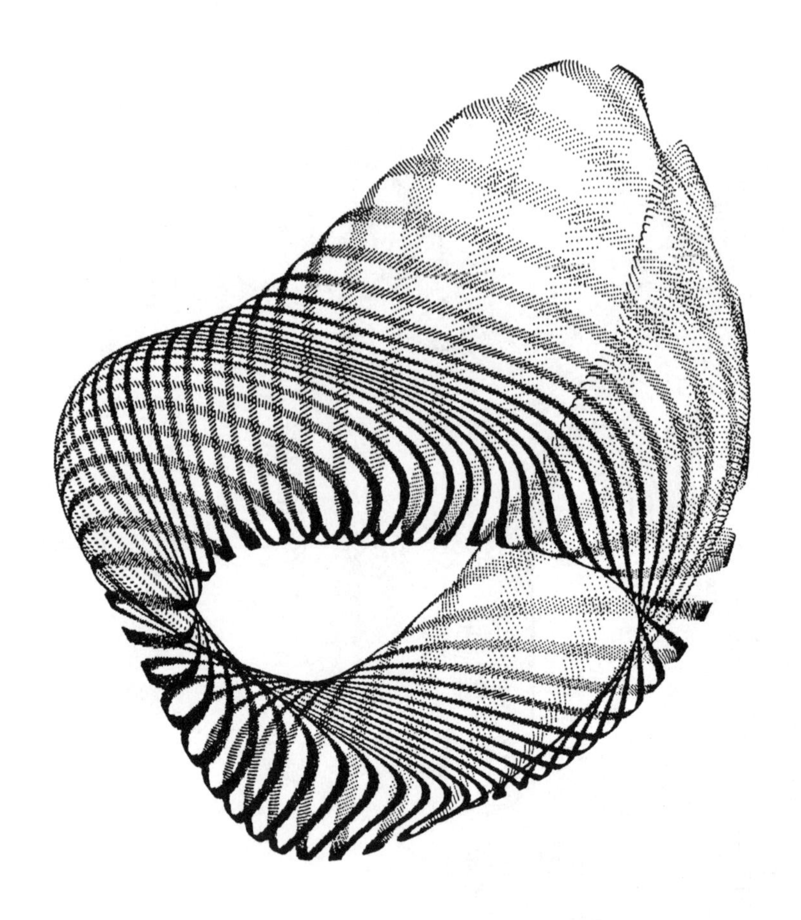

Figure 7

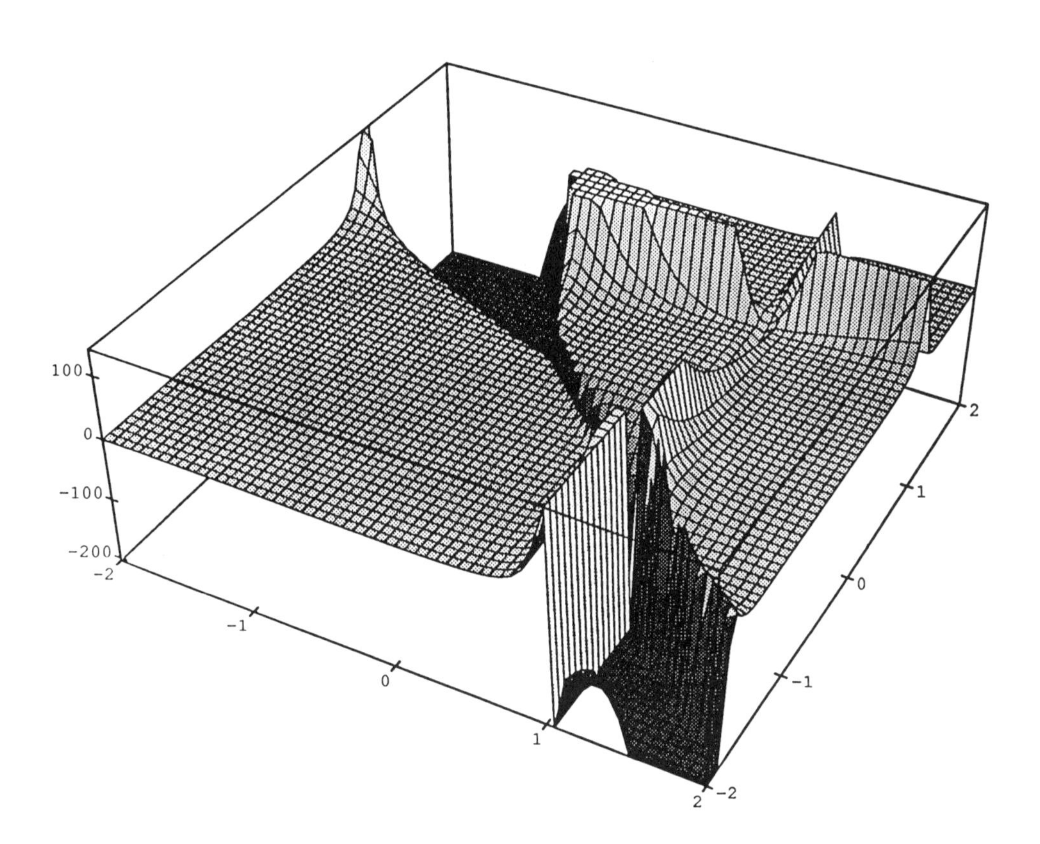

Figure 8: Δ (u,v) as a function of u and v

References

[1] M. Feigenbaum, J. Stat. Phys. **19** (1978) pp. 25-52 and 21 (1979), pp. 669-706 and P. Collet and J-P. Eckmann "Iterated maps on the interval as dynamical systems" Birkäuser (1980) and R.B. May Nature **261** (1976), pp. 459-467.

[2] M. Hénon, Commun. Math. Phys. **50** (1976), 69; R.B. May (1976) Nature **261**, pp. 459-467

[3] M. Kohmoto and Y. Oono, *Cantor spectrum for an almost periodic Schrödinger equation and a dynamical map.* Phys. Lett. **A102** (1984), pp. 145-148.
M. Kohmoto, L.P. Kadanoff and C. Tang "Localization problem in one dimension : mapping and escape", Phys. Rev. Lett. **50** (1983), p. 1870.

[4] L. Kadanoff, page 6, Physics Today, February 1986.

[5] M.P. Bellon, J-M. Maillard and C-M. Viallet. *Integrable Coxeter Groups.* Preprint LPTHE 91-4, and Helsinki HU TFT 91-4.

[6] M.P. Bellon, J-M. Maillard and C-M. Viallet. *Higher dimensional mappings.* Preprint LPTHE 91-5 and Helsinki HU TFT 91-5.

[7] M.P. Bellon, J-M. Maillard and C-M. Viallet. *Infinite Discrete Group for the Yang-Baxter Equations : Spin models.* Preprint LPTHE 91-6 and Helsinki HU TFT 91-6.

[8] M.P. Bellon, J-M. Maillard and C-M. Viallet. *Infinite Discrete Group for the Yang-Baxter Equations : Vertex models.* Preprint LPTHE 91-7 and Helsinki HU TFT 91-7, to be published in Phys. Lett. B.

[9] J-M. Maillard, *Automorphisms of algebraic varieties and Yang-Baxter equations.* Journ. Math. Phys. **27** (1986), p. 2776.

[10] M.P. Bellon, J-M. Maillard and C-M. Viallet. *Mappings for the Z_N chiral Potts Model.* in preparation.

[11] M.T. Jaekel and J-M. Maillard, *Inverse functional relations on the Potts model.* J. Phys. **A15** (1982), p. 2241.

[12] P.W. Kasteleyn. Exactly solvable lattice models. In *Proc. of the 1974 Wageningen Summer School: Fundamental problems in statistical mechanics III,*Amsterdam, (1974). North-Holland.

[13] A.B. Zamolodchikov, *Tetrahedron equations and the relativistic S-matrix of straight-strings in 2+1 dimensions.* Comm. Math. Phys. **79** (1981), p. 489.

[14] R.J. Baxter, *On Zamolodchikov's Solution of the Tetrahedron Equations.* Comm. Math. Phys. **88** (1983), p. 185-205.

[15] R.J. Baxter, *The Yang-Baxter equations and the Zamolodchikov model.* Physics **D18** (1986), p. 321-347.

[16] J-M. Maillard. Exactly solvable models in statistical mechanics and automorphisms of algebraic varieties. In Chin-Kun Hu, editor, *Progress in Statistical Mechanics Vol. 3, World Scientific Proceedings of the 1988 Workshop in Statistical Mechanics (Taïwan). World Scientific, (1988).*

SIMULATIONS OF DYNAMICAL ASPECTS OF A NEURAL NETWORK

Hidetoshi Nishimori
Department of Physics, Tokyo Institute of Technology,
Oh-okayama, Meguro-ku, Tokyo 152, Japan

ABSTRACT

A simple framework to treat the problem of dynamical behavior of a neural network of the Hopfield type has been proposed by Amari and Maginu. In their theory, they made an important assumption on the distribution of noise in the incoming signal to a neuron. I have studied by numerical simulations the validity of this assumption. I review the basic ideas of neural networks and present some recent results on the dynamical behavior of a network in the process of memory retrieval.

INTRODUCTION

A neuron is the basic unit of information processing in the brain. A neuron accepts incoming signals from many other neurons, typically of the order of 10^3 in the human brain. If the weighted sum of incoming signals exceeds a certain threshold, the neuron starts to emit a signal which is passed to many other neurons and processed similarly in each of them. This type of successive signal processing is constantly observed to result in the complex functioning of the brain. Collective behavior of neurons through mutual signal transmission is the key ingredient in realizing the highly non-trivial functions of the brain which is made up of relatively simple elements, neurons. It is in this point that the statistical mechanical approach to the analysis of neural networks has proved to be powerful and effective.

In the first part of this paper, I briefly review the modeling of a neural network in terms of the Ising spin variables. Various types of notion, such as memory and its retrieval, will be explained in close analogy with the Ising model. Then two interesting questions are asked: How can one embed various memories in a neural network? How does the network behave in the retrieval process of embedded memories? The first question is related to the thermal equilibrium properties of the Hopfield model.[1] The second one requires an essentially dynamical treatment of the macroscopic system, a highly non-trivial task. The Amari-Maginu approach[2] to this problem will be examined with the aid of Monte Carlo simulations. A generalization of the theory to the finite temperature case will be presented and the results are compared with the corresponding statistical mechanical predictions in the equilibrium limit.

FORMAL NEURONS AND THE ISING MODEL

At any given moment, a neuron is either firing a train of electric impulses or not according as the weighted sum of incoming signals exceeds a threshold or not. These two states can be represented by the Ising spin variable: Let S_i be 1 if the ith neuron is firing and -1 otherwise. The incoming signal from neuron j to neuron i is then expressed as $J_{ij}(S_j+1)$, where J_{ij} is the synaptic efficacy (or the efficiency of signal transmission from neuron j to i). If J_{ij} is positive, the synapse is called excitatory, and if the sign is negative, it is called inhibitory. It is this possibility of the existence of both signs in J_{ij} which leads to the highly non-trivial behavior of the network of connected neurons. The sum of all input signals to neuron i is thus written as

$$h_i = \sum_j J_{ij}(S_j + 1) \quad . \tag{1}$$

The basic relation to describe the behavior of the network is the following threshold dynamics of a single neuron:

$$S_i(t+1) = \mathrm{sgn}(h_i(t) - \theta_i), \tag{2}$$

where sgn denotes the step function returning either +1 or -1 according to the sign of its argument and θ_i is the threshold. By substituting (1) into (2) and setting the threshold to an appropriate value, one finds that the equation describing the dynamics of a single neuron is reduced to a relatively simple form,

$$S_i(t+1) = \mathrm{sgn}(\sum_j J_{ij} S_j(t)). \tag{3}$$

This equation is nothing but the Glauber dynamics[3] of the Ising model at zero temperature, because the argument on the right hand side can be interpreted as the effective field acting on spin i, and Eq. (3) represents an instruction to bring the spin S_i parallel to the effective field. Therefore, the dynamic as well as static behavior of the neural network is determined by the structure of the energy landscape of the Ising model with the given set of interactions $\{J_{ij}\}$.

ENERGY LANDSCAPE AND MEMORY RETRIEVAL

If the energy landscape of the model has a simple structure, such as that of the ferromagnetic Ising model with all $J_{ij} > 0$, the network evolves toward one of the few (typically, two) minima from the given initial state (Fig. 1). This final steady state has a rather trivial structure (such as the all-spin-up state), and hence the network does not serve as a processor of complex information. If, on the other hand, the energy function has many minima, the time evolution strongly depends upon the initial state and the final steady state reflects the

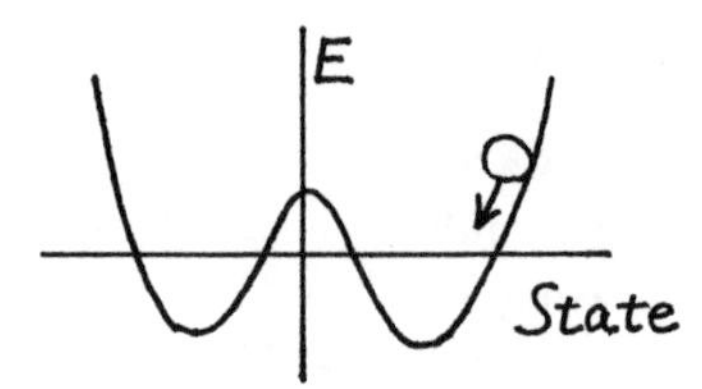

Fig. 1. A simple energy landscape.

Fig. 2. Strong dependence on the initial condition arises for a complex landscape.

initial information given to the network (Fig. 2). This situation arises if the sign of J_{ij} changes rather randomly from a pair (ij) to another, like the exchange interactions of spin glasses. Thus, in order to embed non-trivial memories in a neural network, one has to devise a method to assign minima of the energy to memories. The resulting synaptic efficacies are expected to have both signs.

HOPFIELD MODEL

A simple answer to the question given above is to set J_{ij} as follows:

$$J_{ij} = \frac{1}{N} \sum_{\mu=1}^{p} \xi_i^{(\mu)} \xi_j^{(\mu)} \tag{4}$$

where N is the total number of neurons, and $\xi_i^{(\mu)}$ denotes the state (± 1) of neuron i in the μth embedded spatial pattern. There are p embedded patterns and these are assumed to be uncorrelated (or orthogonal) with each other,

$$\frac{1}{N} \sum_{i=1}^{N} \xi_i^{(\mu)} \xi_i^{(\nu)} = \delta_{\mu\nu} \quad . \tag{5}$$

It is easy to see that the assignment of Eq. (4) assures stability of any embedded pattern under the dynamics of Eq. (3). That is, if $S_i(t)$ is equal to $\xi_i^{(\mu)}$ at all i at a given instant t, then the new state $S_i(t+1)$ is unchanged $\xi_i^{(\mu)}$ at any i as shown by Eqs. (3) to (5). One should note that the synaptic efficacy (4) can be positive or negative depending upon the pair (ij).

The problem of memory embedding in the network has thus been solved. The next natural question as a physicist is what happens if thermal noise is taken into account in the dynamics (3). Amit, Gutfreunt and Sompolinsky[4](AGS) investigated this problem using techniques developed in the theory of spin glasses. The goal is to solve the statistical mechanics of the Ising model specified by the Hamiltonian

$$H = -\sum_{i,j} J_{ij} S_i S_j \tag{6}$$

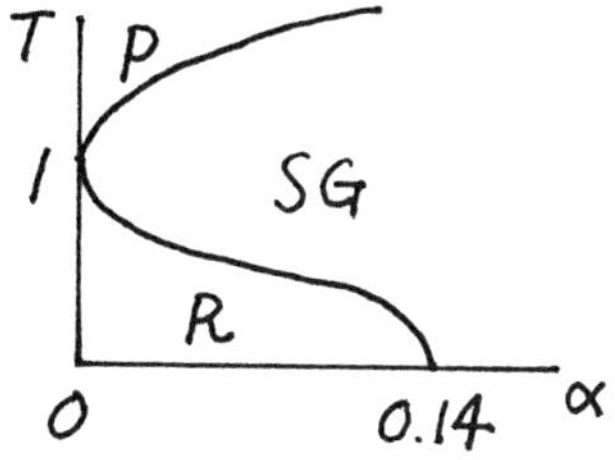

Fig. 3. The phase diagram of the Hopfield model at thermal equlibrium.[4]

with J_{ij} given by Eq. (4). They chose random patterns to carry through the calculations. Note that random patterns ($\xi_i^{(\mu)} = \pm 1$ with equal probability) satisfy the orthogonality condition (5) in the thermodynamic limit $N \to \infty$. The Hamiltonian (6) with Eq. (4) represents an infinite-range model in the sense that J_{ij} is nonvanishing for any pair (ij). Therefore, as is well known in the theory of spin systems, the mean field theory is expected to be exact in the present problem.

AGS thus followed the standard prescription of mean-field theory to disintegrate the many-body problem into a few-body one using the Gaussian integral techniques. The detailed calculations are actually much more complicated than the counterpart of, for instance, the simple ferromagnetic system, since non-trivial averaging procedure over the patterns $\{\xi_i^{(\mu)}\}$ is involved. They made full use of the tricks developed in the study of the Sherrington-Kirkpatrick model of spin glasses[5] in which the average over independent random variables $\{J_{ij}\}$ has to be carried out. The reader is referred to the original paper for details. The results are sketched below.

The system (or the network) has essentially three thermodynamic phases (Fig. 3). The parameters to describe the phase diagram are the temperature T and the pattern ratio $\alpha = p/N$ which represents how much the network is loaded with memory. The first phase is a rather trivial paramagnetic state (P) in which all spins (or neurons) are randomly changing states at very fast rates. No macroscopic order can be observed. This paramagnetic state exists as a stable state in the high-temperature region of the phase diagram. The second one is the spin glass phase (SG) in which spins are frozen randomly; the state of a single spin does not change with time. However, the spatial distribution of the sense of spins is random and has no relation with the embedded patterns. This phase occupies the high-α region of the phase diagram in which the network is heavily loaded with memory resulting in total confusion. The third retrieval phase (R) in the low-T low-α region is the most interesting one in which the state of the network has macroscopic overlap with one (or more) of the embedded memories. The free energy in this phase has a structure similar to Fig. 2. Note that Fig. 2 is a rough sketch of the energy of a complex network in the absence of thermal noise ($T = 0$). In this region of the phase diagram, one can reach an appropriate memorized pattern, by starting within the basin of attraction of a local minimum, many of which correspond to embedded patterns as can be seen from the stability of the patterns in the zero-temperature case explained before. This behavior is nothing but retrieval of memory from incomplete information (see Fig. 4).

Fig. 4. Retrieval of an embedded pattern from noisy initial information.

I should add an explanation of the significance of the temperature in the context of neural networks. The variable T does not represent the actual temperature of the brain. It is simply the parameter to represent the level of uncertainty (or stochasticity) in the functioning of a neuron; instead of Eq. (3), which is for $T = 0$, one assumes that $S_i(t+1)$ is 1 with probability $p = 1/\{1 + \exp(-\beta h_i)\}$, where β is the inverse temperature $1/T$. One can easily check that this prescription reduces to Eq. (3) in the zero-temperature limit.

DYNAMICS OF MEMORY RETRIEVAL IN THE HOPFIELD MODEL

Thermodynamic properties of the Hopfield model have been investigated by the statistical mechanical approach as explained in the preceding section. A next interesting problem is how the network evolves from a given initial state to a final steady state. This question is related to the size and shape of the basin of attraction of an embedded memory of the Hopfield model. It is a problem of dynamics and is generally very difficult to solve even in the mean-field (or the infinite-range) model: One should derive equations to describe the time evolution of macroscopic variables from the microscopic dynamics (3). I first explain the Amari-Maginu approach[2] to this problem, and then present the results of Monte Carlo simulations to clarify the limit of applicability of their theory.

The best macro-variable to represent the network state is the overlap of the current state with the embedded pattern to be recalled,

$$m = \frac{1}{N}\sum_{i=1}^{N} S_i \xi_i^{(1)} \tag{7}$$

where I assumed that the network is in the process of retrieval of an arbitrarily chosen (this case, the first) pattern. m is 1 if the network perfectly retrieves the pattern and is 0 if it is in an uncorrelated state with $\{\xi_i^{(1)}\}$. Our task is to derive the time evolution of m from the microscopic dynamics (3).

Amari and Maginu proceeded by first assuming that the dynamics (3) is realized synchronously at all neurons. This means that all neurons update their states simultaneously using the information of states of all neurons at the previous time step. Strictly speaking, this prescription yields a slightly different thermodynamic state[6] from the conventional asynchronous dynamics of the Glauber model in which each neuron updates its state at a randomly chosen timing independently from that of other neurons. Although Amari and Maginu used this synchronous dynamics to simplify the treatment, it turns out that the resulting phase diagram obtained by applying their method to finite temperatures has very close similarity with that of AGS. (The AGS theory is for the asynchronous system).

The essence of the Amari-Maginu theory is to separate the incoming signal into real signal and noise. Substituting Eq. (4) into the expression of the incoming signal

$$h_i(t) = \sum_j J_{ij} S_j(t),$$

one obtains

$$h_i(t) = \frac{1}{N} \sum_j \sum_\mu \xi_i^{(\mu)} \xi_j^{(\mu)} S_j(t) \quad . \tag{8}$$

If the network is in the retrieval process of the first pattern, contribution from the term corresponding to $\mu = 1$ would be dominant in Eq. (8). Let us thus regard this term as the true signal and the rest as noise. The true signal is easily seen to have a form of the overlap defined by Eq. (7),

$$h_i = \xi_i^{(1)} m(t) + n \quad . \tag{9}$$

Since the noise term

$$n = \frac{1}{N} \sum_{\mu=2}^{p} \sum_j \xi_i^{(\mu)} \xi_j^{(\mu)} S_j(t) \tag{10}$$

is a sum of quite many ± 1's, it may obey the Gaussian distribution with vanishing mean and variance $\sigma(t)^2$ according to the central limit theorem. If one accepts this assumption, it is rather straightforward (if long and tedious) to calculate the expectation value of the new overlap and the variance. The result is[2]

$$\begin{aligned} m(t+1) &= \langle \frac{1}{N} \sum_i \xi_i^{(1)} S_i(t+1) \rangle \\ &= \mathrm{Erf}[m(t)/\sigma(t)] \end{aligned} \tag{11}$$

and

$$\sigma(t+1)^2 = \alpha + 4P[m(t)/\sigma(t)]^2 + 4\alpha m(t) m(t+1) P[m(t)/\sigma(t)]/\sigma(t), \tag{12}$$

where

$$P(x) = \exp(-x^2/2)/\sqrt{2\pi}.$$

The brackets in Eq. (11) represent an average over the stochastic variable n.

By inspecting the time evolution equations (11) and (12), they found that the network is capable of memory retrieval if the pattern ratio $\alpha = p/N$ is smaller than a critical value of 0.15 and the initial overlap $m(0)$ is larger than a threshold (which depends on α). The critical value 0.15 is quite close to that of AGS who derived 0.14 at $T = 0$ by means of statistical mechanics applied to the Hamiltonian (6) with asynchronous dynamics. Amari and Maginu also found that the overlap m initially increases and then turns to decrease even in the case of retrieval failure in which the limiting value of m as $t \to \infty$ is 0. This latter fact implies that the basin of attraction does not have a simple structure determined only by the value of m; m alone is not sufficient to describe the macroscopic state of the network.

They have further derived relations to describe the time evolution of the network by assuming that the distribution of the noise n is Gaussian with non-vanishing mean. This latter assumption is naively expected to yield more reliable results because it is more general than the previous assumption of Gaussian distribution with vanishing mean. The result was a disappointment in that the threshold phenomenon observed in the zero-vanishing results did not exist in the apparently improved version: The critical value 0.15 could be predicted only under the less general assumption of vanishing mean. Thus it is necessary to clarify what part of their theory is acceptable.

MONTE CARLO SIMULATIONS

There are two basic assumptions in the Amari-Maginu theory. The first one is that the incoming signal is separated into real signal and noise, the latter of which can be treated by stochastic methods. The second is that the noise distribution is Gaussian. This second assumption at first seems very reasonable since the noise term n, Eq. (10), is the sum of many ±1's. However, the current state $S_j(t)$ in Eq. (10) is a result of updating processes from the initial state and each update reflects the values of embedded patterns $\{\xi_i^{(\mu)}\}$. Consequently, various terms in Eq. (10) depend upon each other quite strongly, and there is no *a priori* reason for the central limit theorem to hold in this situation. I have carried out Monte Carlo simulations to see under what conditions their assumptions are actually valid.

First I discuss the case of light memory loading $\alpha = 0.08$. Fig. 5 shows the time dependence of the overlap $m(t)$ and cummulants of the noise n when the initial $m(0)$ is 0.5. It is observed that the memory retrieval succeeds ($m(t)$ readily saturates to a value of almost 1) and all cummulants except the second one (which is nothing but the variance σ^2) vanish. Therefore, the distribution is actually Gaussian in the present case. The time evolution of m and σ^2 is seen

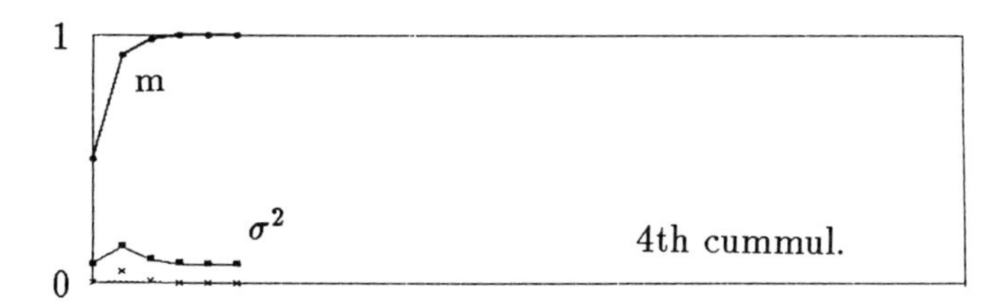

Fig. 5. Simulation results (lines) and predictions of Eqs.(11) and (12) (discrete symbols). $N = 9000$, $p = 720$ and $m(0) = 0.5$.

to be well described by the Amari-Maginu prediction, Eqs. (11) and (12). The data have quite small size dependence in the range from $N = 5000$ to $N = 9000$.

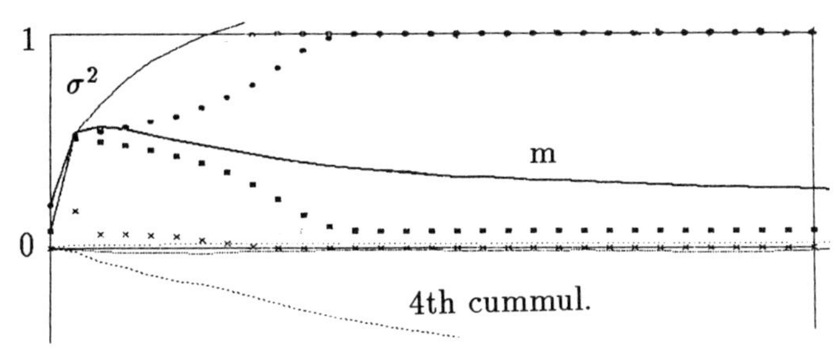

Fig. 6. $N = 9000$, $p = 720$ and $m(0) = 0.2$.

When the initial value of m is small and memory retrieval fails, the noise distribution immediately deviates from Gaussian (Fig. 6). The fourth cummulant increases rapidly with time, and the Amari-Maginu predictions on m and σ^2 do not agree with simulation data. Again, the size dependence of data seems negligible within the range of $N = 5000$ and 9000.

If the memory loading is heavy, $\alpha = 0.20$ (which is above the retrieval critical value 0.15), and the initial m is small, the distribution of n deviates from Gaussian in a quite strong manner, Fig. 7. The situation is somewhat more subtle when $\alpha = 0.20$ and $m(0)$ is rather large. As seen in Fig. 8 which is for $m(0) = 0.9$, higher order cummulants than the second one remain rather small until a certain intermediate time step $t(N)$, and then suddenly start to increase. This tendency has been observed in all data for sizes ranging from $N = 5000$ to 9000. The deviation time $t(N)$ showed only very small, if any, dependence on N in the above range. Amari and Maginu actually accepted the Gaussian assumption of the noise distribution by stating that this assumption would hold up to a certain time step $t(N)$ which would tend to infinity as the system size N increases indefinitely. My tentative conclusion concerning this issue is that the N dependence of $t(N)$ is quite weak, say logarithmic, if their claim is correct in the case of $\alpha = 0.20$ and $m(0) = 0.9$. Another noteworthy character of the data in Fig. 8 is that the simulation values of m and σ^2 do not agree with those of Eqs. (11) and (12). A possible reason for this discrepancy is again the smallness of the system size although it would require inhibitingly larger sizes to confirm it.

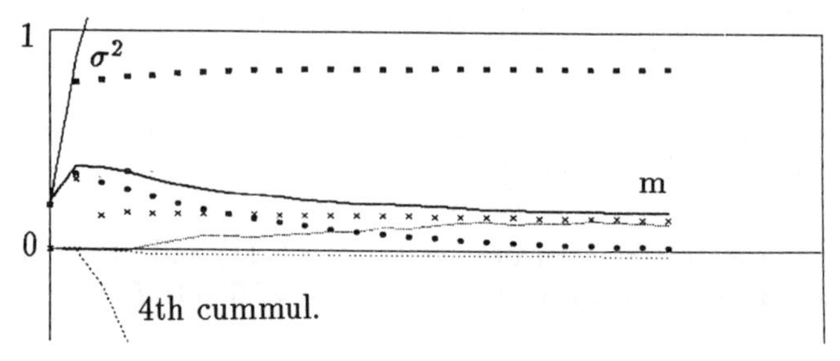

Fig. 7. $N = 9000$, $m = 1800$ and $m(0) = 0.2$.

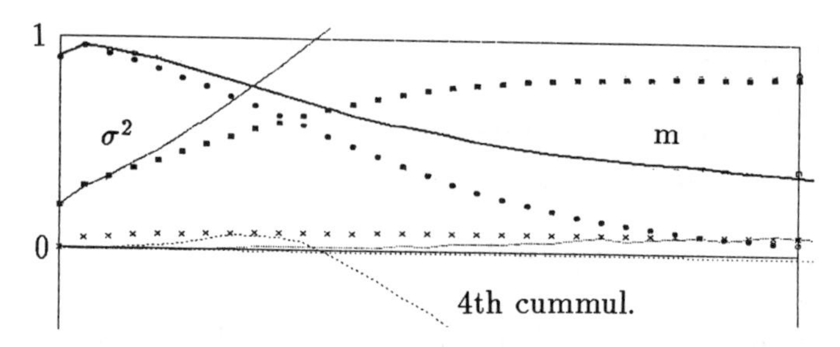

Fig. 8. $N = 9000$, $m = 1800$ and $m(0) = 0.9$

The assumption of Gaussian noise distribution thus seems plausible if memory retrieval is successful, namely for small α and large $m(0)$. There is also a possibility of this assumption to be valid when the initial overlap $m(0)$ is rather large even when the network fails to retrieve an embedded memory. In other cases (in which $m(0)$ is small), the distribution of noise is far from Gaussian, irrespective of α. This latter observation may be accounted for by considering that the separation of the incoming signal into real signal and noise, Eq. (9), is not justified when the initial overlap is not large; the noise term may be larger than the 'true signal' contribution, and consequently more elaborate treatment is required than a simple the stochastic approach. This point is now under investigation.

FINITE TEMPERATURE DYNAMICS

Since the Gaussian assumption is likely to hold at least in the case of successful retrieval, it is natural to ask how the equilibrium phase diagram looks like if the Amari-Maginu method is applied to finite temperature problems. Ozeki and I have obtained the following evolution equations of m and σ^2 at finite temperatures assuming Gaussian distribution of the noise term:

$$m(t+1) = f[\tilde{m}(t), \tilde{\sigma}(t)] \tag{13}$$

and

$$\sigma(t+1)^2 = \alpha + \tilde{\sigma}(t)^2 h[\tilde{a}(t), \tilde{\sigma}(t)]^2 + 2\alpha\tilde{m}(t)m(t+1)h[\tilde{m}(t), \tilde{\sigma}(t)] \ . \tag{14}$$

Here $\tilde{m}$ denotes βm and $\tilde{\sigma}$ is $\beta\sigma$. The function f is defined as

$$f(x,y) = \int_{-\infty}^{\infty} \frac{1}{\sqrt{2\pi}} \exp(-\frac{u^2}{2}) \tanh(x+yu)du$$

and $h(x,y)$ is given by replacing $\tanh(\cdot)$ above by $\mathrm{sech}(\cdot)^2$.

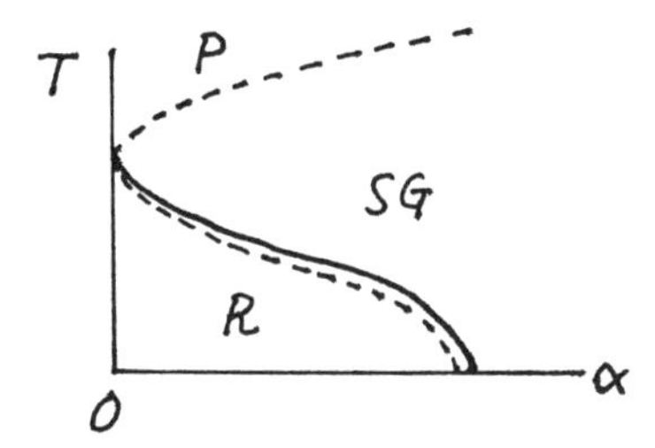

Fig. 9. Equilibrium phase diagram obtained by the signal/noise analysis. The results of AGS are drawn by dotted lines for comparison.

Equations (13) and (14) describe how the network approaches (or deviates from) an embedded pattern in the presence of thermal disturbance. The equilibrium phase diagram is drawn by inspecting the limiting behavior of the solution of these equations. The result is given in Fig. 9. The network succeeds in retrieval in the low-α low-T region. In Fig. 9, I have also drawn the phase boundaries obtained by AGS. It is immediately clear that the retrieval phase is common in both theories and the phase boundary separating the retrieval and other phases agrees well, even quantitatively. The main difference is that the dynamics approach does not distinguish the spin glass phase from the paramagnetic phase. This point can be easily understood by recalling that the Gaussian noise assumption does not hold in the case of unsuccessful retrieval, and Eqs. (13) and (14) should not be applied in the spin glass and paramagnetic phases. Therefore, the dynamical behavior of the network is predicted by Eqs. (13) and (14) only in the low-α low-T phase. In other ranges of parameters, the network behaves in a more complicated manner which will require other macroscopic parameters to describe than m and σ^2.

SUMMARY AND CONCLUSION

I have given brief accounts of the basic functioning and modeling of a neural network of the Hopfield type. Each neuron may be represented by an Ising spin, and the connection (synaptic efficacy) between neurons has strong resemblance to exchange interactions. Hence the problem is that of the Ising model, and the full statistical mechanical apparatus can be used to investigate the macroscopic properties. Ingenious methods developed for equilibrium statistical mechanics of spin glasses have been actually applied by Amit, Gutfreunt and Sompolinsky, and the result showed how the network behaves in thermal equilibrium as a function of various parameters.

It is not easy to analyze the dynamical behavior of a complicated system like the present neural network. Amari and Maginu set a bold assumption on the properties of incoming signal to a neuron, and derived time evolution equations for macroscopic parameters. The results were attractive, in particular in that the equilibrium limits were in close agreement with the predictions of the Amit-Gutfreunt-Sompolinsky theory applied to $T = 0$. I have investigated the validity of the Amari-Maginu theory by Monte Carlo simulations to find that their assumption seems to hold if the parameters of the network are in the range where memory retrieval is successful. Their theory has been generalized to take account of finite temperature properties, and the equilibrium phase diagram of AGS has been partly reproduced. Of course, the present dynamical treatment yields much more information than the equilibrium statistical mechanical approach because the time evolution of the network in the intermediate stage of memory retrieval is explicitly given.

We are carrying out more detailed analysis of the dynamics, Eqs. (13) and (14), to clarify the basin of attraction and other problems. Further Monte Carlo simulations are also being carried out to get information on the distribution of noise in the finite temperature updating processes.

ACKNOWLEDGMENT

This research was supported by the Grant-in-Aid for Priority Area by the Ministry of Education, Science and Culture.

REFERENCES

1. J.J. Hopfield, Proc. Natl. Acad. Sci. USA. **79**, 2445 (1982).
2. S. Amari and K. Maginu, Neural Network **1**, 63 (1988).
3. R.J. Glauber, J. Math. Phys. **4**, 294 (1963).
4. D.J. Amit, H. Gutfreunt and H. Sompolinsky, Phys. Rev. Lett. **55**,1530 (1985).
5. D. Sherrington and S. Kirkpatrick, Phys. Rev. Lett. **35**, 1972 (1975).
6. P. Peretto, Biol. Cybern. **50**, 51 (1984).

MASTER EQUATION APPROACH TO NEURAL NETWORKS

M. Y. Choi
Department of Physics, Seoul National University, Seoul 151–742, Korea

ABSTRACT

The neural network model, which explicitly takes into account the existence of several time scales without discretizing the time, is characterized by its dynamics. It is described by a non–Markov master equation from which static properties can be obtained. Dynamic behavior of the network is also discussed.

I. INTRODUCTION

Neural network models attempt to explain intriguing features of the brain such as memory, learning, fault tolerance, and information storage and retrieval in terms of collective properties of the networks.[1-6] In particular, two such models, one proposed by Hopfield[2] and a closely related model, proposed by Little,[1] have attracted much interest mainly because they are relatively simple and facilitate the use of statistical mechanics in their analysis. In these models of neural networks, each neuron is viewed as an Ising spin s_i $(i = 1, \ldots, N)$ which takes on the value $+1$ if the neuron is firing, and -1 if the neuron is inactive.

In the Little model the existence of the refractory period is effectively taken care of by discretizing time. It uses random sequential updating, where at each time step all neurons update their states simultaneously. Thus the dynamics of the Little model is essentially synchronous, allowing multiple neuron flips at each time step. On the other hand, although time is still discretized, the Hopfield model, originally proposed as a zero-temperature model and subsequently generalized to finite temperatures, uses completely asynchronous dynamics: Only a single neuron is updated at each time step, which is essentially a Monte Carlo process.

Both models have proved quite successful in exhibiting a number of desirable features.[3] However, they rely upon several assumptions, some of which pose questions in view of the biological situation. One assumption in particular is the synchronous or asynchronous character of the dynamics used in the two models. It is obvious that neither dynamics provides a very realistic description of the dynamics in real networks, which presumably lies in between.

On the other hand, the dynamic model proposed recently[6] deals with usual continuous time rather than digital time, but still takes into account the existence of relevant time scales in the nervous system such as the refractory period, time duration of the action potential and retardation of the signal propagation. The dynamics of the model is neither totally synchronous nor totally asynchronous, and may be regarded as more realistic. The price of this is, of course, the lack of a Hamiltonian, which requires a dynamic analysis of the model. The static properties of the dynamic model have been studied via the recently introduced path integral formulation,[7] which facilitates performing the quenched average over the random patterns. This approach is, on the one hand, unavoidable since there is no self-averaging property in the case of storing the infinite number of patterns and is, on the other hand, desirable since it allows a rigorous description of the dynamics. In particular, it reduces the dynamics to a mean–field equation of a single neuron, from which dynamic as well as static properties can be obtained.

II. DYNAMIC MODEL OF NEURAL NETWORKS

We consider a neural network consisting of N neurons with two possible states. The state of the network may be represented by the configuration of all the neurons, $\mathbf{s} = (s_1, s_2, \ldots, s_N)$, where $s_i = +1$ or -1 according to whether the ith neuron fires or not. The neurons are interconnected by synaptic junctions of strength $2J_{ij}(J_{ii} \equiv 0)$, which determine the contribution of a signal fired by the jth neuron to the postsynaptic potential acting on the ith neuron. The threshold behavior of the ith neuron at time t is described by a probability that depends on the difference between the total potential on the ith neuron V_i and its threshold value V_0,

$$V_i - V_0 = \sum_j J_{ij} s_j(t - t_d) + h_i \equiv E_i(t - t_d), \tag{2.1}$$

where it is usually assumed that the external local field $h_i \equiv \sum_j J_{ij} - V_0$ vanishes. In Eq. (2.1) $s_j(t - t_d)$ denotes the state of the jth neuron at time $t - t_d$, where t_d is the delay in the interactions including the synaptic delay. The couplings are assigned according to

$$J_{ij} = \begin{cases} \frac{1}{N} \sum_{\mu=1}^{p} \xi_i^\mu \xi_j^\mu, & i \neq j, \\ 0, & i = j, \end{cases} \tag{2.2}$$

which corresponds to the situation that p patterns $\{\xi_i^\mu\}(\mu = 1, 2, \ldots, p)$ are learned. Here every ξ_i^μ is an independently distributed stochastic variable, taking the values ± 1 with equal probabilities.

We start with consideration of the conditional probability that the ith neuron does not fire at time $t + \delta t$ given that it fires at time t. For sufficiently small δt, the probability on the average over all neurons may be written in the form

$$p[s_i(t + \delta t) = -1 | s_i(t) = 1] = \delta t / t_0, \tag{2.3}$$

where t_0 is the time duration of the action potential, usually of the order of a few milliseconds. Thus the time average has been essentially incorporated in the above expression. Similarly, we write the conditional probability that the ith neuron fires at time $t + \delta t$ given that it does not fire at time t:

$$p[s_i(t+\delta t) = 1 | s_i(t) = -1; \mathbf{s}_{-i}(t - t_d)] = [1 + \tanh \beta E_i(t - t_d)]\delta t / 2t_r, \quad (2.4)$$

where $\mathbf{s}_{-i}(t)$ denotes the set

$$\{s_1(t), \ldots, s_{i-1}(t), s_{i+1}(t), \ldots, s_N(t)\}$$

and the inverse *temperature* $\beta \equiv T^{-1}$ measures the width of the threshold region, i.e., a measure of the level of synaptic noise. Except for the factor $\delta t / t_r$, which takes into account the existence of the refractory period t_r, again of the order of a few milliseconds, the above expression has been chosen essentially following Little.[1] Equations (2.3) and (2.4), together with their counterparts, can be combined to give a general expression for the conditional probability

$$p[s_i(t+\delta t) = s_i' | s_i(t) = s_i; \mathbf{s}_{-i}(t - t_d)],$$

which, in the limit $\delta t \to 0$, can be expressed in terms of the transition rate,

$$p[s_i(t+\delta t) = s_i' | s_i(t) = s_i; \mathbf{s}_{-i}(t - t_d)]$$

$$= \begin{cases} w_i[s_i; \mathbf{s}_{-i}(t - t_d)]\delta t & \text{for } s_i' = -s_i, \\ 1 - w_i[s_i; \mathbf{s}_{-i}(t - t_d)]\delta t & \text{for } s_i' = s_i, \end{cases} \quad (2.5)$$

where the transition rate is given by

$$w_i[s_i; \mathbf{s}_{-i}(t - t_d)] = \frac{1}{2t_r}\left\{(a + \frac{1}{2}) + (a - \frac{1}{2})s_i + \frac{1}{2}(1 - s_i) \tanh \beta E_i(t - t_d)\right\} \quad (2.6)$$

with $a \equiv t_r / t_0$. In the above equations, the dependence of the conditional probabilities on $\mathbf{s}_{-i}(t)$ is implicit, through $E_i(t)$.

The behavior of the neural network is then governed by the master equation for the joint probability $P[\mathbf{s}(t) = \mathbf{s}; \mathbf{s}(t - t_d) = \mathbf{s}'']$ that the system is in its state $\mathbf{s}''$ at time $t - t_d$ and in state $\mathbf{s}$ at time t:

$$P[\mathbf{s}(t+\delta t) = \mathbf{s}; \mathbf{s}(t - t_d) = \mathbf{s}''] - P[\mathbf{s}(t) = \mathbf{s}; \mathbf{s}(t - t_d) = \mathbf{s}'']$$

$$= -\sum_{\mathbf{s}'}\Big\{ p[\mathbf{s}(t+\delta t) = \mathbf{s}' | \mathbf{s}(t) = \mathbf{s}; \mathbf{s}(t - t_d) = \mathbf{s}'']P[\mathbf{s}(t) = \mathbf{s}; \mathbf{s}(t - t_d) = \mathbf{s}'']$$

$$- p[\mathbf{s}(t+\delta t) = \mathbf{s} | \mathbf{s}(t) = \mathbf{s}'; \mathbf{s}(t - t_d) = \mathbf{s}'']P[\mathbf{s}(t) = \mathbf{s}'; \mathbf{s}(t - t_d) = \mathbf{s}''] \Big\} \quad (2.7)$$

with

$$p[\mathbf{s}(t+\delta t)=\mathbf{s}'|\mathbf{s}(t)=\mathbf{s};\mathbf{s}(t-t_d)=\mathbf{s}'']$$
$$\equiv \prod_{i=1}^{N} p[s_i(t+\delta t)=s_i'|s_i(t)=s_i;\mathbf{s}_{-i}(t-t_d)=\mathbf{s}''_{-i}].$$

Thus we obtain a non-Markov master equation, which has also been suggested to describe many-body systems with retarded interactions.[8] In limit $\delta t \to 0$, Eq. (2.7) can be written in the differential form through the use of the transition rate defined by Eq. (2.6):

$$\frac{\partial}{\partial t}P[\mathbf{s}(t)=\mathbf{s};\mathbf{s}(t-1)=\mathbf{s}'']$$
$$=-\sum_k \Big\{\omega_k[s_k;\mathbf{s}_{-k}(t-1)]P[\mathbf{s}(t)=\mathbf{s};\mathbf{s}(t-1)=\mathbf{s}'']$$
$$-\,\omega_k[-s_k;\mathbf{s}_{-k}(t-1)]P[\mathbf{s}(t)=\mathsf{F}_k\mathbf{s};\mathbf{s}(t-1)=\mathbf{s}'']\Big\}, \qquad (2.8)$$

where time t has been rescaled in units of the delay time t_d, $\omega_k[s_k;\mathbf{s}_{-k}(t-1)] \equiv bw_k[s_k;\mathbf{s}_{-k}(t-1)]$ with $b \equiv t_d/t_r$, and $\mathsf{F}_k\,\mathbf{s} \equiv (s_1,\ldots,s_{k-1},-s_k,s_{k+1},\ldots,s_N)$. Then equations describing the time evolution of relevant physical quantities in general assume the form of differential-difference equations due to the retardation in interactions. In particular, the activity of the kth neuron

$$\sigma_k(t) \equiv \langle s_k\rangle_t \equiv \sum_{\mathbf{s},\mathbf{s}'} s_k P[\mathbf{s}(t)=\mathbf{s};\mathbf{s}(t-1)=\mathbf{s}']$$

is determined by the differential-difference equation

$$b^{-1}\frac{d}{dt}\sigma_k(t)=(\frac{1}{2}-a)-(\frac{1}{2}+a)\sigma_k(t)+\frac{1}{2}\langle(1-s_k)\tanh\beta E_k(t-1)\rangle_t. \qquad (2.9)$$

We introduce the vector notation $\boldsymbol{\xi}_k\cdot\mathbf{m} \equiv \sum_\mu \xi_k^\mu m^\mu$, where $m^\mu \equiv \frac{1}{N}\sum_i \xi_i^\mu \sigma_i$ describes the overlap between the neurons and the memory μ. Multiplying Eq. (2.9) by $N^{-1}\xi_k^\mu$ and summing over k, we obtain the equation for the *order parameter* $\mathbf{m}$:

$$b^{-1}\frac{d}{dt}\mathbf{m}(t)=-(\frac{1}{2}+a)\mathbf{m}(t)+\frac{1}{2N}\sum_k \boldsymbol{\xi}_k\langle(1-s_k(t))\tanh\beta E_k(t-1)\rangle_t. \qquad (2.10)$$

For finite p, we may replace the average over the neurons by the average over the distribution of the memories, and the corresponding asymptotic behavior has been studied in some detail.[6] In particular, the possibility of the static Mattis-state solution of the form $\mathbf{m}=(m,0,\ldots,0)$ corresponding to the fully correlated states with just one of the quenched memories has been pointed out. The stability of these solutions has been also examined, revealing that when

$a > a_c \equiv (\sqrt{3}-1)/2$, the network undergoes a continuous phase transition at the critical temperature $T_2^0 = 4a/(1+2a)^2$ from the disordered state (with no memory) to the Mattis state (with memories). In case that $a < a_c$, the network exhibits successive transitions as the temperature is lowered; a first-order transition from the disordered state to the *mixed* state where order and disorder coexist, and then a continuous transition to the ordered (Mattis) state.

III. DYNAMIC MEAN–FIELD THEORY

In the case of storing infinite number of patterns, i.e., for finite $\alpha \equiv p/N$ with $N \to \infty$, there is no self-averaging property which allows us to reduce the degrees of freedom from N to p. Since the simultaneous retrieval of the infinite number of the stored patterns is not plausible in the biological sense, we investigate a situation where only a finite number l of patterns, say, $(\xi^1, \ldots, \xi^l)$ to be condensed in the network. Then the remaining patterns have overlap at most of order $O(\frac{1}{\sqrt{N}})$, and it is needed to consider corresponding correlation and response functions for the condensed patterns and for the random patterns. One way of handling this problem is by resorting to the path integral formulation introduced recently.[7] The application of the path integral formulation to the dynamic model has been discussed in Ref. 9, which concludes that the original problem can be essentially reduced to a single neuron problem. In particular the variables carrying noncondensed pattern indices are decoupled and separated from those carrying spatial indices; the integration over variables carrying indices $i = 1, 2, \ldots, N$ and $\nu = l+1, \ldots, \alpha N$ can be performed explicitly. Thus the local field $E_i(t)$ in Eq. (2.1) can be replaced by the averaged local field

$$\begin{aligned} H_i(t) &= h + \sqrt{\alpha}\Phi(t) + \sum_{\mu=1}^{l} m^\mu(t)\xi_i^\mu \\ &\quad + \alpha \int_{t'<t} dt'\, \tilde{S}(t,t')\sigma_i(t'), \end{aligned} \tag{3.1}$$

where Φ is a Gaussian random field with zero mean and correlation[9]

$$\langle \Phi(t)\Phi(t')\rangle_\Phi = R(t,t'). \tag{3.2}$$

Here $R(t,t')$ and $\tilde{S}(t,t')$ are the random overlap correlation and response functions, respectively, and related to the autocorrelation function $C(t,t') \equiv \langle\!\langle\, \langle s_i(t)s_i(t')\rangle \,\rangle\!\rangle$ and the local response function $G(t,t') \equiv \langle\!\langle \partial\langle s_i(t)\rangle/\partial h_i(t')\rangle\!\rangle$ via

$$\tilde{S}(t,t') = G(t-1,t'+1) + \int_{\tau<t-1} d\tau\, G(t-1,\tau)\tilde{S}(\tau,t'), \tag{3.3}$$

$$\begin{aligned} R(t,t') &= C(t,t') + \int_{\tau<t} d\tau G(t,\tau)R(\tau-1,t') + \int_{\tau'<t'} d\tau' G(t',\tau')R(t,\tau'-1) \\ &\quad - \int_{\tau<t} d\tau \int_{\tau'<t'} d\tau'\, G(t,\tau)R(\tau-1,\tau'-1)G(t',\tau'). \end{aligned} \tag{3.4}$$

Equations (2.9) and (2.10) then take the form of single–neuron equations ($b \equiv 1$)

$$\frac{d}{dt}\sigma(t) = (\frac{1}{2} - a) - (\frac{1}{2} + a)\sigma(t) + \frac{1-\sigma(t)}{2}\tanh\beta H(t-1), \quad (3.5a)$$

$$\frac{d}{dt}\mathbf{m}(t) = -(\frac{1}{2} + a)\mathbf{m}(t) + \frac{1}{2}\langle\langle \boldsymbol{\xi}[1-\sigma(t)]\tanh\beta H(t-1)\rangle\rangle, \quad (3.5b)$$

where $\langle\langle\ \rangle\rangle$ denotes both the Gaussian average over Φ and the quenched average over the patterns $\{\xi^\mu\}_{\mu=1,\ldots,l}$. In equilibrium, various correlation and response functions are expected to depend only on time difference, e.g., $C(t,t') = C(t-t')$. Thus the autocorrelation function satisfies ($t \geq 0$)

$$\frac{d}{dt}C(t) = -\frac{1}{2}[1+2a+\tanh\beta H(t-1)]C(t) + \frac{1}{2}[1-2a+\tanh\beta H(t-1)]\sigma(0), \quad (3.5c)$$

where the average is to be taken.

We first consider the statics of the dynamic model, which is described by the long-time limit of the variables characterizing the state of the network. In addition to the static macroscopic overlaps, we define the static order parameters for the autocorrelation and the random overlap correlation by

$$q \equiv \lim_{t\to\infty} C(t), \qquad r \equiv \lim_{t\to\infty} R(t), \quad (3.6)$$

which are the Edward-Anderson(EA) order parameter and the mean square random overlap, respectively. Meanwhile, G and $\tilde{S}$ should vanish in the long time limit, as long as we consider the case of a finite relaxation time. Also the statics is expected to be independent of the short-time parts of the fluctuations for the noncondensed patterns, in which case $\tilde{S}(t)$ and $R(t) - r$ can be set equal to zero. We further note that the average of a function over Φ can be reduced to the average over a Gaussian variable z with zero mean and unit variance:

$$\langle\langle f(\Phi(t))\rangle\rangle_\Phi = \int \frac{dz}{\sqrt{2\pi}} e^{-z^2/2} f(\sqrt{r}z) \equiv \langle\langle f(\sqrt{r}z)\rangle\rangle_z \quad (3.7a)$$

$$\langle\langle f(\Phi(0))f(\Phi(t))\rangle\rangle_\Phi = \langle\langle f(\sqrt{r}z)f(\sqrt{r}z)\rangle\rangle_z. \quad (3.7b)$$

This allows us to write the static order parameter equations in the form

$$m^\mu = \langle\langle \xi^\mu M_z\rangle\rangle, \quad (3.8a)$$

$$q = \langle\langle (M_z)^2\rangle\rangle, \quad (3.8b)$$

where

$$M_z \equiv \frac{1-2a+\tanh\beta H_z}{1+2a+\tanh\beta H_z}$$

stands for the activity of the neuron with local field $H_z \equiv h + \sqrt{\alpha r}z + \sum_{\mu=1}^{l}\xi^\mu m^\mu$, and $\langle\langle\ \rangle\rangle$ denotes both the Gaussian average over variable z and the quenched average over the random patterns $\{\xi^\mu\}_{\mu=1,\dots,l}$. Note that the local field H_z consists of three parts: an external field or the threshold value for the synaptic potential, a ferromagnetic part $\sum_{\mu=1}^{l} \xi^\mu m^\mu$ resulting from the l condensed overlaps, and a Gaussian noise $\sqrt{\alpha r}z$ generated by the instantaneous fluctuations of the random overlaps with the rest of the patterns.

Since the long-time limit results in the delta peak in the Fourier component $\omega = 0$, the Fourier transform of Eq. (3.4) yields

$$r = \frac{q}{(1-G_0)^2} = \frac{q}{(1-\langle\langle \frac{\partial}{\partial H_z} M_z \rangle\rangle)^2} \tag{3.8c}$$

with $G_0 \equiv \int dt\, G(t)$. In Ref. 9, Eqs. (3.8) have been derived through the use of a generating functional and analyzed in detail. When α is zero, Eqs. (3.8) are decoupled and Eq. (3.8a) with no external field ($h = 0$) reduces to that for the case of storing a finite number of patterns in Sec. II.

At zero temperature ($\beta \to \infty$), the network undergoes a first-order transition at $\alpha = \alpha_c$ from the disordered state to the mixed state as the number of stored patterns is lowered. The capacity of the network at zero temperature is given by $\alpha_c = 0.138/(1+a^2)$. For finite α, the transition temperature below which the ordered (Mattis) state appears can be computed numerically, and the overall phase diagram in the (T, α, a) space has been also obtained.

Next we investigate the dynamic behavior in equilibrium at high temperatures ($\beta^2\alpha \ll 1$). Equation (3.5c) can be integrated to give the autocorrelation function ($t \geq 0$)

$$\begin{aligned} C(t) &= \exp\Big(-\frac{1}{2}\int_0^t d\tau[1+2a+\tanh\beta H(\tau-1)]\Big) \\ &\quad +\frac{1}{2}\int_0^t dt' \exp\Big(-\frac{1}{2}\int_{t'}^t d\tau[1+2a+\tanh\beta H(\tau-1)]\Big) \\ &\qquad \times\sigma(0)[1-2a+\tanh\beta H(t'-1)], \end{aligned} \tag{3.9}$$

which is to be averaged over Φ and $\{\xi^\mu\}$. To the zeroth order (in $\beta^2\alpha$), Eq. (3.9) yields the ordinary exponential relaxation

$$C(t) = q + (1-q)\exp(-\frac{1+2a}{2}t), \tag{3.10}$$

where $q = \frac{1-2a}{1+2a}$ to the zeroth order. it is tedious but straightforward to calculate the first order contribution. Upon Fourier transforming, we obtain

$$\begin{aligned} \tilde{C}(\omega) &\equiv C(\omega) - q\delta(\omega) \\ &= \frac{32a}{(1+2a)[(1+2a)^2+4\omega^2]} + \beta^2\alpha\frac{16a^2}{(1+2a)^2[(1+2a)^2+4\omega^2]}\tilde{R}(\omega) \\ &\quad +\beta^2\alpha F(\omega;a), \end{aligned} \tag{3.11}$$

where $F(\omega; a)$ is a complicated function of ω and a. (Its detailed form is not important.) In the limit $\omega \to 0$ the *fluctuation–dissipation theorem* reads

$$\mathrm{Im}G(\omega) = \frac{1+a}{2}\beta\omega\tilde{C}(\omega). \tag{3.12}$$

For low frequencies, therefore, Eq. (3.11) becomes

$$\tilde{C}(\omega) = \frac{32a}{(1+2a)^3} + \beta^2\alpha\frac{16a^2}{(1+2a)^4}\frac{\tilde{C}(\omega)}{(1-G_0)^2} + \beta^2\alpha F(\omega=0;a), \tag{3.13}$$

which displays the breakdown of expansion as $\beta^2\alpha$ approaches $(1+2a)^4(1-G_0)^2/16a^2$. For small β, Eq. (3.8c) leads to

$$(1-G_0)^2 = \frac{q}{r} \approx \left[1 - \frac{4a\beta}{(1+2a)^2}\right]^2,$$

and the generalized *Almeida–Thouless* (AT) *line* below which the expansion breaks down is given by

$$T_c \equiv \beta_c^{-1} = \frac{4a}{(1+2a)^2}(1+\sqrt{\alpha}). \tag{3.14}$$

For $a = 1/2$, Eq. (3.14) exactly corresponds to the spin–glass transition temperature found in the static analysis. For $a \neq 1/2$, on the other hand, Eq. (3.14) is to be contrasted with the static analysis which indicates the absence of spin–glass transition due to lack of the spin up–down symmetry.[9] In this case it would be of interest to investigate the nature of the dynamic AT line. Below the AT line, higher–order contributions to $\tilde{C}(\omega)$ must be included, and the dynamics for $\alpha \neq 0$ is expected to be governed by algebraic rather than exponential relaxation. Exactly on the AT line, preliminary analysis indeed indicates algebraic decay of the correlation function: $\tilde{C}(t) \sim t^{-\nu}$ with $\nu = 1/3$. Dynamic properties below the transition temperature and their implications to the memory are also of interest and for further investigation.

ACKNOWLEDGEMENTS

I have benefited from my collaborators, G.M. Shim and D. Kim and thanks to the organizers of the 1991 Taipei International Symposium on Statistical Physics for the invitation and hospitality. This work was supported in part by the Seoul National University-Daewoo Research Fund and in part by the Basic Science Research Institute Program, Ministry of Education of Korea.

REFERENCES

1. W. A. Little, Math. Biosci. **19**, 101 (1974); W. A. Little and G. A. Shaw, Math. Biosci. **39**, 281 (1978).

2. J. J. Hopfield, Proc. Natl. Acad. Sci. USA **79**, 2554 (1982).

3. P. Peretto, Biol. Cybernet. **50**, 51 (1984); W. Kinzel, Z. Phys. B **60**, 205 (1985); D. J. Amit, H. Gutfreund, and H. Sompolinsky, Phys. Rev. A **32**, 1007 (1985).

4. For a recent review and an extensive list of references, see, e.g., D. J. Amit, *Modeling Brain Function* (Cambridge University Press, Cambridge, 1989) and the articles in J. Phys. A **22** (1989).

5. B. Derrida, E. Gardner, and A. Zippelius, Europhys. Lett. **4**, 167 (1987); A. Crisanti and H. Sompolinsky, Phys. Rev. A **37**, 4865 (1988); M. Shiino and T. Fukai, J. Phys. A **23**, L1009 (1990).

6. M. Y. Choi, Phys. Rev. Lett. **61**, 2809 (1988).

7. H. J. Sommers, Phys. Rev. Lett. **58**, 1268 (1987); H. Rieger, M. Schreckenberg, and J. Zittartz, Z. Phys. B **72**, 523 (1988).

8. M.Y. Choi and B. A. Huberman, Phys. Rev. B **31**, 2862 (1985).

9. G. M. Shim, M. Y. Choi, and D. Kim, Phys. Rev. A **43**, 1079 (1991).

The Abelian Sandpile Model Of Self-organized Criticality

DEEPAK DHAR

Tata Institute of Fundamental Research,
Homi Bhabha Road, Bombay 400005, India

ABSTRACT

Recent results about the abelian sandpile model of self-organized criticality are briefly reviewed.

The abelian sandpile model has attracted a lot of attention recently as an example of self-organized critical[1-13] system which shows interesting mathematical structure and this enables calculation of several properties of the model analytically. This talk reviewed what is known about the model at present, focussing mainly on my own investigations in this field undertaken in collaboration with R. Ramaswamy of Jawaharlal Neheru University (New Delhi) and S.N. Majumdar at Tata Institute(Bombay). Since most of this work has already appeared in print elsewhere, only a brief summary with reference to original articles is given here.

The sandpile model of self-organized criticality was defined by Bak, Tang and Wiesenfeld.[1-2]The model is defined as follows: The configuration of a sandpile is defined by specifying the 'height of the sandpile' at each site of a lattice. We take the heights to be non negative integers. Sand is added to the pile one grain at a time, at random. adding a grain at a site, causes the height there to increase by 1. If the height of the sandpile at any site exceeds a critical value, that site becomes unstable, causing z grains of sand to drop out of that site. These grains drop at nearby sites, increasing the height of each of them by 1. Sand particles leave the system if a toppling occurs at the boundary of the lattice.

As the number of particles added tends to infinity, the sandpile evolves to a steady state. In the steady state, the size of disturbance (avalanche), caused by the addition of a particle, is a random variable with a probability distribution that decreases as a power-law for large avalanches. The precise value of the exponent depends on the measure of avalanche size used. These powers have been estimated by numerical simulations,[1,2,9-11]though as yet there does not seem to be a consensus about the precise values of the exponents even for the two dimensional model.

This model, clearly, does not describe the properties of real sand very well. In particular, one expects that a toppling criterion based on gradients of heights would be more realistic. Also, since topplings are forced by gravity, one expects that more particles will drop downward than upward. However, it turns out that the model, where the criticality condition depends on the height variable itself, has particularly simple mathematical structure, which allows an analytical calculation of several of its properties.

Let C denote any stable configuration of the sandpile. We define operators a_i which act on the space of stable configurations by the equation

$$a_i C = C' \tag{01}$$

for all C, where C' is the stable configuration obtained from C by adding a paricle at site i, and allowing the system to relax. Then one notices that the operators a_i's commute with each other

$$[a_i, a_j] = 0 \qquad \text{for all} \quad i \quad and \quad j. \tag{02}$$

Models where Eq. (2) holds, are particularly simple, and would be refered to as abelian sandpile model (ASM) in the following.

In addition, there are other relations satisfied by these operators. For the 2-dimensional square lattice, these relations are of the form[13]

$$a_i{}^4 = a_j a_k a_l a_m \tag{03}$$

where the sites j, k, l and m are the four neighbours of i.

Using the commutativity of the a_i's, one can show[13]that in the critical state, most stable configurations do not occur at all. The configurations that do occur, do so with equal probability. There exists a simple recursive algorithm called the burning algorithm to decide whether a given configuration is allowed or not in the steady state. The total number of allowed configurations is given by the determinant of an $N \times N$ integer matrix (N=number of sites in the lattice) specifying the toppling rules. For the square lattice, the number of allowed states varies as $\exp(\mu N)$ for large N, where

$$\mu = \frac{1}{(2\pi)^2} \int_0^{2\pi} d\theta \int_0^{2\pi} d\phi \, \ln(4 - 2\cos\theta - 2\cos\phi) \tag{04}$$

Eq. (4) is easily recognized as arising in the asymptotic enumeration of spanning trees on the square lattice[14] . In fact, one can establish a direct one to one correspondence between the recurrent configurations in the critical state of ASM, and spanning trees on the same lattice. [15] . The spanning tree problem is known to correspond to the q-state Potts model in the $q \to 0$ limit. We thus have an intrigueing correspondence between the non-equilibrium ASM and an equlibrium statistical mechanical model.

Let G_{ij} be the average number of topplings at site i if one adds a particle at site j. Then it can be shown that G_{ij} satisfies the Laplace's equation

$$\nabla^2 G_{ij} = \delta_{ij} \tag{05}$$

which has a simple solution $G(r) \sim r^{2-d}$ in d-dimensions. This implies that in this problem $\eta = 0$, a result which also follows from the Potts model equivalence.

For relaxation to the critical state, the spectrum of relaxation times can also be determined completely[13]. We find that in d-dimensions, the largest relaxation time varies as L^d, where L is the linear size of the system.

The heights of different sites in the critical state are correlated. For example, two adjacent sites both cannot have minimum height 1. It turns out that one can calculate exactly[16] the fraction of sites f_1 having height 1. For a square lattice we get

$$f_1 = \frac{2}{\pi^2}(1 - \frac{2}{\pi}) \tag{06}$$

We also showed that the correlations in height at two sites separated by distance R vary as R^{-2d} in d-dimensions. No analytical calculation of f_2, f_3 or f_4, the concentration of sites having heights 2, 3 or 4 respectively, has been possible so far, though some numerical estimates are available[9].

The only other analytical result about the undirected 2-dimensional ASM is related to the known chemical distance exponent for spanning trees $z = 5/4$.[17] This implies a scaling relation between exponents characterizing the distribution of sizes of avalanches. Let the probability that the duration of an avalanche, resulting from adding a single particle, exceeds T vary as $T^{-\tau_t+1}$, and the probability that the number of sites affected exceeds S vary as $S^{-\tau_s+1}$, then scaling arguments give[15]

$$\frac{\tau_t - 1}{\tau_s - 1} = \frac{8}{5} \tag{07}$$

In special cases, the ASM can be solved exactly. One case is the ASM with preferred direction. This model becomes equivalent to the voter model.[18] The upper critical dimension for this model is 3. In $d \geq 3$, we get

$$\tau_t = 2, \qquad \tau_s = 3/2$$

In the two dimensional case, we find the problem is equivalent to annihilating random walkers, and one gets

$$\tau_t = 3/2, \qquad \tau_s = 4/3$$

Another special case is the ASM on the Bethe lattice. In this case, explicit expressions for various avalanche distribution functions have been obtained.[19] The exponents agree with the mean field percolation exponents

$$\tau_t = 2, \qquad \tau_s = 3/2$$

Another quantity which has evoked much interest is the power spectrum of the fluctuations. Originally, Bak et al had argued that this system shows '$1/f$' type power spectra. Numerical experiments of Kertesz and Kiss[20] have shown fairly convincingly that the power spectra of ASM is '$1/f^2$' type. Numerical simulations of the directed 2-dimensional ASM by us[21] show that power spectra are '$1/f^2$' type even for finite rate of sand addition, except at very small frequencies.

I believe that a clearer understanding of the critical exponents of the ASM (eg. power laws characterizing the avalanche distributions) will help in understanding the wider problem self-organized criticality in general. This seems to be a promising area for further study.

REFERENCES

1. P. Bak, C. Tang and K. Wiesenfeld, *Phys. Rev. Lett,* **59,** 381 (1987).
2. P. Bak, C. Tang and K. Wiesenfeld, *Phys. Rev.,* **A38,** 364 (1988).
3. C. Tang and P. Bak, *J. Stat. Phys,* **51,** 797 (1988).
4. L.P. Kadanoff, S.R. Nagel, L. Wu and SM. Zhou, *Phys. Rev.,* **A39,** 6524 (1989).
5. L.P. Kadanoff, *Physica,* **A163,** 1 (1990).
6. T. Hwa and M. Kardar, *Phys. Rev. Lett,* **62,** 1813 (1989).
7. T. Hwa, Ph.D. Thesis (MIT), unpublished.
8. P. Bak and K. Chen, *Physica,* **D38,** 5 (1989).
9. S.S. Manna, *J. Stat. Phys,* **59,** 509 (1990).
10. P. Grassberger and S.S. Manna, *J. Physique,* **51,** 1077 (1990).
11. S.S. Manna, HLRZ preprint, 1991.
12. K. Wiesenfeld, J. Theiler and B. McNamara, *Phys. Rev. Lett,* **65,** 949 (1990).
13. D. Dhar, *Phys. Rev. Lett,* **64,** 1613 (1990).
14. F.Y. Wu, *Rev. Mod. Phys,* **54,** 235 (1982).
15. S.N. Majumdar and D. Dhar, to appear in Physica **A**, 1992.
16. S.N. Majumdar and D. Dhar, *J. Phys.,* **A24,** L357 (1991).
17. A. Coniglio, *Phys. Rev. Lett,* **62,** 3054 (1989).
18. D. Dhar and R. Ramaswamy, *Phys. Rev. Lett,* **63,** 1659 (1989).
19. D. Dhar and S.N. Majumdar, *J. Phys.,* **A23,** 4333 (1990).
20. J. Kertesz and L.B. Kiss, *J. Phys.,* **A23,** L433 (1990).
21. R. Ramaswamy and D. Dhar, unpublished.

QUANTUM SYSTEMS

NUMERICAL STUDIES OF CONJUGATED POLYMERS

W. P. Su
Department of Physics, University of Houston
Houston, Texas 77204-5504, U. S. A.

ABSTRACT

Conjugated polymers are well described by a simple one dimensional electron-phonon model. The characteristic one dimensional lattice instability leads to the formation of localized excitations such as solitons and polarons. Many important solutions of the Hamiltonian can be obtained by iterating the Hartree equations. But there are some subtle questions such as the effect of Coulomb interactions and quantum lattice fluctuations for which one needs to go beyond the mean field treatment. A numerical scheme for carrying out such studies is discussed.

I. INTRODUCTION

Many important and interesting molecules are conjugated. The heme group of myoglobin, chlorophyll and β-carotene etc. all derive their characteristic colors from the conjugated structures[1]. The *cis* to *trans* conformational change of the rhodopsin molecule constitutes the primary vision process[1]. Conjugated polymers are the simplest conjugated systems-long polyenes. As such they offer a convenient starting point for investigating the general properties of conjugated systems in general.

Since the successful synthesis of polyacetylene in the late seventies, there has been an impressive amount of experimental and theoretical work on conjugated polymers[2]. It is fair to say that we have reached a good theoretical understanding of those systems. Some of the ideas proposed for the polymers such as the soliton excitations with anomalous charge and spin quantum numbers have had a strong impact on other areas of physics. Here we attempt to give a brief review of the theoretical results. The model Hamiltonian and a simple numerical scheme to solve the mean field equations are discussed in Section II. Some mean field solutions are discussed in Section III. We also discuss the need to go beyond the mean field theory and the problem of including the electron-electron interactions and quantum lattice fluctuations. In Section IV a powerful path integral Monte Carlo method is introduced to handle this problem. Section V contains a summary.

II. MODEL HAMILTONIAN AND METHODOLOGY

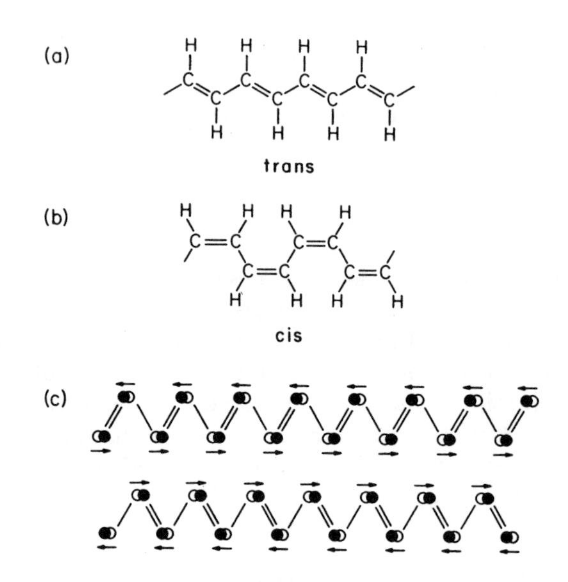

Fig. 1. Structural diagrams for polyacetylene: (a) *cis*-$(CH)_x$; (b) *trans*-$(CH)_x$; (c) the two degenerate ground states of *trans*-$(CH)_x$.

Polyacetylene $(CH)_x$ is a prototype conjugated polymer. It comes in two forms, the *trans* and *cis*, as shown in Figure 1. We will focus on the trans form first. To simplify the presentation we consider only the one dimensional motion of the carbons along the chain direction. The coordinate u_i denotes the displacement of the i-th carbon away from the equilibrium position(Figure 2).

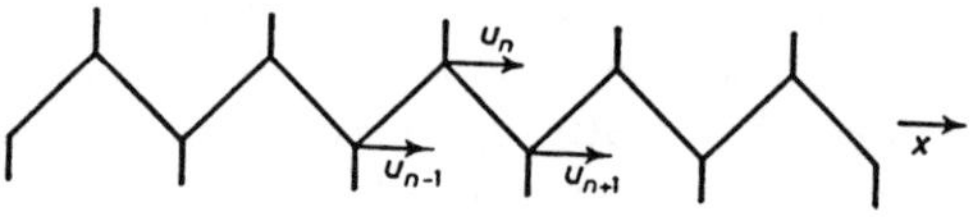

Fig. 2. Dimerization coordinate u_n defined for *trans*-$(CH)_x$.

$(CH)_x$ assumes a planar backbone structure due to the sp^2 hybridized orbitals. The σ-bonds between the carbons are hard to excite and for our purpose they act like harmonic springs connecting the carbons. The p_z orbitals perpendicular to the plane are partially occupied. The substantial overlap between adjacent p_z orbitals leads to a wide π-band of about 10 eV. Naively, one would expect pristine $(CH)_x$ to be a metal. Surprisingly, it turns out that a one dimensional metal is unstable. Any slight electron-phonon coupling would lead to a lattice distortion, which then turns the metal into a semiconductor or an insulator. For a half-filled system like $(CH)_x$ the lattice dimerizes, resulting in a bond alternation. For quantitative calculations Su, Schrieffer and Heeger[3] have proposed the following simple Hamiltonian

$$H = -\sum_{n,s} t_{n,n+1}(c^{\dagger}_{n+1,s}c_{n,s} + \text{H.c.}) + \frac{M}{2}\sum_n \dot{u}_n^2 + \frac{K}{2}\sum_n (u_n - u_{n+1})^2 \qquad (1)$$

where the hopping intergral of the π-electrons is usually linearized to

$$t_{n,n+1} = t_0 + \alpha(u_n - u_{n+1}). \qquad (2)$$

This is justified because the u's are of the order of 0.05Å compared to the average lattice spacing of 1.2Å.

Under ordinary circumstances one would assume that in the ground state, the expectation value of the u's all equal zero. Thus in the zeroth order approximation the π-electrons move in a uniform lattice with a constant hopping integral $-t_0$. To treat the phonons one quantizes the last two terms of the SSH Hamiltonian. Substituting the u dependent $t_{n,n+1}(2)$ into (1) results in a linear electron-phonon coupling. By using the standard many-body perturbation theory one can in principle carry out straightforward calculations of various physical quantities. However, the renormalized energies of some phonons turn out to be negative. In other words, the system can gain energy by emitting an infinite number of phonons with a particular momentum $2k_F$ (k_F is the Fermi momentum). So it seems starting the theory with a metallic phase is not a proper one. Instead a different approach in which the ground state lattice configuration is self-consistently determined rather than assumed has been more fruitful. In such an approach, the lattice field is first treated classically, quantum lattice fluctuations are left for more elaborate treatments. Specifically, for any given lattice configuration u_i , one can diagonalize the fermion part of the SSH Hamiltonian (the first term in (1)). By summing over the occupied states, one obtains the total electronic energy. Combining this with the lattice strain energy (the last term in (1)) gives the total enengy. Minimizing the total energy with respect to the lattice configuration then leads to static classical solutions. A convenient way of achieving this is to take the partial derivative of the Hamiltonian with respect to u_i and set the expectation value of the derivative to zero. In this way, the following set of Hartree equations are obtained

$$u_i = \frac{1}{2}(u_{i+1}+u_{i-1}) + \frac{\alpha}{2K}\sum_s < c^{\dagger}_{i+1,s}c_{i,s} + \text{H.c.} > - \frac{\alpha}{2K}\sum_s < c^{\dagger}_{i-1,s}c_{i,s} + \text{H.c.} > . \qquad (3)$$

Starting with any initial lattice configuration u_i one diagonalizes the fermion part of (1) and uses the single particle wave functions to evaluate the expectation value in (3). The left hand side of equation (3) yields updated value of u_i. This procedure is repeated until u_i converges. The calculation is usually carried out for a finite chain with a given number of electrons. The self-consistent lattice configuration depends sensitively on the average number of electrons per site ν per spin. The system dimerizes for $\nu = 1/2$ and trimerizes for $\nu = 1/3$ etc. By removing one electron from a half-filled system a polaron solution is found. Removing another electron, two solitons are found in the ground state of *trans*-polyacetylene and a bipolaron is found in *cis*-polyacetylene.

III. STATIC MEAN-FIELD SOLUTIONS

In this section we examine some mean field solutions:
(a) Half-filled case-the system dimerizes spontaneously in the ground state, with $u_n = u_0(-1)^n$. In *trans*-polyacetylene $u_0 = \pm|u_0|$ corresponding to two equivalent ground states which will be called the A and B phases respectively as shown in Figure 1(c). For each dimerized ground state configuration there is a gap in the electronic energy spectrum separating the occupied states from the empty states(Figure 3).

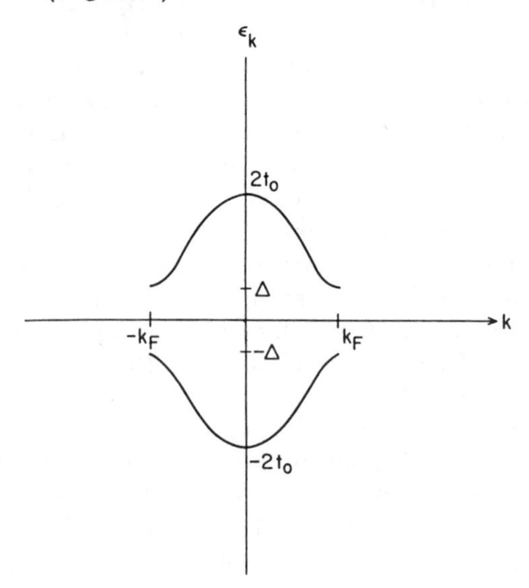

Fig. 3. π-bands for a dimerized chain in the reduced zone scheme.

Compared to the undimerized state(Figure 4) this represents a lower total electronic energy.

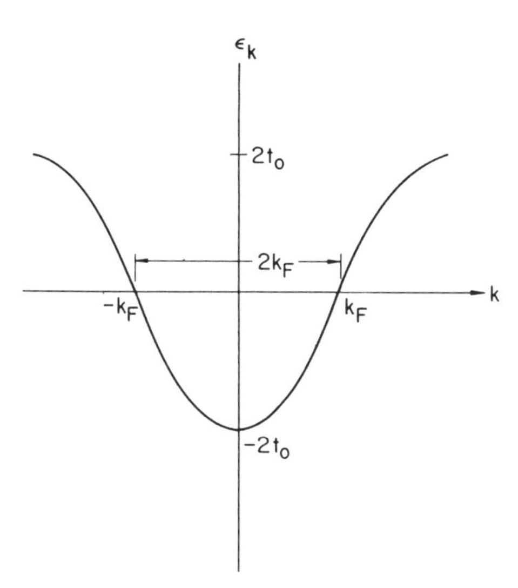

Fig. 4. π-band for undimerized $(CH)_x$.

This is the origin of the spontaneous lattice dimerization (the Peierls' distortion). The system is therefore a semiconductor. From an approximate analytical caculation the energy gap is found to be

$$E_g = 2\Delta = We^{-\frac{1}{\lambda}} \tag{4}$$

where W is the bandwidth $4t_0$ and $\lambda = \frac{4}{\pi}\frac{\alpha^2}{Kt_0}$ is the dimensionless electron-phonon coupling constant. It is clear from this expression that the dimerized state can never be reached from the metallic state through any perturbative approach.

In a nondegenerate polymer such as *cis*-polyacetylene the symmetry between the A and B phases is lifted by introducing a small symmetry-breaking term to the Hamiltonian in (1)

$$\Delta H = t_e \sum_{n,s} (-1)^n (c^{\dagger}_{n+1,s} c_{n,s} + \text{H.c.}). \tag{5}$$

This renders one of the conjugated states, A say, as the true ground state whereas B is only metastable.

(b) By removing an electron from an exactly half-filled system, a polaron configuration is found in the ground state. Removing another one yields two solitons. A soliton is a domain wall separating the A phase from the B phase(Figure 5). Because of topological constraint, the two solitons thus formed is a soliton-antisoliton pair. There is an effective repulsive interaction between the soliton and the antisoliton in trans-polyacetylene, whereas they attract each other and form a bound state(a bipolaron) in *cis*-polyacetylene.

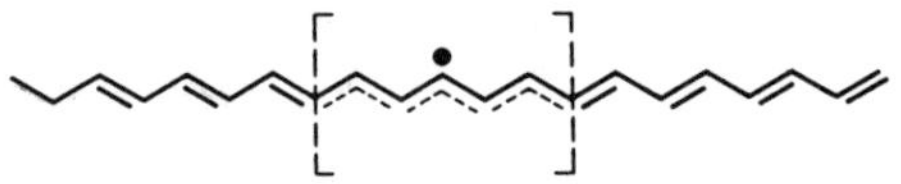

Fig. 5. Solitons (or bond-alternation domain walls in polyacetylene): schematic form of a neutral soliton on a *trans*-$(CH)_x$ chain.

Each of the polaron and soliton configurations is associated with a characteristic electronic structure within the energy gap. This can be experimentally detected as subgap absorption. Qualitative agreement between the mean field theory and experiment has been achieved. To improve on this, at least two things are required. The first is to consider quantum lattice effect. Some progress has been made along these lines within the Born-Oppenheimer approximation. The phonon effect usually manifests itself in the broadening of an absorption peak[4] or in the appearance of vibronic side peaks[5].

Another thing is to include the electron-electron interactions. Usually this is done by adding the following type of terms to the original Hamiltonian (1)

$$U\sum_{n} c^{\dagger}_{n\uparrow}c_{n\uparrow}c^{\dagger}_{n\downarrow}c_{n\downarrow} + \sum_{n,i,s,s'} V_i c^{\dagger}_{n,s}c_{n,s}c^{\dagger}_{n+i,s'}c_{n+i,s'} \tag{6}$$

The actual value of U and V are comparable to the energy gap. Thus the interactions are not negligible in general. Even a perturbative treatment of (6) leads to a substantial renormalization of the energy gap, the subgap absorption energy and the effective electron-phonon coupling constant. As another instance of the necessity to include the quantum lattice fluctuations and electron-electron interactions we mention the formation of a bipolaron versus two separated polarons in a nondegenerate polymer. The electron-phonon interaction favors bipolaron, but the electron-electron repulsion might favor separate polarons. Since the energy difference of these two competing effects is comparable to the phonon energy, quantum lattice fluctuations can be important.

IV. WORLD-LINE MONTE CARLO METHOD

The problem of quantum lattice fluctuations and electron-electron interactions can be rigorously treated by a powerful path integral Monte Carlo method. It has been employed by Hirsch *et al.*[6] to attack several one dimensional models. In this method one can follow the motion of the lattice and the electrons in imaginary time. The probability distribution of the lattice and the electronic configurations can be readily visualized. Various dynamical correlation functions can be calculated. For example the current-current correlation function

$$< J(\tau)J(0) > \tag{7}$$

contains information about the absorption intensity $\alpha(\omega)$. To extract spectral functions from the dynamical correlation functions in general requires inverse Laplace tranform, which is ill-defined. Fortunately good progress[7] on this problem has recently been made by adopting the maximum entropy method employed in the image reconstruction.

V. CONCLUSION

Conjugate polymers are interesting one dimensional materials which seem to be adequately modelled by some tight-binding Hamiltonians. Those types of electron-phonon Hamiltonians are not amenable to completely analytic treatments. Numerical calculation will remain a very powerful tool for many years ahead.

REFERENCES

1. L. Stryer, Biochemistry (Freeman, N. Y., 1981).
2. A. J. Heeger, S. Kivelson, J. R. Schrieffer, and W. P. Su, Rev. Mod. Phys. 60, 781 (1988).
3. W. P. Su, J. R. Schrieffer, and A. J. Heeger, Phys. Rev. B22, 2099 (1980).
4. J. Yu, H. Matsuoka, and W. P. Su, Phys. Rev. B37, 10367 (1988).
5. B. Friedman and W. P. Su, Phys. Rev. B39, 5152 (1989).
6. J. E. Hirsch, R. L. Sugar, D. J. Scalapino, and R. Blankenbecler, Phys. Rev. B26, 5033 (1982).
7. R. N. Silver, D. S. Sivia, and J. E. Gubernatis, Phys. Rev. B41, 2380 (1990).

PHASE TRANSITIONS IN TWO- AND THREE-DIMENSIONAL FLUIDS WITH INTERNAL QUANTUM STATES: COMPUTER SIMULATIONS AND THEORY

P. Nielaba
Institut für Physik, Universität Mainz, Staudingerweg 7, D-6500 Mainz, F.R.G.

ABSTRACT

The properties of model fluids are investigated, whose particles have classical degrees of freedom in two and three dimensions and two internal quantum states. Attractive interactions are "turned on", when the internal states are hybridized, corresponding to molecules acquiring a "dipole" moment. The phase diagram of this system in the temperature- density plane as well as the static and imaginary time correlations at various densities are investigated by path integral Monte Carlo simulations. These are compared with mean field theory predictions.

INTRODUCTION

Phase transitions in two dimensional systems have become an interesting field of research in the last years. One part of the motivation comes from new experimental facts [1-4], an other from model computations [5-7] as well as from exact solutions of certain statistical models in two dimensions. Here we report on path integral Monte Carlo simulation results of particular phase transitions in two- [8] and in three[9] dimensions, finite size scaling[10] block analysis techniques[11, 12] are utilized.

BLOCK ANALYSIS METHOD

The idea of the block size analysis[11, 12] is briefly sketched. Consider a fluid system in a volume $\Lambda = S^d$ (d is the dimension of space). Further we imagine that the volume is divided into cells of linear dimension b. We can ask now for the the idealized distribution of densities which we find inside the cells or blocks with a linear dimension b with ξ (correlation length) $\ll b \ll S$. In a one phase fluid situation the underlying system is homogeneous yielding an unimodal gaussian density distribution function

$$P_b \propto \exp(-\frac{\beta(\rho-\bar{\rho})^2 b^d}{2\bar{\rho}^2 K^{(b)}}) \tag{1}$$

centered around the average density $\bar{\rho}$ of the overall system, β^{-1} is the temperature. The width of the gaussian depends on the block size dependent *effective* isothermal compressibility $K^{(b)}$ of the fluid phase. Using scaling arguments[11, 12] the *physical* compressibility in the thermodynamic limit $K := K^{(\infty)}$ is accessible in an NVT - ensemble using

$$K^{(b)} = K(1 - K^{bc}(\xi/b)) \tag{2}$$

where K^{bc} stands for a boundary correction term. In a two phase situation with coexistence of a gas and a liquid phase the distribution function is bimodal

$$P_b \propto C_{gas} \exp(-\frac{\beta(\rho - \rho_{gas})^2 b^d}{2\rho_{gas}^2 K_{gas}^{(b)}}) + C_{liq} \exp(-\frac{\beta(\rho - \rho_{liq})^2 b^d}{2\rho_{liq}^2 K_{liq}^{(b)}}) \quad (3)$$

peaked around the densities ρ_{gas} and ρ_{liq}, respectively; C_{gas} and C_{liq} are weight factors. The shape of the density distribution is distinctly non gaussian for densities in between ρ_{gas} and ρ_{liq} due to interfacial effects of the coexisting phases. Nevertheless the widths and positions of these peaks can be used to extract the phase densities and compressibilities in the thermodynamic limit.

FLUIDS WITH INTERNAL QUANTUM STATES

We study a system of particles (molecules) whose relevant internal states can be represented by a two-level tunneling system, while their translations can be treated classically. We ignore all other degrees of freedom. The N-particle Hamiltonian of the system is

$$H = \sum_{i=1}^{N} \mathbf{p}_i^2/2M - \frac{1}{2}\omega_0 \sum_{i=1}^{N} \sigma_i^x + \sum_{i<j} U(r_{ij}) - \sum_{i<j} J(r_{ij})\sigma_i^z \sigma_j^z \quad (4)$$

$$H = K + V; \qquad r_{ij} = \mathbf{r}_i - \mathbf{r}_j \ ;$$

where $\mathbf{p}_i$ and $\mathbf{r}_i$ are the momentum and position in two or three dimensions of particle i, M is the mass of the particles, σ^x and σ^z are the usual Pauli spin - 1/2 matrices, and K and V are, respectively, the kinetic and potential energy; the latter consists of a one-particle (two-level) part and two pair interaction terms $U(r)$ and $J(r)$, which will be specified later.

We are interested in the equilibrium properties of this system when the mass M is sufficiently large for the translational degrees of freedom to be treated classically. The classical-quantum canonical distribution function is then a product of a purely classical momentum part $\propto exp(-\beta K)$ (which is trivial) and an N-particle density matrix μ; μ is classical, e.g., diagonal, in the coordinates $\{\mathbf{r}_i\}$ and a quantum-mechanical operator in the spin variables $\sigma = \{\sigma_i\}$.

Examples of systems which can be modeled to a greater or lesser extent in this way, particularly when the positions are frozen in a regular or (quenched) disordered array, can be found in Refs.[13–19,9,8]. Our model corresponds to annealed disorder in which the positions $\{\mathbf{r}_i\}$ take on continuous values in some box Λ : we think of the particles as two state molecules with an internal Hamiltonian $-\omega_o\sigma^x/2$, interacting via a pair potential depending on their internal state[9]. The aim here is, however, not to mimic any particular real system but to establish reliable methods for dealing with the effects of strong interactions on the internal structure of molecules or atoms in two dimensions when there are no obvious collective coordinates in terms of which the description is simple. (We are particularly interested in being able to eventually treat systems such as molecules adsorbed on substrates at very low temperatures, which show very interesting phase diagrams [1–4]). For this reason we shall take $U(r)$ and $J(r)$ to have very simple forms, U is a hard sphere potential with diameter R and $J(r) = J$ for $R < r < 1.5R$ and zero elsewhere.

The important feature of the Hamiltonian (4) is that the interaction term will tend to lift particles out of their internal ground state corresponding to $\sigma^x = 1$ into a hybrid state, i.e., the eigenstates of σ^z. We shall study this phenomenon as a function of β and the particle density ρ.

The limit $\omega_0 \to 0$ corresponds to a two-component classical system with attractive interactions $J(r)$ between particles of the same species, which favors segregation. The opposite limit, $\omega_0 \to \infty$, gives a one - component classical system with only hard core interaction $U(r)$.

We study the properties of this system by path integral Monte Carlo methods (cf. Ref.20 for an overview), and compare the results of the simulations with mean field theory.

The path integral MC formalism is most easily described by considering the partition function [20]

$$Z(\beta, N, \Lambda) = \lambda^{-3N} \frac{1}{N!} \int \cdots \int d\mathbf{r}_1 \cdots d\mathbf{r}_N tr_\sigma exp(-\beta V), \tag{5}$$

where λ is the de Broglie wavelength. Following Suzuki [18], we use the Trotter formula to write Z as [17, 9, 8]

$$Z = \lim_{L \to \infty} \frac{A_L^{NL}}{\lambda^{3N} N!} \int d\mathbf{r}_1 \cdots d\mathbf{r}_N e^{-\beta \sum_{i<j} U(r_{ij})} \sum_{\{S\}} e^{-\beta \tilde{V}_L(\{S\})}, \tag{6}$$

where

$$-\tilde{V}_L(\{S\}) = \sum_{i=1}^{N} \sum_{l=1}^{L} \left(K_L S_{i,l} S_{i,l+1} + \frac{1}{L} \sum_{j=i+1}^{N} J(r_{ij}) S_{i,l} S_{j,l} \right), \tag{7}$$

$$S_{i,l} = \pm 1.$$

and $A_L = [\frac{1}{2} sinh(\beta\omega_0/L)]^{1/2}$, $K_L = \frac{1}{2\beta} ln[coth(\beta\omega_0/2L)]$. The properties of the system can now be obtained as thermal averages over the classical canonical distribution of the $N \times L$ particles, $exp\left(-\beta \sum_{i<j} U(r_{ij}) - \beta\tilde{V}_L(\{S\})\right)$.

In order to study quantitatively the features of the model described by eq.(4), for different densities and temperatures we[8, 9] performed Monte Carlo simulations for fixed $J = 1$ and $R = 1$, at $\omega_0/J = 4$ and several different temperatures. The number of classical particles N was 200 and the number of "monomers" in the polymer L was chosen such that $L/\beta = 40$ in most cases, these values of L appeared sufficient for points investigated, i.e., increasing them did not seem to affect the results. A typical run with 10^6 Monte Carlo steps took about 4 CPUh on a CRAY-YMP.

Using standard Monte Carlo techniques [20, 21], we computed the internal energy [17, 9, 8], the interaction energy, the magnetization and the susceptibility. In the two dimensional case we studied[8] in detail the block size distributions in this system as well as the compressibility and the specific heat[22].

We observed the following general behavior: as the density is increased there is a continuous increase of $-\beta u^z$, and a decrease of $-\beta u^x$, indicating a changeover from occupation of eigenstates of σ^x to that of σ^z,i.e., hybridization. The degree of this hybridization at a given ρ depends strongly on ω_0. The second order transition is visible by a large increase in the susceptibility. Beyond the phase transition density the magnetization takes nonzero values approaching one as ρ is increased, indicating the dominance of cooperative effects. Caused by the smaller coordination number in two dimensions, the phase transition density in three dimensions[9] is smaller than in two dimensions[8].

For small βJ the system undergoes, as ρ is increased, a second - order phase transition from the paramagnetic to the ferromagnetic phase. At larger values of βJ, a formation of clusters sets in, and the system undergoes a first order phase transition from the paramagnetic to the ferromagnetic phase. A mean field study (see below), shows the transition densities only poorly and in principle only contains the critical exponents of the system in three dimensions, which are different from the exponents in two dimensions, in two dimensions the mean field tricritical temperature is off by a factor of two as compared to the simulation results. The deviations of the mean field predictions for the location of the phase transition from the values obtained with the path integral Monte Carlo simulations are larger in two dimensions[8] compared to the three dimensional[9] case.

Histograms of the magnetization at temperatures above the tricritical point show a broad distribution about the magnetization $M = 0$ in the paramagnetic region and about $M \neq 0$ in the ferromagnetic region. At temperatures below the tricritical point the magnetizations are sharply peaked about $M \neq 0$ because of the presence of a high density ferromagnetic liquid phase in coexistence with a low density paramagnetic gas phase. The different behavior of the system at high and at low temperatures can be visualized in "snapshot" pictures, which show the two coexisting phases at low temperatures and the phase of the ferromagnetic fluid at high densities and temperatures. With an analysis of the density distributions in subsystems [11, 12, 8] the coexistence densities in the two phase region can be obtained. In this region we obtain density distributions peaked about two densities which are significantly shifted away from the average density to a high and a low density. At temperatures above the tricritical point we obtain only gaussian density distributions about the average density in the subsystems, in this case we have no two phase coexistence. Applying the sketched finite size block analysis technique in conjunction with the density cumulant intersection method[11, 12, 8, 22] we are able to locate this tricritical point at the end of the critical line at $((\beta_{tri} J)^{-1}, \rho_{tri} R^2) = (0.57 \pm 0.02, 0.45 \pm 0.01)$ in two dimensions.

For $(\beta J)^{-1} = 1$ in two dimensions we are above the tricritical temperature. At this temperature we computed the fluid phase isothermal compressibility[8] in the thermodynamic limit extracted via finite size scaling from the width of $P_b(\rho)$ as indicated in eq. (2). Near the para- ferromagnetic transition the fluctuations in the system are large, resulting in a small cusp in the compressibility which cannot be predicted by the mean- field Hamiltonian, whereas off criticality simulation and mean field are in satisfactory agreement as in the case of the susceptibility.

In three dimensions we studied[9] the behavior of the imaginary - time correlation functions $C_z(\tau, \rho)$ and $C_x(\tau, \rho)$. These give information about the degree to which the system is in "eigenstates" of σ^z or σ^x. At low densities the particles are in eigenstates of σ^x, $C_x(\tau, \rho)$ is nearly one, and $C_z(\tau, \rho)$ shows a pronounced minimum at $\tau = \beta/2$, indicating that the eigenstates of σ^z are least correlated at opposite points in the "quantum- polymer". With increasing density the system goes into hybrid states: the $\sigma^x$are now less correlated and the tendency to more correlated σ^z eigenstates increases. $C_x(\tau, \rho)$ starts building a minimum at $\beta/2$, and the value of $C_z(\beta/2, \rho)$ increases. As the density crosses the phase transition between the paramagnetic and ferromagnetic phases, $C_z(\tau, \rho)$ and $C_x(\tau, \rho)$ interchange their form. $C_x(\beta/2, \rho)$ is now smaller than $C_z(\beta/2, \rho)$, corresponding to the ordering tendency of the $\sigma_i^z \sigma_j^z$ interaction.

The pair correlation functions for higher densities show maxima at distances near multiples of the hard sphere diameter R with intermediate minima, representing the layering structure of the classical fluid. The probabilities for finding particles with parallel spin are enhanced at close distances, compared to the uncorrelated case. The integral of the difference $\beta J\rho[g_+(r) - g_-(r)]$ over the interaction shell gives the interaction energy.

For large distances the expectation values of $< \sigma_i^z \sigma_j^z >$ factorize. For densities beyond the transition, i.e., in the ferromagnetic phase, the probability for finding two parallel spins at large distances is higher than for antiparallel spins, $g_+(r) > g_-(r)$, the value of the difference, for $r \to \infty$, is given by m^2.

At $r = \frac{3}{2}R$ the correlation functions are discontinuous. The deviations of the ratios of $g_\pm(\frac{3}{2}R^-)/g_\pm(\frac{3}{2}R^+)$ from the classical ratios, given by the Boltzmann factors $exp(\pm\beta J)$, may be thought of in terms of a quantum screening effect. The two-particle approximation for $g_\pm(r)$ already shows this effect of the weakened effective interaction. The agreement of the Monte Carlo results with those obtained from the lowest terms in a virial expansion[9] is good for low densities.

The Hamiltonian of the mean-field version of our model is given by [13, 14, 15, 9]

$$H_{MF}^N = \sum_{i=1}^{N} \mathbf{p}_i^2/2M + \sum_{i<j} U(r_{ij}) - (J_0/N)\sum_{i<j} \sigma_i^z \sigma_j^z - (\omega_0/2)\sum_{i=1}^{N} \sigma_i^x, \tag{8}$$

i.e., the interaction between the internal degrees of freedom of two particles is distance independent; it decreases as $1/N$ with increasing number of particles to get a sensible thermodynamic limit.

In order to compare the results of mean-field theory with those of our model we need to choose the value of J_0 in eq.(8). Stratt [15] takes $J_0 = \rho \int d\mathbf{r} J(r) g(r)$, where $g(r)$ is the two-point correlation function of the underlying classical model. This is motivated as follows: In the mean-field model the magnetic field felt by, say, the first particle is $(J_0/N)\sum_{i>1} \sigma_i^z$. Replacing σ_i^z by m and approximating the two-particle distribution by the classical one, we find that the effective field on any particle can be approximated by $m\rho \int d\mathbf{r} J(r) g(r)$. Ideally, one should of course try to take the actual pair-correlation function $g(r)$ of the model. Since this is, however, not known, one takes in practice the $g(r)$ from the Percus-Yevick approximation for hard spheres. In the mean-field model, positional correlations are fully determined by the distance correlation functions of the underlying classical model.

We obtain $m = 0$ for $\beta \leq \beta_c$, where β_c is given by the solution of $(2J_0/\omega_0)$ $tanh(\beta_c\omega_0/2) = 1$. If $J_0 < \omega_0/2, \beta_c = \infty$. In this case no phase transition occurs. If $J_0 > \omega_0/2$, then for $\beta > \beta_c$ a phase is possible with $m \neq 0$.

In order to get the mean field phase diagram, it is important to remember that the free energy of any physical model always needs to be a convex function of the density. This is not the case for our mean-field model. However, if the mean-field free energy contains a concave part as a function of the density, we take the convex envelope. The result of introducing the convex envelope is that our system now also can show first order transitions.

By means of path integral Monte Carlo simulation we computed the phase diagram in the temperature - density plane of a liquid whose particles have internal quantum states. We evaluated the correlation functions in real space and imaginary time. For low density we found good agreement with a low density expansion. In our aim to understand more about complex phase transitions of two dimensional fluids with internal quantum states we combined block size analysis techniques successfully with path integral Monte Carlo simulation techniques. Thus additional studies of quantum effects in two dimensional systems at low temperatures are possible and will be presented in the near future [22].

Acknowledgements: The work presented is based on a cooperation with D. Marx, K. Binder, Ph. de Smedt, L. Dooms, J.L. Lebowitz and J. Talbot, The partial support from the *SFB 262* is gratefully acknowledged, many of the computations presented here were done partly on the CRAY-YMP of the HLRZ Jülich and at the RHRK in Kaiserslautern.

References

1.) S.K. Sinha (ed.) *Ordering in Two Dimensions* (North- Holland, Amsterdam 1980).

2.) A.D. Migone, H.K. Kim, M.H.W. Chan, J. Talbot, D.J. Tildesley, W.A. Steele, Phys. Rev. Lett. **51**, 192 (1983); N.S. Sullivan, J.M. Vaissiere, Phys. Rev. Lett. **51**, 658 (1983).

3.) H. Freimuth, H. Wiechert, H.J. Lauter, Surf. Sci **189/190**, 548 (1987).

4.) J. Chui, S.C. Fain, H. Freimuth, H. Wiechert, H.P. Schildberg, H.J. Lauter, Phys. Rev. Lett. **60**, 1848 (1988); ibid. **60**, 2704 (1988).

5.) S.F.O'Shea, M.L. Klein, Chem. Phys. Lett. **66**, 381 (1979); Phys. Rev. **B25**, 5882 (1982).

6.) O.G. Mouritsen, A.J. Berlinsky, Phys. Rev. Lett. **48**, 181 (1982).

7.) D.J. Tildesley, W.B. Street, Mol. Phys. **41**, 85 (1980); G. Fiorese, J. Chem. Phys. **73**, 6308 (1980); **75**, 1427 (1981); **75** 4747 (1981); B. Jönsson, G. Karlstrom, S. Romano, J. Chem. Phys. **74**, 2896 (1981).

8.) D. Marx, P. Nielaba, K. Binder, preprint.

9.) P. de Smedt, P. Nielaba, J.L. Lebowitz, J. Talbot, L. Dooms, Phys. Rev. **A38**, 1381 (1988).

10.) V. Privman (ed.) *Finite Size Scaling and Numerical Simulation* (World Scientific, Singapore 1990).

11.) K. Binder, Z. Phys. **B43**, 119 (1981); Ferroelectrics **73**, 43 (1987).

12.) M. Rovere, D.W. Heermann, and K. Binder, Europhys. Lett. **6**, 585 (1988); M. Rovere, D.W. Heermann, K. Binder, J. Phys. C: Condensed Matter **2**, 7009 (1990).

13.) D. Chandler, P.G. Wolynes, J. Chem. Phys. **74**, 4078 (1981); K.S. Schweizer, R.M Stratt, D. Chandler, P.G. Wolynes, *ibid* **75**, 1347 (1981); D. Chandler, K.S. Schweizer, P.G. Wolynes, Phys. Rev. Lett **49**, 1100 (1982).

14.) E. Martina, G. Stell, Phys. Rev. **B 26**, 1389 (1982).

15.) R.M. Stratt, J. Chem. Phys. **80**, 5764 (1984); Phys. Rev. Lett. **53**, 1305 (1984); S.G. Desjardins, R.M. Stratt, J. Chem. Phys. **81**, 6232 (1984).

16.) R.W. Hall, P.G. Wolynes, J. Stat. Phys. **43**, 935 (1986); Phys. Rev. **B 33**, 7879 (1986).

17.) P. Ballone, Ph. de Smedt, J.L. Lebowitz, J. Talbot, E. Waisman, Phys. Rev. **A 35**, 942 (1987).

18.) M. Suzuki, Prog. Theor. Phys. **46**, 1337 (1971); Commun. Math. Phys. **51**, 183 (1976); Prog. Theor. Phys. **56**, 1454 (1976).

19.) D.F. Coker, B.J. Berne, D. Thirumalai, J. Chem. Phys. **86**, 5689 (1987); M. Parrinello, A. Rahman, *ibid.* **80**, 860 (1984).

20.) Proceedings of the Conference of Frontiers of Quantum Monte Carlo, Los Alamos, 1985 [J. Stat. Phys. **43**, (1986)]; *Quantum Monte Carlo Methods*, edited by M. Suzuki (Springer- Verlag, Berlin, 1987); *Quantum Simulations of Condensed Matter Phenomena*, (Doll, Gubernatis (eds.)) (World Scientific, Singapur 1990).

21.) K. Binder, D.W. Heermann, *Monte Carlo Simulation in Statistical Physics: An Introduction* (Springer, Berlin 1988).

22.) D. Marx, P. Nielaba, K. Binder, in preparation.

CRYSTALLIZATION OF QUANTUM HARD SPHERES. A DENSITY FUNCTIONAL APPROACH

P. Nielaba
Institut für Physik, Universität Mainz, Staudingerweg 7, D-6500 Mainz, F.R.G.

ABSTRACT

At zero temperature Bose hard sphere systems are in the fluid phase at low densities and in the solid phase at high densities. The fluid- solid transition in this quantum system is described by density functional methods (modified weighted density functional theory). Predictions for solid- phase energies and for freezing parameters are in good agreement with available simulation data.

MODIFIED WEIGHTED DENSITY APPROXIMATION FOR QUANTUM LIQUIDS

A fundamental application of the density functional method[1] is to the freezing transition in simple liquids. Using information on the structure and thermodynamics of the *uniform* liquid, the method leads to predictions for the densities of the coexisting liquid and solid phases, the latent heat of transition, and the Lindemann parameter. Although agreement with simulation results tends to vary with the system studied and with the version of the method used, especially notable success has been obtained in the important case of the *classical* hard-sphere liquid, where the predicted freezing parameters agree with simulation usually within a few per cent.[1] An interesting issue, therefore, is the manner in which the general method may be extended from classical to *quantum* systems, and whether, in particular, it can illuminate the physical nature of the freezing transition in quantum liquids. Recently, one version of the density- functional method, the Ramakrishnan- Yussouff theory[2], has been extended[3, 4, 5] and applied to freezing of a Lennard- Jones model of ^4He at finite temperatures[3], and also to freezing (Wigner crystallization) of the ground- state Fermi one- component plasma[4]. Here we report on a general extension of a quite different version of the method, the modified weighted- density approximation[6, 8] from classical systems to quantum systems at zero temperature and demonstrate its utility in the specific case of freezing of a Bose liquid of *hard spheres.*

At zero temperature the ground state energy $E[\rho]$ of a nonuniform quantum liquid is a unique functional[9] of the one body density $\rho(\mathbf{r})$. For zero external field we can separate this functional into the *ideal- gas* energy $E_{id}[\rho]$, which can be treated exactly, and the *correlation* energy $E_c[\rho]$, containing interatomic interactions and exchange. $E_c[\rho]$ is generally not known for nonuniform systems, and we can approximate it by an extension of the modified weighted density approximation[6] (MWDA) to quantum liquids.[7] The basis of this approximation is that the average correlation energy per particle of the nonuniform system can be equated to its counterpart for the *uniform* liquid evaluated at a *weighted* density assumed to depend on a weighted average over the volume of the system of the (spatially varying) physical density, i.e.,

$$E_c^{MWDA}[\rho]/N = \varepsilon(\hat{\rho}) \tag{1}$$

where N is the number of particles, ε is the uniform- liquid correlation energy per particle and $\hat{\rho}$ is *defined* by

$$\hat{\rho} = \frac{1}{N}\int d^3r\rho(\mathbf{r})\int d^3r'\rho(\mathbf{r}')w(\mathbf{r}-\mathbf{r}';\hat{\rho}) \tag{2}$$

A *self- consistent* choice of the density argument of the "weight function" w in Eq.2 is essential[6, 8]. In order to ensure that the approximation is exact in the limit $[\rho(\mathbf{r}) \to \rho]$, w must be normalized to unity.
A *unique* determination of w follows from requiring that $E_c^{MDWA}[\rho]$ satisfies

$$\lim_{\rho(\mathbf{r})\to\rho} \frac{\delta^2 E_c^{MWDA}[\rho]}{\delta\rho(\mathbf{r})\delta\rho(\mathbf{r}')} = v(|\mathbf{r}-\mathbf{r}'|;\rho) \tag{3}$$

where $v(|\mathbf{r}-\mathbf{r}'|;\rho)$ is to be interpreted as an extension of the classical direct correlation function. Equation 3 ensures that a functional Taylor- series expansion of $E_c^{MWDA}[\rho]$ about the density of a uniform reference liquid is exact to second order, *and* includes approximate terms to all higher orders.[6, 8]

In Fouricr space the weight function is then determined by

$$w(k;\rho) = \frac{1}{2\varepsilon'(\rho)}[v(k,\rho) - \delta_{k,0}\rho\varepsilon''(\rho)], \tag{4}$$

where primes denote derivatives with respect to density. With the normalization condition for w it follows

$$v(k=0;\rho) = 2\varepsilon'(\rho) + \rho\varepsilon''(\rho) \tag{5}$$

which may be interpreted as a "quantum compressibility rule".

Equations 1, 2 and 4 now constitute the MDWA for a *nonuniform quantum liquid* at zero temperature. For the Bose hard sphere system, we can approximate the required liquid- state information ε and $v(k)$, *via* the paired phonon analysis[10] (PPA). The PPA gives approximations both for ε and for the structure factor $S(k)$, from which we then obtain $v(k)$ from the relation[11]

$$v(k) = \frac{\hbar^2k^2}{4m}(\frac{1}{S^2(k)} - 1) \tag{6}$$

where m is the mass of the particle. It is important to mention, however, that the PPA does not guarantee consistency between ε and $v(k)$, in the sense that the compressibility rule [Eq. 5] is not satisfied exactly. Therefore, in order to ensure that Eq. 5 *is* satisfied, we *scale* the PPA $v(k)$ by the factor $(2\varepsilon' + \rho\varepsilon'')/v(k=0)$, resulting in an increase in magnitude of about 20%.

GROUND STATE OF THE QUANTUM LIQUID

For the Bose hard sphere system one can approximate the ground state properties in the PPA. The N- body ground state wave function is approximated in the Jastrow form[12]

$$\psi_T(\mathbf{r_1},\cdots,\mathbf{r_N}) = \prod_{(i,j)} e^{-u(r_{ij})}, \tag{7}$$

where u is a variational function chosen to minimize the ground state energy,

$$\frac{\delta}{\delta u(\mathbf{r})}[\frac{<\psi_T, H\psi_T>}{<\psi_T,\psi_T>}] = 0. \tag{8}$$

The probability density $|\psi|^2$ corresponds to a Boltzmann factor of a classical system, thus the correlation function $g(r,[u])$ can be approximated by the hypernetted chain approximation. The correlation energy is given by

$$E_c(\rho) = \frac{\hbar^2}{4m}\rho \int d^3r g(r,[u])\nabla^2 u(r). \tag{9}$$

The change $\Delta u(r)$ is then computed in the random phase approximation and added to $u(r)$, and the procedure is repeated until convergence. $v(k)$ is then given by the structure factor $S(k)$ via Eq.6.

GROUND STATE ENERGY OF THE QUANTUM SOLID

In order to compute the ground state energy for the quantum *solid* we have to choose a crystal structure, to parametrize the one body density and to minimize the total solid energy with respect to the parametrized density. As in previous studies of the *classical* hard sphere system one can assume for the one body density distributions simple Gaussians at a given width α centered on perfect crystal lattice sites,

$$\rho_S(\mathbf{r}) = (\frac{\alpha}{\pi})^{3/2} \sum_{\mathbf{R}} e^{-\alpha(\mathbf{r}-\mathbf{R})^2} \tag{10}$$

It follows then from Eqs.4 and 10 that Eq.2 takes the form

$$\hat{\rho}(\alpha,\rho_S) = \rho_S[1 + \frac{1}{2\varepsilon'(\hat{\rho})} \sum_{\mathbf{G}\neq \mathbf{0}} e^{-G^2/2\alpha} v(G;\hat{\rho})], \tag{11}$$

where ρ_S is the average solid density, and G the magnitude of the reciprocal-lattice vector $\mathbf{G}$ of the solid density. This implicit equation for $\hat{\rho}$ can be solved for fixed α and ρ_S by numerical iteration, and the approximate correlation energy E_c^{MWDA} for the solid is then given by Eq.1. The ideal- gas energy is given by

$$E_{id}[\rho] = \frac{\hbar^2}{2m} \int d^3r |\nabla \sqrt{\rho(\mathbf{r})}|^2 \tag{12}$$

which results in the case for nonoverlapping Gaussians- a good approximation near freezing- in $E_{id}/N \simeq \frac{3}{4}\frac{\hbar^2}{m}\alpha$, identical to the form usually assumed in variational Monte Carlo simulations,[13]. The total solid ground state energy is then given by

$$E^{MWDA}(\alpha,\rho_S)/N \simeq \frac{3}{4}\frac{\hbar^2}{m}\alpha + \varepsilon(\hat{\rho}(\alpha,\rho_S)) \tag{13}$$

which is still to be minimized with respect to α for *fixed* ρ_S. The ideal gas energy E_{id} increases with α, strongly *opposing* localization of the atoms about lattice sites, while E_c falls off rapidly with α, strongly *favoring* localization. The competition between E_{id} and E_c may result- for sufficiently high ρ_S- in a minimum in the *total* energy at nonzero α, implying a mechanically stable solid. *Thermodynamic* stability of the solid relative to the liquid is determined by comparing the liquid and solid energies.

FREEZING OF A QUANTUM HARD SPHERE BOSE SYSTEM AT ZERO TEMPERATURE

By varying ρ_S and repeating the minimization procedure described above, the solid- phase energy curve (E/V vs. ρ_S) is obtained. In comparing this curve for a particular choice of crystal symmetry with the corresponding results for the liquid phase as obtained by the PPA, one observes a crossing of the two curves confirming a freezing transition. The results for the energies and the localization parameter α at a given density are in close agreement with the Monte Carlo results of Hansen, Levesque and Schiff[13]. α increases with the solid density indicating that the stronger localization at higher densities. The Lindemann parameter L_α is related to α at the solid coexistence density, for an fcc crystal we have $L_\alpha = (3/\alpha a^2)^{1/2}$, where a is the lattice constant. The coexistence densities are determined with the Maxwell construction, equating the pressures and chemical potentials of the two phases. The resulting coexistence densities in the case of an fcc lattice are $\rho_l\sigma^3 = 0.246, \rho_S\sigma^3 = 0.284$, resulting on a density change of $\Delta\rho\sigma^3 = 0.038$, in close agreement with the simulation results[13] ($\rho_l\sigma^3 = 0.23 \pm 0.02, \rho_S\sigma^3 = 0.25 \pm 0.02, \Delta\rho\sigma^3 = 0.02$). The Lindemann parameter $L_\alpha = 0.24$ is again in very good agreement with the value from the simulation ($L_\alpha = 0.27$).

The Lindemann parameter for the hard sphere Bose system is about three times the value for a classical hard sphere system, in agreement with simulation[12], the larger quantum value resulting from the strong variation of the energies with the localization.

For the bcc and the hcp crystal structures the freezing parameters are quite similar to the fcc values, these similarities for the close packed structures are known as well from simulations[14, 15] for ^4He.

In total the modified weighted density approximation for quantum hard spheres at zero temperature gives a good description of the freezing transition. However the required liquid state input is difficult to obtain, and even in the PPA the consistency of the results have to be enforced by scaling of the correlations.

Acknowledgements: The work presented is based on a cooperation with N.W. Ashcroft, A.R. Denton and K.J. Runge at the Cornell University. Support by the DFG is gratefully acknowledged.

References

1.) M. Baus, J. Phys.: Condens. Matter **2**, 2111; J. Stat. Phys. **48**, 1129 (1987); A.D.J. Haymet, Prog. Solid State Chem. **17**,1 (1986); R. Evans, Adv. Phys. **28**, 143 (1979).

2.) T.V. Ramakrishnan and M. Yussouff, Phys. Rev. **B19**, 2775 (1979); see also A.D.J. Haymet and D.J. Oxtoby, J. Chem. Phys. **74**, 2559 (1981).

3.) J.D. McCoy, S.W. Rick, and A.D.J. Haymet, J. Chem. Phys. **92**, 3034, 3040 (1990); J. Chem. Phys. **90**, 4662 (1981).

4.) G. Senatore and G. Pastore, Phys. Rev. Lett. **64**, 303 (1990).

5.) S.T. Chui, Phys. Rev. **B 41**, 796 (1990).
6.) A.R. Denton, and N.W. Ashcroft, Phys. Rev. **A 39**, 4701 (1989).
7.) A.R. Denton, P. Nielaba, K.J. Runge, N.W. Ashcroft; Phys. Rev. Lett. **64**, 1529 (1990), A.R. Denton, P. Nielaba, K.J. Runge, N.W. Ashcroft; J. Phys.: Condensed Matter **3**, 593 (1991).
8.) W.A. Curtin and N.W. Ashcroft, Phys. Rev. **A 32**, 2909 (1985); Phys. Rev. Lett. **56**, 2775 (1986).
9.) P. Hohenberg and W. Kohn, Phys. Rev. **136**, B864 (1964).
10.) C.C. Chang and C.E. Campbell, Phys. Rev. **B 15**, 4238 (1977); H.W. Jackson and E. Feenberg, Ann. Phys. (N.Y.) **15**, 266 (1961).
11.) E. Krotscheck, Phys. Rev. **B 33**, 3158 (1986).
12.) W.L. McMillan, Phys. Rev. **A 138**, 442 (1965).
13.) J.-P. Hansen, D. Levesque, and D. Schiff, Phys. Rev. **A 3**, 776 (1971).
14.) P.A. Whitlock, M.H. Kalos, G.V. Chester, and D.M. Ceperley, Phys. Rev. **B21**, 999 (1981).
15.) D. Frenkel and A.J.C. Ladd, J. Chem. Phys. **81**, 3188 (1984).

ROTATOR IMPURITY IN A CRYSTAL. LATTICE DEFORMATIONS AND QUANTUM MONTE CARLO SIMULATIONS

P. Nielaba
Institut für Physik, Universität Mainz, Staudingerweg 7, D-6500 Mainz, F.R.G.

ABSTRACT

At zero temperature the equilibrium structures of a system consisting of a quantum- rotator embedded in a relaxing surrounding are studied with a variational approach. At finite temperatures we investigate the properties of this system with a quantum Monte Carlo simulation. We obtain results for the energies, eigen state occupation numbers and the local lattice structure.

INTRODUCTION

Recent results [1, 2, 3] on experimental investigations of orientational glasses reveal a wealth of very interesting properties. For example, even the structure at low temperatures is very anomalous, as exhibited by a broad but singular lineshape of Bragg peaks [2]. Such results can neither be explained by analytical theories where the coupling between rotational and translational degrees of freedom is treated on a mean-field like level [4] nor by Edwards- Anderson- type models of orientational glasses [5] which omit the translational degree of freedom altogether. A related exciting low temperature property is the anomalous specific heat, which commonly [6] is attributed to two-level systems, which also show up in anomalous dynamical properties. For a detailed understanding of such properties, computer simulations of realistic models would be rather desirable.

A recent molecular dynamics study of a model for $N_2 - Ar$ mixtures has provided a rather satisfactory overall agreement of the simulated phase diagram [7, 8] with experiment [7]. This *purely classical* approach cannot account for the low temperature properties of this system, of course.

As a first step to develop methods for understanding such low temperature properties of disordered solids, we consider a simple limiting case, namely the ground state structural properties of a model for $N_2 - Ar$ mixtures in the diluted limit. We treat only the N_2 molecules as a quantum rotator, while the Ar atoms are treated classically. For a given configuration of the classical particles we solve the eigenvalue problem for the lowest eigenstate[9]. Thus we obtain the ground state equilibrium configurations of the coupled system. This approach is then combined[10] with a Monte Carlo procedure to treat low but nonzero temperatures. Already at the present stage of our approach, the agreement with pertinent experimental results is encouraging.

One of the reasons of our investigation was the study of the contributions of the lattice deformations due to the quantum nature of the rotator at low temperatures. The new positions of the Ar atoms result in a new potential for the rotator, and so in turn the lattice deformations have an effect on the level splittings of the rotator and thus the tunneling behavior.

OPTIMIZED WAVE FUNCTION AND LATTICE DEFORMATION

At zero temperature the Ar atoms occupy well defined lattice positions on a fcc lattice. The substitution of a single Ar atom by a linear molecule (we think of N_2 immersed in an Ar lattice in the diluted limit) results in a lattice deformation around the impurity site. One of the reasons of our investigation was the study of the contributions of the lattice deformations due to the quantum nature of the rotator at low temperatures. The new positions of the Ar atoms result in a new potential for the rotator, and so in turn the lattice deformations have an effect on the level splittings of the rotator and thus the tunneling behavior.

The procedure for the computation is now described briefly. As model parameters for the Lennard-Jones potentials V_{Ar-Ar}, V_{Ar-N} between our particles we take the values obtained previously [8]. We put $M+1$ Ar particles with positions $\{R_i\}$ on the fcc lattice sites $\{R_i^0\}$ and substitute a single Ar particle by a rigid rotator (N_2) with the center of mass located on the lattice site $r_0 = R_0$ and compute the symmetrized potential $V_S(\vartheta, \varphi, \{R_i\})$ (ϑ and φ are polar angles with the C_4 symmetry axes) for the rotator atoms on the sphere with the diameter d (for N_2:[8] $d = \sigma^l_{N-N}$) given by the distance between the rotator atoms at positions r_1 and r_2 , $r_{1,2} = R_0 \pm (\sin\vartheta\cos\varphi, \sin\vartheta\sin\varphi, \cos\vartheta)d/2$.

$$V_S(\vartheta, \varphi, \{R_i\}) = \frac{1}{2}\sum_{i=1}^{M}[V_{N-Ar}(r_1 - R_i) + V_{N-Ar}(r_2 - R_i)] \qquad (1)$$

This potential has 8 maxima in the (111) directions, 6 minima in the (100) directions, and neither minima nor saddle points are in the (110) directions, in contrast to fixed model potentials studied for rigid rotators previously [11, 12, 13].

Next we compute approximatively the energy eigenvalues and eigenfunctions for the rotator with rotational constant B (for N_2: [8] $B = B_{N_2} = 4.00345 \times 10^{-23} J$) in the potential $V_S(\vartheta, \varphi, \{R_i\})$. For fixed positions of the Ar particles we have to solve the eigenvalue problem with the Hamiltonian H_Q:

$$H_Q = BL^2/\hbar^2 + V_S(\vartheta, \varphi, \{R_i\}) \qquad (2)$$

L is the angular momentum operator. The total model Hamiltonian H is given by

$$H = H_Q + \sum_{i<j}^{M} V_{Ar-Ar}(R_i - R_j) \qquad (3)$$

Similar model Hamiltonians with a coupling of a quantum particle to a heat bath have been studied recently [14].

Our potential V_S is of cubic symmetry, thus the Hamiltonian is invariant under operations with elements of the cubic group O_h. At low temperatures we are only interested in the lowest energy levels, in particular of the ten possible eigenfunction symmetry types [15] only the solutions for the irreducible representations A_{1g} and T_{1u} of O_h. The energy eigenvalues $\varepsilon_k^{A_{1g}}$ (k is the angular momentum eigenvalue of the free rotator) are non degenerate, the eigenfunctions are symmetric, $\varepsilon_0 = \varepsilon_0^{A_{1g}}$ denotes the ground state of the rotator. The eigenvalues $\varepsilon_k^{T_{1u}}$ are triply degenerate, an eigenfunction is antisymmetric and has a nodal line along one of the C_4 axes, $\varepsilon_1 = \varepsilon_1^{T_{1u}}$ denotes the lowest possible eigenvalue in this symmetry.

For a particular symmetry type ST we expand the eigenfunctions with symmetry adapted combinations of spherical harmonics [16] $X_{k,n}^{ST}$ (n distinguishes orthonormal functions for the same k and different coefficients for the magnetic numbers m).

$$\psi^{ST}(\vartheta,\varphi,\{R_i\}) = \sum_{k,n} a_{k,n}^{ST}(\{R_i\}) X_{k,n}^{ST}(\vartheta,\varphi) \quad (4)$$

Energy eigenfunctions which minimize the energy should have minimal energy eigenvalues as functions of the $a_{k,n}^{ST}$'s, thus all first partial derivatives of the energy eigenvalues with respect to the a's are zero. Utilizing the Schrödinger equation and differentiation of the expectation value
$< \psi^{ST}, H_Q \psi^{ST} > = \varepsilon \sum (a_{k,n}^{ST})^2$ with respect to the $a_{k,n}^{ST}$ gives a numerical eigenvalue problem, where an eigenvector is given by an array of the a's for a particular eigenvalue ε. These numerical eigenvalues are the possible energy eigenvalues and the corresponding a's optimize the energy eigenfunctions. This method is known as the *Ritz-Galerkin* method [17].

We then vary the distances of the nearest neighbors (NN) in the (110) directions and next nearest neighbors (NNN) in the (100) directions, $R_i^{NN} = \lambda^{NN} R_i^{0,NN}$, $R_i^{NNN} = \lambda^{NNN} R_i^{0,NNN}$, other distances are held fixed. For every combination of λ^{NN} and λ^{NNN} we approximate the ground state wave function for the rotator with the Ritz Galerkin method in the new potential of cubic symmetry [18]. We find a minimum in the total energy of the coupled system for $\lambda^{NN} \neq 1 \neq \lambda^{NNN}$. The wave function for the equilibrium ground state has six maxima representing the enhanced probability for the rotator orientation in one of the (100) directions. The resulting wave function for the next energy level, ε_1, is energetically triply degenerate, because of the cubic symmetry. In this case the probability for the rotator orientation in a particular state is enhanced only in one direction.

These wave functions are needed for a Quantum Monte Carlo calculation at nonzero temperatures[10].

For a density of $\rho^* = \rho\sigma_{Ar-Ar}^3 = 0.967$ we obtain the relative distances $(\lambda^{NN} - 1) = 0.82\%$ and $(\lambda^{NNN} - 1) = 0.14\%$. The values in a classical computation are $(\lambda_c^{NN} - 1) = 0.76\%$ and $(\lambda_c^{NNN} - 1) = 0.14\%$. For $\rho^* = 0.893$ we obtain $(\lambda^{NN} - 1) = 1.39\%$, $(\lambda^{NNN} - 1) = 0.24\%$, $(\lambda_c^{NN} - 1) = 1.35\%$ and $(\lambda_c^{NNN} - 1) = 0.24\%$. Thus the effect of quantum delocalization leads to a larger local lattice deformation compared to the classical case, in particular for the nearest neighbors.

We can compare the equilibrium level splitting $\varepsilon_1 - \varepsilon_0$ in the relaxed surrounding with the value for the non equilibrium configuration, where all Ar particles occupy the fixed fcc lattice sites. At the density $\rho^* = 0.893$ we obtain for the equilibrium case $\varepsilon_1 - \varepsilon_0 = 1.9204B$, which is smaller than the level splitting for the free rotator $(2.B)$, but larger than in the case of the non relaxed surrounding $(1.86B)$. Thus in the equilibrium structures at low temperatures the occupancy of higher levels is suppressed compared to the case of a non relaxed surrounding. This effect seems to enhance the importance of the ground state in the equilibrium structures even at nonzero temperatures, at least within our simple model, which may give the results relevance even at nonzero temperatures.

In addition we varied the rotation constant B and the size d of the rotator for different densities. Within our approach we are able to obtain results for the deformed lattice structures even at densities, where the system is in a metastable (overheated) state, which may be of interest for experimental motivation. The substitution of a single Ar atom by a N_2 molecule (with *two* atoms) causes mainly additional repulsion between the N and Ar atoms. Generally with smaller densities the average distance between the Ar atoms is larger as well as the part in configuration space, where non repulsive forces between the Ar atoms dominate. Thus the Ar atoms nearby the rotator in their tendency to relax away from the rotator have equilibrium distances from the rotator which are relatively larger compared to the values at higher densities.

For very large molecules ($d = 2\sigma^l_{N-N}$) the repulsive interactions between N and Ar are very large, such that the quantum contribution to the lattice deformation is negligible. For molecules in the size of the N_2 molecule ($d = \sigma^l_{N-N}$) however we obtain generally a larger lattice deformation in a quantum mechanical treatment than in a classical. This is mainly due to the fact that in the ground state the rotator favors to delocalize and to optimize the wave function by minimizing its curvature, which can only be achieved by removing structure in the potential for the rotator, a large fraction of these structures resulting from interactions with the nearest neighbors. In order to minimize the difference of the ground state energy for the rotator to the classical value the nearest neighbors have to move further away compared to the classical case, and thus the equilibrium lattice deformations are larger in a quantum mechanical treatment.

For a given density and fixed rotator size we varied the rotation constant. Only little difference for the lattice deformation was obtained for varying rotation constants, indicating that the influence of the level splitting of the free rotator on the ground state wave function is small, the lattice deformations approaching the classical values with decreasing rotation constant. At finite temperatures however the higher levels of the rotator are occupied as well and may give different equilibrium structures.

These results were obtained in the diluted limit and at constant volume. Outside the dilute limit interactions between the N_2 molecules will enter and possibly lead to rather different results. However, experimental results [19] give evidence for a linear dependency of the lattice deformations with the N_2 concentrations. This empirical fact (which is surprising because one would expect that interactions V_{N-N} significantly change the results) motivated us to make the following simplifying Ansatz for the presumably very complicated equilibrium structures. We take the vectors from the N_2 center of mass to the nearest neighbors as defining vectors for a new unit cell, thus assuming periodic positions of the N_2 molecules in the lattice at a concentration of 25% of N_2 in Ar. The lattice structure is now different from the structure in the dilute limit and the distances and interactions beyond the nearest neighbors do not correspond to the values obtained in the previous computation, of course. In this crude Ansatz the implicit symmetry property of the new lattice structure would stabilize the nearest neighbor distances at values obtained from the above computation of the dilute limit. Let us now also assume that the lattice spacing of this regular crystal at a concentration of 25% N_2 and that of a random crystal of the same concentration are the same, and that the change in the lattice parameter with concentration is linear for concentrations below 25%. Although the correctness of these assumptions is not at all obvious, they can perhaps be motivated by considering that it is most important to have energetically favorable nearest neighbor distances in the equilibrium structure, distances between other neighbors being much less important. Thus, comparing this crude approximation with experimental data [19], we find good agreement for the density of $\rho^* = 0.967$ for the three smallest lattice deformations, for which experimentally the fcc lattice structure was observed.

At higher concentrations the experimental findings indicate an equilibrium structure of the *hcp* type. In our work we investigated lattice deformations in the case of cubic symmetry, thus for higher concentrations a direct comparison of our results to experiment certainly is not possible. Nevertheless it is interesting to note that our fcc results at a (metastable) lower density $\rho^* = 0.893$ would yield a deformation similar to experiment (assuming again a linear density dependence as above, as is suggested also by the data). Of course, for a definitive comparison to experiment (e.g. to diffuse scattering experiments from single crystals [20]) at higher densities one must wait for an extension of the present approach to systems containing many N_2 rotators in the hexagonal structure.

The method employed for the computation of the ground state can be utilized for higher levels as well.

QUANTUM MONTE CARLO TREATMENT OF A ROTATOR IMPURITY IN A CRYSTAL

At nonzero temperature the Ar - atoms fluctuate around their equilibrium positions. The mass of the rotator is small compared to the mass of the inherent phonon masses such that the time scale for the quantum dynamics is much smaller than the time scale for the lattice vibrations. In this sense we treat the full quantum problem in an adiabatic approximation, where the velocities of the environment atoms can be integrated out in the partition function. The atoms may now be described by a classical translational degree of freedom coupled to the quantum mechanical orientational degree of freedom of the fixed rotator on a site.

For small temperatures the classicaly described environment of the rotator will fluctuate around this groundstate configuration and the dynamical variations of the particles modify the potential for the rotator in such a way, that we can treat the difference to the groundstate potential as a small perturbation. We start with the Hamiltonian for the whole system :

$$\hat{\mathcal{H}} = \frac{\hat{L}^2}{2I} + \hat{V}_{qm} + V_{kl} + E_{kl} \tag{5}$$

which consists of the kinetic part of the rotator

$$\frac{\hat{L}^2}{2I},$$

his interaction with the environment

$$\hat{V}_{qm} = 4\varepsilon_{NAr} \sum_i \left(\frac{\sigma_{NAr}}{r_{Ri}}\right)^6 \left(\left(\frac{\sigma_{NAr}}{r_{Ri}}\right)^6 - 1\right),$$

the classical potential between the Ar particles

$$V_{kl} = 4\varepsilon_{ArAr} \frac{1}{2} \sum_{i,j} \left(\frac{\sigma_{ArAr}}{r_{ij}}\right)^6 \left(\left(\frac{\sigma_{ArAr}}{r_{ij}}\right)^6 - 1\right),$$

and their kinetic energy

$$E_{kl} = \sum_i \frac{\vec{p}_i^{\,2}}{2m_{Ar}}.$$

The operator attribute of the interaction potential between the rotator and his environment is manifested through the angle dependence of $r_{Ri} = r_{Ri}(\Omega)$, the distance between the $i-th$ particle and a point in the rotator sphere, generated through all orientations. The rotator Hamiltonian at $T = 0K$ is:

$$\hat{\mathcal{H}}_R^0 = \frac{\hat{L}^2}{2I} + \hat{V}_{qm}^0 \tag{6}$$

with the eigenvalues ε_n^0 and eigen functions ψ_n^0. The interaction potential $\hat{V}_{qm}^0$ in the static groundstate configuration of the environment which deviates from the fcc - lattice described above. Small dynamical variations of the particles give rise to a perturbation of the potential for the rotator :

$$\hat{V}_1 = \hat{V}_{qm} - \hat{V}_{qm}^0 \ , \tag{7}$$

which modify the rotator Hamiltonian

$$\hat{\mathcal{H}}_R = \hat{\mathcal{H}}_R^0 + \hat{V}_1 \ . \tag{8}$$

We evaluate the system properties[10] in such a temperature range that only the ground state and the next higher state, which is triply degenerate, are occupied. Thus at thermal energies below the energy difference between the lowest levels, we neglect the contributions of the higher levels in the sense of a two level approximation. Thus we restrict the spectrum to $n_{max} = 4$ states. Taking into account that the classical interaction of the particles have no operator attribute, the trace over the Boltzmann factor takes the form:

$$tr(\exp(-\beta(\hat{\mathcal{H}} - E_{kl}))) = \exp(-\beta V_{kl} \sum_{n=1}^{n_{max}} < \psi_n^0, \exp(-\beta(\hat{\mathcal{H}}_R^0 + \hat{V}_1))\psi_n^0 > \tag{9}$$

Introducing the configuration dependent density operator:

$$\hat{\varrho} = \frac{\exp(-\beta(\hat{\mathcal{H}}_R^0 + \hat{V}_1))}{\sum_{n=1}^{n_{max}} < \psi_n^0, \exp(-\beta(\hat{\mathcal{H}}_R^0 + \hat{V}_1))\psi_n^0 >}, \tag{10}$$

and defining the quantum mechanical average through

$$< \hat{A} >_{qm} = tr(\hat{\varrho}\hat{A}), \tag{11}$$

we can describe the thermal average as

$$< \hat{A} >_{th} = \frac{1}{Z'} \int d\{r_i\} \exp(-\beta\hat{\mathcal{H}}_{eff}) < \hat{A} >_{qm}, \tag{12}$$

where

$$\hat{\mathcal{H}}_{eff} = V_{kl} - \frac{1}{\beta} \ln \sum_{n=1}^{n_{max}} < \psi_n^0, \exp(-\beta(\hat{\mathcal{H}}_R^0 + \hat{V}_1))\psi_n^0 > \tag{13}$$

is the temperature and configuration dependent effective energy, which gives the statistical weight for the quantum mechanical expectation value of an observable at a given configuration. For calculating the density matrix $< \psi_i^0, \hat{\varrho}\psi_k^0 >$ we used the Trotter formula ($\hat{V}_1$ does not commute with $\hat{\mathcal{H}}_R^0$),

$$\exp(-\beta(\hat{\mathcal{H}}_R^0 + \hat{V}_1)) = lim_{L\to\infty}(\exp(-\beta\frac{\beta}{L}\hat{\mathcal{H}}_R^0)(\exp(-\beta\frac{\beta}{L}\hat{V}_1))^L. \quad (14)$$

Because of the low Hilbert space dimension of our model we can choose the upper bound for the Trotter dimension extremely high ($L \leq 10^{30}$). The pseudo dynamics of the system is generated with the Metropolis procedure.

Monte Carlo simulations at constant density $\rho^* = \rho\sigma_{ArAr}^3 = 0.893$ and different temperatures show[10] that with increasing temperature the probability to find the rotator in the second level increases and the mean square displacement of a nearest neighbor of the rotator increases and thus thc mean distance between them. We conjecture that the rotator in the upper state has mainly repulsive interaction with this nearest neighbors because of the strong orientation in this state. The nearly spherically symmetric groundstate is more sensitive to the attractive part of the potential. This may result in a decrease of the lowest energy eigenvalue and an increase of the next energy eigenvalue with increasing temperature, as obtained in the simulations.

An analysis of the influence of the mixed state of the rotator on the local structure of the surrounding particle positions shows the center of the nearest neighbor distributions is in a quantum simulation more remote from the fcc-lattice sites than in the fully classical case[10].

With our method we can on the one hand describe the quantum mechanical attributes of the rotator at low temperatures, and on the other hand, we are able to study the dynamical influence of the mixed state onto the local lattice structure which we obviously have to distinguish from the classical treatment of the orientational degree of freedom. In summary we found with this Quanten Monte Carlo Method an interesting access to the microscopic mutual influence between the quantum mechanical nature of a rotator and his classical environment. We have used the wavefunctions at temperature $T = 0K$ to formulate the formal frame for a Quantum Monte Carlo Simulation and calculated the perturbated energy eigenvalues and state probabilities for the rotator at finite temperatures. Interesting differences between this system and a fully classical system were found.

Acknowledgements: The work presented is based on a cooperation with K. Binder and W. Helbing. This research was carried out in the framework of the Sonderforschungsbereich 262 der Deutschen Forschungsgemeinschaft. The computations were carried out on the *CRAY YMP* of the *HLRZ* at Jülich.

References

1.) A. Loidl, Ann. Rev. Phys. Chem. **40**, 29 (1989).
2.) K. Knorr, A. Loidl, Phys. Rev. Lett. **57**, 460 (1986).
3.) A. Heidemann, A. Magerl, M. Prager, D. Richter, T. Springer (eds.), *Quantum Aspects of Molecular Motions in Solids*, Proceedings on an ILL-IFF workshop, (Springer, Berlin 1987).
4.) K.H. Michel, Phys. Rev. **B 35**, 1405, 1414 (1987).

5.) H.-O. Carmesin and K. Binder, Europhys. Lett. **4**, 269 (1987); J. Phys. **A** **21**, 4053 (1988); D. Hammes, H.-O. Carmesin, K. Binder, Z. Phys. **B 76**, 115 (1989)
6.) W.A. Phillips (ed.) *Amorphous Solids, Low Temperature Properties* (Springer, Berlin 1980)
7.) H. Klee, H.-O. Carmesin, K. Knorr, Phys. Rev. Lett. **61**, 1855 (1988)
8.) H.-O. Carmesin, Thesis, Mainz (1988). Lennard - Jones potential parameters are $\sigma_{Ar-Ar} = 3.4\AA$, $\sigma_{N-N} = 3.31\AA$, $\sigma_{N-Ar} = 3.35\AA$, $\varepsilon_{Ar-Ar} = 1.67 \times 10^{-21}J$, $\varepsilon_{N-Ar} = 0.927 \times 10^{-21}J$, $\varepsilon_{N-N} = 0.515 \times 10^{-21}J$. The distance between the N atoms is $\sigma^l_{N-N} = 2 \times 0.1646 \times 3.31\AA$.
9.) P. Nielaba, K. Binder, Europhys. Lett. **13**, 327 (1990).
10.) W. Helbing, P. Nielaba, K. Binder, Phys. Rev **B**, in press.
11.) A.F. Devonshire, Proc. Roy. Soc. (London) **A 153**, 601 (1936).
12.) P. Sauer, Z. Phys. **194**, 360 (1966).
13.) H.U. Beyeler, Phys. Stat. Sol. **B 52**, 419 (1972).
14.) P. Nielaba, J.L.Lebowitz, H. Spohn, J.L. Valles, J. Stat. Phys. **55**, 745 (1989).
15.) M. Tinkham, *Group theory and quantum mechanics* (New York, McGraw-Hill 1964).
16.) S.L. Altmann, A.P. Cracknell, Rev. Mod. Phys. **37**, 19 (1965).
17.) in S.G. Michlin, *Variationsmethoden der Mathematischen Physik*, (Akademie- Verlag, Berlin 1962).
18.) We work with $M + 1$ particles in a cubic box with periodic boundary conditions, $M = 4 \times 12^3 - 1$; change of M to $M = 4 \times 16^3 - 1$ did not effect the resulting NN and NNN equilibrium positions. The expectation values for the quantum particle were obtained with a numerical integration over a grid of 120×120 points on the surface of a sphere with diameter σ^l_{N-N}. For the Ritz - Galerkin method we expanded our wave functions up to order 6 and 12 with no significant changes in the results. The solution of the eigenvalue problem requires a treatment of matrices in the typical size of 39×39 for each of the different combinations of λ^{NN} and λ^{NNN} .
19.) C.S. Barrett, L. Meyer, J. Chem. Phys. **42**, 107 (1965); H. Klee, K. Knorr, preprint.
20.) S. Dietrich, W. Fenzl, Phys. Rev.**B39**, 8873, 8900 (1989).

A NEW TRIAL WAVE FUNCTION FOR THE ONE DIMENSIONAL $t - J$ MODEL

T.K. Lee
Center for Stochastic Processes in Science and Engineering
and
Department of Physics
Virginia Polytechnic Institute and State University
Blacksburg, VA 24061 USA

ABSTRACT

Using the property that charge and spin are decoupled, a new variational wave function is constructed for the one dimensional $t - J$ model. In this wave function the spins in this reduced-length Heisenberg chain interact as if the holes are absent. For $J = 0$, the exact ground state energy is reproduced. The spin and charge structure factors, the momentum distribution $n(k)$ calculated by Monte Carlo method agree very well with exact results obtained by Bethe Ansatz.

Recently a lot of effort has been devoted to understanding the $t - J$ model in one dimension. The Hamiltonian of this model written in the subspace of no doubly occupied sites has two terms,

$$H_t = -t \sum_{i\sigma} C^+_{i\sigma} C_{i+1\sigma} + h.c. \tag{1}$$

and

$$H_J = J \sum_i (\vec{s}_i \cdot \vec{s}_{i+1} - \frac{1}{4} n_i n_{i+1}) \tag{2}$$

The model has been solved exactly at[1] $J = 0$ and[2] $J = 2t$ by using Bethe Ansatz. In both of these cases, the ground states belongs to an unconventional class of interacting Fermi systems known as Tomonaga-Luttinger (TL) liquids[3]; which exhibit power law singularities in correlation functions and the momentum distribution at the Fermi surface. A more interesting property of the TL liquid is the separation of spin and charge excitations. Ren and Anderson[4] have proposed that such a property will also survive in two dimensions.

Although a substantial amount of analytical work[4,5,6,7] has been carried out for this $1D$ model, the phase diagram and physical properties of this model is much more clarified by numerical work[8]. But the method of exact diagonalization is useful only for small-size systems. A very successful method of treating these strongly correlated systems with large sizes is to use the variational Monte Carlo (VMC) method. The $t-J$ model in two dimensions has been studied extensively by using many different variational wave functions[9].

Most earlier work on $1D\, t-J$ model by VMC method uses the Gutzwiller wave function (GW)[10,11]. This wave function is just the projected Fermi liquid state,

$$|GW\rangle = Pd|FL\rangle = Pd\Pi_{k<k_F} C^+_{k\uparrow} C^+_{k\downarrow} |0\rangle \tag{3}$$

where the projection operator $Pd = \Pi(1-n_{i\uparrow}n_{i\downarrow})$. Alternatively we may rewrite $|GW\rangle$ as a sum of all possible configurations with coefficient as a product of two determinants $Det[\phi_{k_i}(r_i^\uparrow)]\ Det[\phi_k(r_j^\downarrow)]$, where $\phi_k(r_i) = exp(ikr_i)$.

Recently Hellberg and Mele[12] proposed two new wave functions, one of them is the so called slave-Boson (SB) wave function. The electron operator is written as a product of two operators, $C^+_{i\sigma} = d^+_{i\sigma}e_i$, where the spin is assosciated with the Fermi opertor d^+ and the charge with the Bose operator e. In contrast to reference 12, here we have chosen the annihilation operator e instead of the creation operator e^+ to relate to C^+, but same results are expected. The wave function for the spin is chosen as an ideal Fermi liquid similar to GW. The hole wave function will be also written as a Fermion Slater determinant by using the Jordan-Wigner transformation. In the SB state the coefficient of each configuration is of the form

$$H(r_\ell^0)Det[\phi_k(r_i^\uparrow)]Det[\phi_k(r_j^\downarrow)] \tag{4}$$

where the wave function of the hole is

$$H(r_\ell^0) = \Pi_{\ell<m} sgn(r_\ell^0 - r_m^0)Det[\phi_{k^0}(r_\ell^0)] \tag{5}$$

For $J=0$, which corresponds to $U=\infty$ Hubbard model, the ground state energy is known to be given by the ideal spinless fermions: $E_0 = -t\frac{2}{\pi}\sin k_F^0$. k_F^0 is the Fermi wave vector of the hole and it is related to the Fermi wave vector k_F of the spins, $k_F^0 = \pi - 2k_F$. Hellberg and Mele[12] show that both GW and SB have higher energy than the exact vale E_0.

To obtain this exact energy E_0, we must construct a wave function such that the holes are completely decoupled from the spins, hence the holes form an ideal spinless Fermi gas. In both GW and SB any change of the hole positions would also alter the spin wave function, in other words, hole and spin interact and scatter. To rid of this interaction we propose the squeezed Heisenberg chain (SHC) state where each configuration of spins and holes has coefficient of the form,

$$H(r_\ell^0)Det[\phi_{\bar{k}}(\bar{r}_i^\uparrow)]Det[\phi_{\bar{k}}(\bar{r}_j^\downarrow)] \tag{6}$$

SHC is quite similar to SB except that the spin coordinates in SB are replaced by the effective coordinates $\bar{r}_i^{\uparrow} = r_i^{\uparrow} - \sum_{\ell<i} n_h(\ell)$, where the last term denotes the total number of holes on the left of site i. The lattice constant is taken to be unity in this paper. The effective spin coordinates changes only when up and down spins are exchanged. Hence the spins are not influenced by the holes and they form a squeezed Heisenberg chain with a reduced length $N_s = L - N_h$, where N_s is the total number of spins, and N_h the number of holes. The effective wave vector $\bar{k}$ is also rescaled by a factor L/N_s.

As the holes are completely decoupled from the spin wave function in SHC, it is straight forward to show that the expectation value of H_t is the exact ground state energy E_0. It should be noticed that SHC is not an eigenstate of H_t. The ground state of $U = \infty$ Hubbard model as shown in reference 1 is much more complicated and it is difficult to use it to calculate correlation functions for a reasonably large size lattice. On the other hand the VMC method can be easily used to calculate any correlation functions for all of the trial functions discussed above.

A brief discussion of the VMC method will be given here. Readers please refer to reference 10 for more details. A trial wave function may be expanded in terms of a set of complete orthonormal states, i.e. $|\psi\rangle = \sum a(\alpha)\,|\alpha\rangle$. The state $|\alpha\rangle$ is usually chosen to be in the particle-occupation representation in the configuration space. The expectation value of an operator O using this trial state $|\psi\rangle$ can be written in the form

$$\langle O \rangle = \sum_{\alpha} \Pr(\alpha) \sum_{\beta} \frac{a(\beta)}{a(\alpha)} \langle \alpha | O | \beta \rangle \tag{7}$$

where the function

$$\Pr(\alpha) = |a(\alpha)|^2 / \sum_i |a(i)|^2.$$

Important sampling is achieved by using Metroplis method to generate the state $|\alpha\rangle$ according to the probability $\Pr(\alpha)$. The function $a(\alpha)$ is either given by equation (6) for the SHC state or equation (4) for the SB state.

The TL liquid is well known[3] to have very unusual spin-spin and charge-charge correlation functions. The momentum distribution $n(k)$ has a power law singularity near k_F obeying the relation

$$n(k) = n(k_F) - C sgn(k - k_F)|k - k_F|^{\alpha} \tag{8}$$

It is essential to examine these correlation functions obtained by the SHC state to understand its relation with the exact ground state.

Using the exact ground state wave function of the $U = \infty$ Hubbard model, Ogata and Shiba[1] have calculated the spin structure factor

$$S(k) = \frac{1}{L}\sum_{i,j}\left\langle S_z^i S_z^j\right\rangle e^{ik(r_i - r_j)} \tag{9}$$

A cusp is found at $k = 2k_F$. In figure 1, $S(k)$ for a lattice of 100 sites and 50 holes is plotted as a function of k for the three states, SHC, SB and GW. In this case $2k_F = \pi/2$.

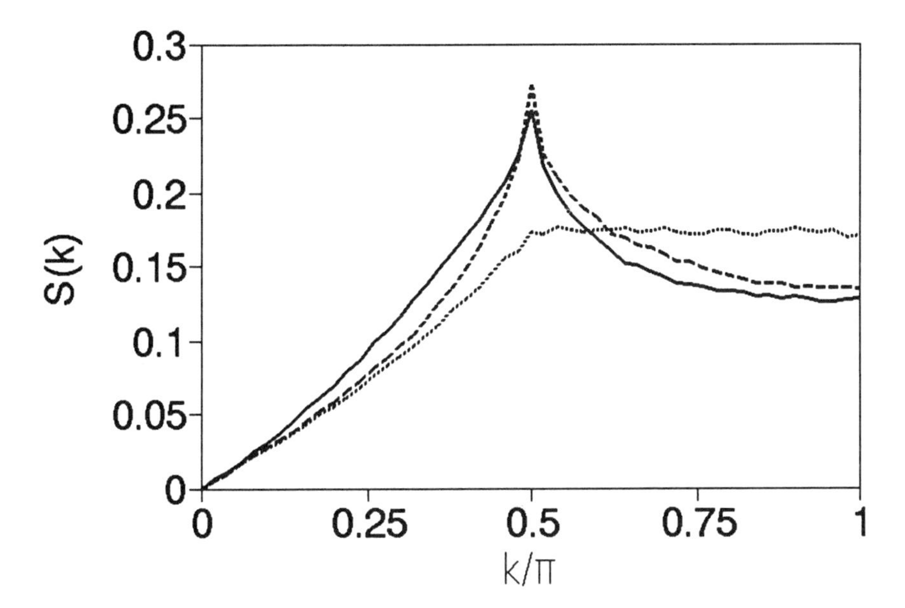

Fig. 1. The spin structure factor $S(k)$ plotted as a function of k for the three trial states: GW, the dotted line; SB, the dashed line; and SHC, the solid line. There are $N_h = 50$ holes in a lattice of $L = 100$ sites.

$S(k)$ of GW is very similar to what we expect from an ideal Fermi gas where $S(k)$ equals to $k/4\pi$ for $k < 2k_F$, and $k_F/2\pi$ otherwise. The cusp of SHC state at $2k_F$ is less than that of SB state. Within the numerical accuracy, the value of the cusp for 16 holes in 32 sites is almost identical with the exact results reported by Ogata and Shiba[1].

The charge structure factor is defined as

$$D(k) = \frac{1}{L}\sum_{i,j}\langle n_i n_j\rangle e^{ik(r_i - r_j)} \tag{10}$$

Because the charges behave as ideal spinless Fermions, we expect $D(k)$ to have the simple form for an ideal gas where $D(k)$ equals to $k/2\pi$ for $k < 2k_F^0$, and

k_F^0/π otherwise. In figure 2, $D(k)$ calculated by VMC method is plotted as a function of k for the three trial states.

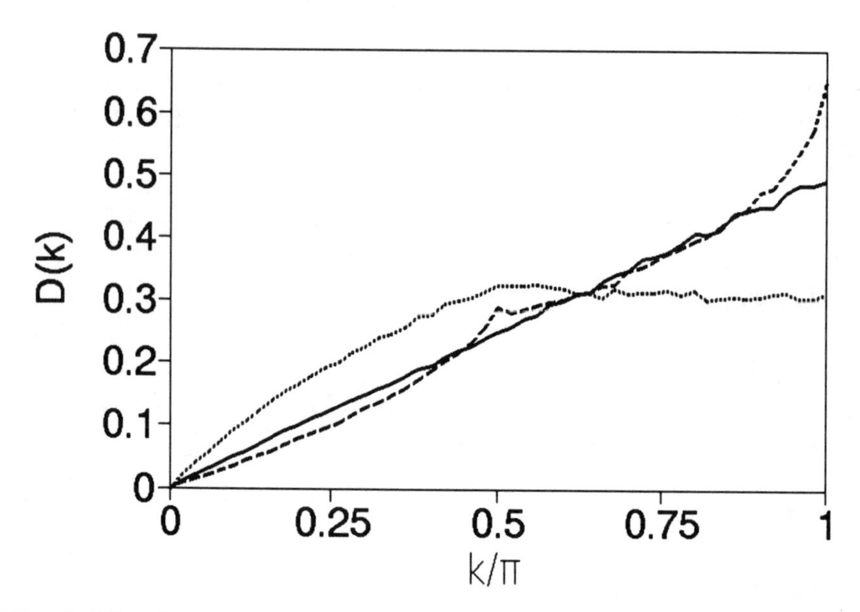

Fig. 2. The charge structure factor $D(k)$ plotted as a function of k for the trial states:GW, the dotted line; SB, the dashed line; and SHC, the solid line.

Notice that $k_F^0 = \pi/2$, hence the result of SHC is exactly what is predicted by an ideal spinless Fermi gas as discussed above. Although $D(k)$ of the SB state is quite close to the exact result but it has incorrectly builded in two peaks at $k = k_F^0$ and $k = 2k_F^0$.

Two very different functional forms for spin and charge correlation functions as shown by the SHC state in figures 1 and 2 is the unique characteristics of the TL liquid. As discussed above the power law singularity of the momentum distribution $n(k)$ at $k = k_F$ provides another important characteristic. In figure 3 we plotted $n(k)$ for the three different trial states. GW shows a very clear discontinuity at $k = k_F$ as the Fermi liquid. Both SHC and SB states show power law singularity. The exponent α in equation (8) is 0.27[12] for the SB state, for the SHC state α is very close to the exact result of 0.125[12].

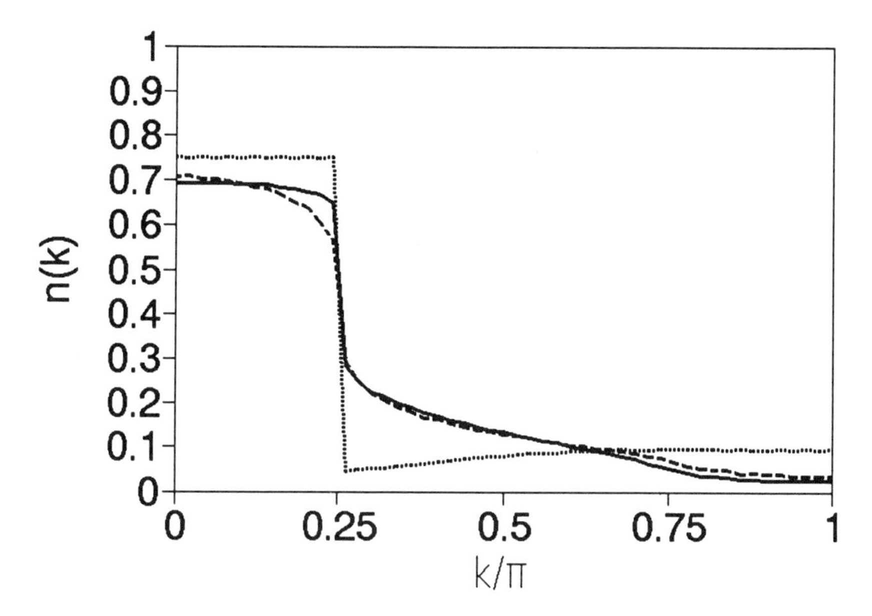

Fig. 3. The momentum distribution $n(k)$ plotted as a function of k for the three trial states:GW, the dotted line; SB, the dashed line; and SHC, the solid line.

In summary, we have proposed a simple wave function that completely separates the charge and spin. This state is like a Heisenberg chain (SHC) with the holes squeesed out. The results of ground state energy and the various correlation functions show that the SHC state has many properties of the exact ground state even though it is not the same.

So far our discussion has been restricted to $J = 0$ or $U = \infty$. Presence of H_J would cause interaction between charge and spin. As J increases, the SHC state is no longer a good variational wave function. In fact, $\langle H_J \rangle / L$ is -0.103(1) for SHC, -0.120(1) for SB and -0.164(1) for GW. For $J < t$, of these three states, SHC has the lowest energy. At $J = 2t$, GW is very close to the ground state. Thus the cusp of $S(k)$ at $k = 2k_F$ decreases with increasing value of J/t, this agrees with the finite temperature Monte Carlo results[13]. The behavior of $D(k)$ also agrees with their result that the maximum shifts to small k value as J/t increases. More work is needed to find variational wave functions that would describe the complete phase diagram[8] of the $t - J$ model.

ACKNOWLEDGEMENTS

The author is grateful to the hospitality of the Material Science Center and Department of Physics of Tsinhua University where most of the numerical work reported in this paper was carried out. The author also acknowledges invaluable discussion with Drs. Shoudan Liang and Z.Y. Weng. This work was partially supported by the National Science Council, Rep. of China and by the Thomas F. Jeffress and Kate Miller Jeffress Memorial Trust.

REFERENCES

1. M. Ogata and H. Shiba, Phys. Rev. B41, 2326 (1990).
2. N. Kawakami and S.K. Yang, Phys. Rev. Lett. 65, 2309 (1990); P.A. Bares and G. Blatter, Phys. Rev. Lett. 64, 2567 (1990).
3. J. Solyom, Adv. Phys. 28, 201 (1979).
4. Y. Ren and P.W. Anderson, preprint.
5. H. Frahm and V.E. Korpin, Phys. Rev. B42, 10553 (1990).
6. T.K. Lee and Z.Q. Wang, preprint.
7. Z.Y. Weng, D.N. Sheng, C.S. Ting and Z.B. Su, preprint.
8. M. Ogata, M. Luchini, S. Sorella and F.F. Assaad, Phys. Rev. Lett. 66, 2388 (1991).
9. T.K. Lee and L.N. Chang, Phys. Rev. B42, 8720 (1990) and references therein.
10. C. Gros., R. Joynt and T.M. Rice, Phys. Rev. B36, 381 (1987).
11. T.K. Lee, W. vonder Linden and P. Horsch, Physica C153, 1265 (1988).
12. C. Stephen Hellberg and E.J. Mele, preprint.
13. F.F. Assaad and D. Wuertz, preprint.

TITPACK
— NUMERICAL DIAGONALIZATION ROUTINES OF QUANTUM SPIN HAMILTONIANS —

Hidetoshi Nishimori
Department of Physics, Tokyo Institute of Technology,
Oh-okayama, Meguro-ku, Tokyo 152, Japan

ABSTRACT

The computer program package TITPACK Ver.2 has been designed to facilitate numerical calculations of the low-lying energy levels of quantum spin systems. For spin-1/2 models, it diagonalizes the Hamiltonian numerically for the specified values of parameters (system size, anisotropy, lattice structure etc.) as long as the memory and CPU time requirements are satisfied on the user's computer. Several low-lying energy eigenvalues and eigenvectors are returned automatically by calling a few subroutines. The basic algorithm is the well-known Lanczos method.

INTRODUCTION

The ground states of quantum spin systems do not have trivial structures if the exchange interactions are antiferromagnetic. A simple example can be found in the two-spin system with spin size 1/2, in which the singlet ground state is quite different from the classical Neel state. The situation is much more complicated in thermodynamically large systems. Exact solutions are found for the one-dimensional case with uniform antiferromagnetic interactions. Most of the other systems escape rigorous analysis in spite of recent high interest in the two-dimensional Heisenberg antiferromagnet.

Numerical approaches often show their power when reliable results are not easily obtained by analytic techniques. One of the main numerical methods in investigating quantum spin systems is the quantum Monte Carlo simulations pioneered by the Suzuki group.[1] This method has an advantage that any system can be treated in principle and main memory requirement is not generally large. However, as is usual for any Monte Carlo simulations, the problem of statistical errors is sometimes serious, in particular when one treats frustrated systems. Very large, often inhibitingly large, CPU time is required to obtain significant data for frustrated cases such as the triangular lattice model and the antiferromagnet with next nearest neighbor interactions.

Another technique is the direct numerical diagonalization of spin Hamiltonian matrices, which became rather popular recently partly because of the fast development of hardware technology of computers. The leading advantage of the diagonalization method is the absence of statistical errors. The method is useful also because the necessary CPU time is not large, often negligible, as compared with Monte Carlo simulations. One does not have to apply for a large grant to carry out a research program using numerical diagonalization. However, the memory requirement is not small, which sets a strong limit on the size of systems to be treated by numerical diagonalization. Therefore one has to determine carefully what method to use in investigating quantum spin systems by numerical calculations. TITPACK is a useful tool for those who wish to check their physical intuition by numerical diagonalization of finite-size systems.

HISTORICAL BACKGROUND

The system to be treated by the package TITPACK is specified by the Hamiltonian

$$H = -2\sum_{i,j} J_{ij}(S_i^x S_j^x + S_i^y S_j^y + \Delta_{ij} S_i^z S_j^z) \ . \tag{1}$$

The spin size S is 1/2. The parameters J_{ij} and Δ_{ij} can be arbitrarily set by the user. In particular, the lattice structure is specified by appropriately setting the values of exchange interactions.

The first version of the package, written by Y. Taguchi and the present author, was released in 1986.[2] The program was intended to be used on a Hitachi's supercomputer HITAC S-810/20 at the Computer Center of the University of Tokyo. This machine had the theoretical peek speed of 800 MFLOPS and the maximum main memory space accessible for a general user was 32 MBytes at that time. This specification enabled us to diagonalize a 21-spin system without using translational symmetries; we were investigating the problem of triangular lattice antiferromagnet by the railroad trestle extrapolation method,[3] which did not allow lattice symmetry other than the trivial conservation of total S_z.

This first version has since been used by many researchers in the field of quantum spin systems in Japan. The main reasons for not having been used abroad were that the program used specific library calls of HITAC and that the specifications for automatic vectorization of FORTRAN codes were different on Japanese supercomputers from those of foreign machines. We therefore did not write a manual in English.

Hardware of computers has advanced quite significantly in these several years. In 1988, a new supercomputer HITAC S-820/80 was introduced to the University of Tokyo Computer Center. This machine has the peek speed of 3GFLOPS and the current maximum memory per user is 480 MBytes. A new supercomputer, Fujitsu's FACOM VP-2600, was installed in 1991 at the Data Processing Center of Kyoto University. Its speed is 5 GFLOPS at the theoretical peek performance, and the main memory for a general user is available up to 200 MBytes. Situations in smaller machines have also changed, and it is now rather easy to install a workstation in one's own office.

The new version of TITPACK has increased portability to catch up with these changes of situations. It runs without library calls of a specific machine since it is written using only FORTRAN 77 statements with the single exception of bit-wise logical operations of two integers, IAND (logical And of corresponding bits) and IEOR (logical Exclusive Or of bits). Since these two functions are now supported by most FORTRAN compilers, in this form or another, this would not set a serious difficulty in using TITPACK on various machines.

USAGE

To use TITPACK Ver. 2, the user has first to specify various parameters in the Hamiltonian (1). With these parameters given in data or other statements, he/she calls several routines successively to get eigenvalues, eigenvectors and correlation functions. An example is given below in which one calculates lowest eigenvalues, an eigenvector and nearest neighbor correlation functions of the one-dimensional Heisenberg antiferromagnet with `n`(the number of spins)=16. The reader is referred to the manual, which is available from the present author upon request, for detailed explanation of routines; I only give a short account here. The routine `sz` generates spin configurations in the space of given `n` and total S_z. The four lowest eigenvalues are calculated in `lnc1`, and the eigenvector in `lncv1`. The precision of the eigenvalues and eigenvector is checked in `check1`, and the correlation functions are evaluated in `xcorr` and `zcorr`.

```
c************* Sample program #1 ***************
c   Eigenvalues and an eigenvector / lnc1, lncv1
c    Precision check and correlation functions
c***********************************************
      parameter (n=16,idim=12870,ibond=n)
      implicit real*8 (a-h,o-z)
      dimension E(4)
      dimension list1(idim),list2(2,0:2**15)
      dimension bondwt(ibond),ipair(2*ibond),zrtio(ibond)
      dimension npair(2)
      dimension wk(idim,2)
      dimension x(idim)
c
      data bondwt/ibond*-1.0d0/
      data zrtio/ibond*1.0d0/
      data ipair/1,2, 2,3,  3,4,  4,5,  5,6,  6,7,  7,8, 8,9,
     &      9,10, 10,11, 11,12, 12,13, 13,14, 14,15, 15,16, 16,1/
      nvec=1
      iv=idim/3
      call sz(n,idim,0.0d0,list1,list2)
c
c*** Eigenvalues
      call lnc1(n,idim,ipair,bondwt,zrtio,ibond,
     &                 nvec,iv,E,itr,wk,idim,list1,list2)
      print 100,e,itr
 100  format(/' [Eigenvalues]  '/2x,4f14.8
     &       /' [Iteration number]'/i8)
c
c*** Ground-state eigenvector
      call lncv1(n,idim,ipair,bondwt,zrtio,ibond,
     &             nvec,iv,x,itr,wk,idim,list1,list2)
```

```
      print *,'[Eigenvector components (selected)]'
      print 120,(x(j),j=13,idim,idim/20)
 120  format(4d18.9)
c
c*** Precision check and correlation functions
      call check1(n,idim,ipair,bondwt,zrtio,ibond,
     &            x,wk,Hexpec,list1,list2)
      npair(1)=1
      npair(2)=2
      call xcorr(n,idim,npair,1,x,sxx,list1,list2)
      call zcorr(n,idim,npair,1,x,szz,list1)
      print 130,sxx,szz
 130  format(/' [Nearest neighbor correlation functions]'/
     &       '     sxx :',d18.10,',     szz :',d18.10)
      end
```

```
==================== RESULT 1 ====================

[Eigenvalues]
   -14.28459272   -13.74421336   -13.39309484   -13.04681410
[Iteration number]
     50
[Eigenvector components (selected)]
  0.343792928D-05   0.125778312D-01   0.892952949D-04  -0.216465016D-02
 -0.341393531D-02   0.717748830D-04  -0.346965348D-02   0.124963114D-03
 -0.121707616D-02  -0.341393530D-02   0.338263590D-07   0.391932543D-03
  0.266317018D-02  -0.431510412D-03   0.392280336D-02  -0.113573424D-03
  0.578613958D-02   0.448821353D-02   0.316356482D-06   0.169426150D-03

---------------------------- Information from check1
<x*H*x> = -1.42845927D+01
H*x(j)/x(j) (j=min(idim/3,13),idim,max(1,idim/20))
 -0.142845090D+02  -0.142845927D+02  -0.142845925D+02  -0.142845927D+02
 -0.142845927D+02  -0.142845924D+02  -0.142845927D+02  -0.142845936D+02
 -0.142845927D+02  -0.142845927D+02  -0.142825155D+02  -0.142845931D+02
 -0.142845927D+02  -0.142845932D+02  -0.142845927D+02  -0.142845920D+02
 -0.142845927D+02  -0.142845927D+02  -0.142844308D+02  -0.142845933D+02
-----------------------------------------------------

[Nearest neighbor correlation functions]
   sxx : -0.1487978408D+00,     szz : -0.1487978407D+00
```

Three routine groups are included in TITPACK Ver. 2, L, M and S. The user chooses an appropriate group for his/her purposes. A rough criterion is the size of the system to be diagonalized. The group L is for large systems, M for medium and S for small. Optimal performance is expected by using the best group for the size. The above sample program uses group L although group M is usually appropriate for this size on most computers. The reader is referred to the manual for details. Examples of CPU time and memory requirements are presented in the following section for each of these groups.

I have to add here that no reduction of matrix size by lattice symmetry is implemented, except for the trivial classification by total S_z, because this package is intended for general use including random system. One therefore has to rewrite relevant parts of the source code if he/she wishes to take full advantage of symmetries of a specific lattice. I think that this simple program without elaboration on symmetry reduction is usually sufficient to get a first-hand test of one's physical intuition, which is actually how the first version has been used.

BENCHMARK TEST

Tables I to III list the necessary main memory space and CPU time on HITAC S-820/80 and NEC EWS-4800/220 (a work station). The Hamiltonian is the one-dimensional Heisenberg antiferromagnet with nearest neighbor interactions. The four lowest eigenvalues in the space $S_z^{\rm total} = 0$ and the ground-state eigenvector have been calculated. All inner do loops have been automatically vectorized by the compiler on HITAC S-820/80.

Table I Main memory requirement
(in MB; – : less than 1MB; * : exceeding 1GB)

n	matrix size	I	II	III	Iv	Iv′	IIv	IIv′	IIIv
12	924	–	–	6.6	–	–	–	–	6.6
16	12,870	–	2.7	*	–	–	2.8	3.0	*
20	184,756	3.5	48	*	4.9	7.8	49	51	*
24	2,704,156	52	701	*	72	113	836	877	*
26	10,400,600	198	*	*	277	436	*	*	*

Table II CPU time on S-820/80 (sec)
(– : under 1 sec; * : unexecutable for insufficient memory)

n	sz	I	II	III	Iv	Iv′	IIv	IIv′	IIIv
12	–	–	–	2	–	–	–	–	2
16	–	–	–	*	–	1	–	–	*
20	5	10	1	*	20	38	2	4	*
24	80	225	*	*	448	839	*	*	*
26	332	883	*	*	1758	2780	*	*	*

Table III CPU time on EWS-4800/220 (sec)
(– : under 1 sec; * : unexecutable for insufficient memory)

n	sz	I	II	III	Iv	Iv′	IIv	IIv′	IIIv
12	–	1	1	816	2	4	2	3	816
16	5	34	21	*	68	128	41	95	*
20	104	852	*	*	1710	2988	*	*	*

I : Eigenvalues by Group L
II : Eigenvalues by Group M
III : Eigenvalues by Group S
Iv : Eigenvalues and an eigenvector by Group L
Iv′ : Eigenvalues and an eigenvector by Group L
IIv : Eigenvalues and an eigenvector by Group M
IIv′ : Eigenvalues and an eigenvector by Group M
IIIv : Eigenvalues and an eigenvector by Group S

BASIC ALGORITHM

Routines in groups L and M use the Lanczos method to tridiagonalize a matrix and the inverse iteration method combined with the Conjugate Gradient method to calculate eigenvectors. I give short accounts of these methods. The reader is referred to Ref. 4 and other textbooks of numerical techniques for details.

It is instructive first to explain the Lanczos method as an improvement of the power method. The power method is a simple way to calculate the eigenvalue of a matrix with the largest absolute value; one starts from an arbitrary initial vector $\mathbf{v}_0$ and multiplies the matrix H repeatedly until the resulting vector converges to the desired eigenvector. More precisely, when the eigenvalues and eigenvectors of an m-dimensional matrix H are $E_j, \psi_j (j = 1, \cdots, m)$, the expansion of the initial vector reads

$$\mathbf{v}_0 = \sum_{j=1}^{m} a_j \psi_j$$

Multiplying this initial vector k times by H, one obtains

$$\mathbf{v}_k \equiv H^k \mathbf{v}_0 = \sum_{j=1}^{m} a_j E_j^k \psi_j$$

It is apparent that the relative weight of the eigenvector corresponding to the eigenvalue with the largest absolute value increases exponentially with k among terms appearing in the above sum.

Acceleration of convergence over the simple power method is achieved by subtracting components of previous vectors $(\mathbf{v}_{k-1}, \mathbf{v}_{k-2}, \cdots)$ from $\mathbf{v}_k$ so that one can eliminate the effects of the arbitrarily chosen initial vector as rapidly as possible. This subtraction of components of previous vectors is incidentally equivalent to tridiagonalization. Let us explain this point.

If the tridiagonal matrix T is obtained from the original matrix H by a transformation matrix V, one has the relation $T = V^{-1}HV$, or $VT = HV$. Let the column vectors of V be $\mathbf{v}_1, \mathbf{v}_2, \cdots$, and the diagonal elements of T be $\alpha_1, \alpha_2, \cdots$, and the subdiagonal elements $\beta_1, \beta_2, \cdots$. Then the relation $VT = HV$ is written as

$$\begin{aligned}
H\mathbf{v}_1 &= \alpha_1 \mathbf{v}_1 + \beta_1 \mathbf{v}_2 \\
H\mathbf{v}_2 &= \beta_1 \mathbf{v}_1 + \alpha_2 \mathbf{v}_2 + \beta_2 \mathbf{v}_3 \\
H\mathbf{v}_3 &= \beta_2 \mathbf{v}_2 + \alpha_3 \mathbf{v}_3 + \beta_3 \mathbf{v}_4 \\
&\cdots \\
H\mathbf{v}_{m-1} &= \beta_{m-2} \mathbf{v}_{m-2} + \alpha_{m-1} \mathbf{v}_{m-1} + \beta_{m-1} \mathbf{v}_m \\
H\mathbf{v}_m &= \beta_{m-1} \mathbf{v}_{m-1} + \alpha_m \mathbf{v}_m
\end{aligned} \tag{2}$$

If one rewrites (2) into a form to calculate $\mathbf{v}_k$ successively,

$$\begin{aligned}
\mathbf{v}_2 &= (H\mathbf{v}_1 - \alpha_1\mathbf{v}_1)/\beta_1 \\
\mathbf{v}_3 &= (H\mathbf{v}_2 - \beta_1\mathbf{v}_1 - \alpha_2\mathbf{v}_2)/\beta_2 \\
\mathbf{v}_4 &= (H\mathbf{v}_3 - \beta_2\mathbf{v}_2 - \alpha_3\mathbf{v}_3)/\beta_3 \\
&\cdots \\
\mathbf{v}_m &= (H\mathbf{v}_{m-1} - \beta_{m-2}\mathbf{v}_{m-2} - \alpha_{m-1}\mathbf{v}_{m-1})/\beta_{m-1}
\end{aligned} \tag{3}$$

The last equation in (3) corresponds the second last of (2). The requirement that the last relation of (2) is compatible with that of (3) leads to the vanishing value of $\mathbf{v}_{m+1}$ when (3) is formally extended to $m+1$. In order to have a vanishing $(m+1)$th vector in the series of $\mathbf{v}_k$, it is sufficient to choose $\mathbf{v}_k$ so that it is orthogonal to all previous vectors $\mathbf{v}_{k-1}, \mathbf{v}_{k-2}, \cdots$, since $(m+1)$ vectors cannot be orthogonal to each other in the m-dimensional space. It is useful here to regard (3) as an iterative orthogonalization process by subtracting components of previous vectors. Actually, it turns out that[4] by choosing

$$\begin{aligned}
\alpha_i &= \mathbf{v}_i^T H\mathbf{v}_i \\
\beta_i &= \| H\mathbf{v}_i - \beta_{i-1}\mathbf{v}_{i-1} - \alpha_i\mathbf{v}_i \|
\end{aligned} \tag{4}$$

all $\mathbf{v}_k$ are orthogonal to each other. Since this process (3) and (4) can be regarded as an improvement of the power method, it is not necessary to execute all of $(m-1)$ steps in (3) if one wishes to evaluate only several low-lying eigenvalues; one may calculate the eigenvalues of the intermediate tridiagonal matrix of dimension $l(< m)$ by the bisection method to see whether or not the low-lying eigenvalues have reached sufficiently converged values.

As for the eigenvectors, one first calculates the eigenvectors of the tridiagonal matrix when convergence of eigenvalues is confirmed. One then transforms the eigenvectors of the tridiagonal matrix into the original representation by use of the transformation matrix V. That is, the eigenvectors in the original representation are obtained by summing up the products of the components $c_1, c_2, \cdots$ of the eigenvector in the tridiagonal representation and the column vectors $\mathbf{v}_1, \mathbf{v}_2, \cdots$ of V: $c_1\mathbf{v}_1 + c_2\mathbf{v}_2 \cdots$. However, it is difficult to store the sequence generated by (3) $\mathbf{v}_1, \mathbf{v}_2, \cdots$ until convergence since convergence usually results after tens of iterations (and tens of the vectors should be stored in memory). A simple way out is to store $c_1, c_2, \cdots$ (real numbers) in the first run, and repeat the Lanczos process to generate $\mathbf{v}_1, \mathbf{v}_2, \cdots$ and sum up the products of the vectors and $c_1, c_2, \cdots$.[5] This method is used in some routines.

Inverse iteration is a method to calculate the eigenvector corresponding to a given approximate eigenvalue E_a. One starts from an arbitrary initial vector $\mathbf{v}_0$ and repeatedly multiplies $(H - E_a)^{-1}$. Let the expansion of the initial vector by the eigenvectors of H be

$$\mathbf{v}_0 = \sum_{j=1}^{m} a_j \psi_j$$

Then, k multiplications of $(H - E_a)^{-1}$ lead to

$$\mathbf{v}_k \equiv (H - E_a)^{-k}\mathbf{v}_0 = \sum_{j=1}^{m} a_j (E_j - E_a)^{-k} \psi_j$$

This equation indicates that the contribution of the eigenvector corresponding to the eigenvalue closest to E_a becomes exponentially large as k increases. Convergence is faster for a better approximate eigenvalue; in TITPACK Ver. 2, the iteration usually converges after a few steps.

If one tries to directly execute multiplication of $(H - E_a)^{-1}$, one has to obtain the inverse matrix, which is in general a difficult task. However, the relation

$$\mathbf{v}_k = (H - E_a)^{-1}\mathbf{v}_{k-1}$$

is equivalent to

$$(H - E_a)\mathbf{v}_k = \mathbf{v}_{k-1}$$

which may be regarded as a system of linear equations to calculate $\mathbf{v}_k$, given $\mathbf{v}_{k-1}$. Thus standard techniques can be used.

The Conjugate Gradient method solves linear equations of large scale by regarding the equation $M\mathbf{x} - \mathbf{b} = 0$ as the extremization of a bilinear form $f(\mathbf{x}') \equiv (\mathbf{r}, M^{-1}\mathbf{r})$ of the residual $\mathbf{r} \equiv \mathbf{b} - M\mathbf{x}'$, where $\mathbf{x}'$ is an approximate solution. Here (,) denotes inner product. To extremize $f(\mathbf{x}')$, one iteratively improves the approximate solution $\mathbf{x}'$ by changing $\mathbf{x}'$ to the direction with the steepest gradient on the landscape of $f(\mathbf{x}')$. The point where $f(\mathbf{x}')$ is extremum along this direction may be chosen as the next improved approximation. This is the Steepest Decent method. The Conjugate Gradient method is its improvement in that the direction along which a new approximation is sought is chosen within a space orthogonal to all previous modification vectors.

CONCLUSION

TITPACK Ver. 2 is a package of computer programs to numerically diagonalize quantum spin systems. It enables a researcher to obtain low-lying energy levels of any quantum spin system of the form in Eq. (1), as long as the computer to be used accommodates the necessary memory. It is sufficient to simply call several subroutines. It is an automatic vendor of eigenstates (Fig. 1); it requires no knowledge of algorithm of numerical diagonalization of large matrices. The user just enters the Hamiltonian and pays for CPU time (some people may be able to skip this last part) to get results.

The source code is available from the author by e-mail. The address is A85085@JPNKUDPC.BITNET. The source may also be obtained by mail in an IBM-PC formatted 5-inch diskette. The manual will be mailed upon request. There is no restriction on personally copying and using TITPACK Ver. 2 for academic purposes. I only ask the user to acknowledge its use, as it is or

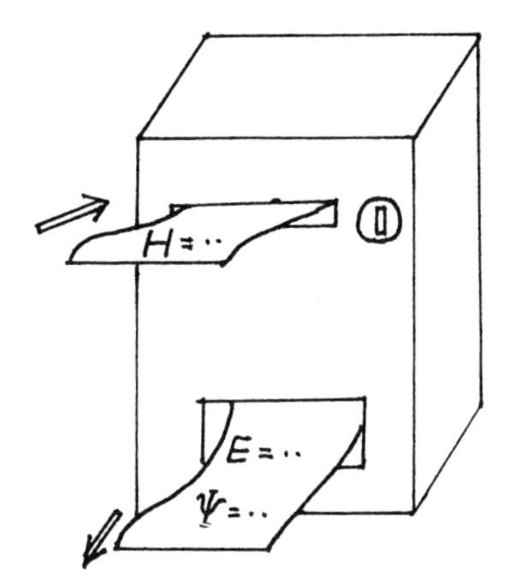

Fig. 1. TITPACK Ver. 2 as a vendor of eigenstates.

after modifications, with the present author's name explicitly indicated when publishing results.

ACKNOWLEDGMENT

I express my sincere thanks to Dr. Y. Taguchi, the co-author of Ver. 1. Useful suggestions on coding techniques by Dr. M. Ogata and Mr. M. Matsushita are gratefully acknowledged. This research was supported by the Grant-in-Aid for Priority Area by the Ministry of Education, Science and Culture.

REFERENCES

1. M. Suzuki, in *Quantum Monte Carlo Methods in Equilibrium and Nonequilibrium Systems* (Springer, Berlin, 1987).
2. Y. Taguchi and H. Nishimori, Bussei Kenkyu **45**, 299 (1986) (in Japanese). H. Nishimori and Y. Taguchi, Prog. Theor. Phys. Suppl. No. 87, 247 (1986).
3. T. Oguchi, H. Nishimori and Y. Taguchi, J. Phys. Soc. Jpn. **55**, 323 (1986).
4. H. Togawa, *Numerical Calculations Of Matrices* (in Japanese) (Ohm Sha, Tokyo, 1971).
5. E.R. Gagliano, E. Dagotto, A. Moreo and F.C. Alcaraz, Phys. Rev. B **34**, 1677 (1986). See also *Encyclopedia of Mathematical Sciences,* Ed. H. Hironaka, (Maruzen, Tokyo, 1991) (in Japanese), p. 909 for detailed analysis of the precision of eigenvectors.

AUTHOR INDEX

AIP Conference Proceedings

		L.C. Number	ISBN
No. 235	Physics and Chemistry of MCT and Novel IR Detector Materials (San Francisco, CA, 1990)	91-55493	0-88318-931-3
No. 236	Vacuum Design of Synchrotron Light Sources (Argonne, IL, 1990)	91-55527	0-88318-873-2
No. 237	Kent M. Terwilliger Memorial Symposium (Ann Arbor, MI, 1989)	91-55576	0-88318-788-4
No. 238	Capture Gamma-Ray Spectroscopy (Pacific Grove, CA, 1990)	91-57923	0-88318-830-9
No. 239	Advances in Biomolecular Simulations (Obernai, France, 1991)	91-58106	0-88318-940-2
No. 240	Joint Soviet-American Workshop on the Physics of Semiconductor Lasers (Leningrad, USSR, 1991)	91-58537	0-88318-936-4
No. 241	Scanned Probe Microscopy (Santa Barbara, CA, 1991)	91-76758	0-88318-816-3
No. 242	Strong, Weak, and Electromagnetic Interactions in Nuclei, Atoms, and Astrophysics: A Workshop in Honor of Stewart D. Bloom's Retirement (Livermore, CA, 1991)	91-76876	0-88318-943-7
No. 243	Intersections Between Particle and Nuclear Physics (Tucson, AZ 1991)	91-77580	0-88318-950-X
No. 244	Radio Frequency Power in Plasmas (Charleston, SC 1991)	91-77853	0-88318-937-2
No. 245	Basic Space Science (Bangalore, India 1991)	91-78379	0-88318-951-8
No. 246	Proceedings of the Ninth Symposium on Space Nuclear Power Systems	91-58793	Casebound Pt. 1: 1-56396-021-4 Pt. 2: 1-56396-023-2 Pt. 3: 1-56396-025-7 Set: 1-56396-027-3 Paperback Pt. 1: 1-56396-020-6 Pt. 2: 1-56396-022-2 Pt. 3: 1-56396-024-9 Set: 1-56396-026-5
No. 247	Global Warming: Physics and Facts (Washington, DC 1991)	91-78423	0-88318-932-1
No. 248	Computer-Aided Statistical Physics (Taipei, Taiwan 1991)	91-78378	0-88318-942-9